AF352205

International Exploration of Technology Equity and the Digital Divide:
Critical, Historical and Social Perspectives

Patricia Randolph Leigh
Iowa State University, USA

INFORMATION SCIENCE REFERENCE

Hershey · New York

Director of Editorial Content:	Kristin Klinger
Director of Book Publications:	Julia Mosemann
Acquisitions Editor:	Lindsay Johnston
Development Editor:	Christine Bufton
Publishing Assistant:	Milan Vracarich, Jr.
Typesetter:	Milan Vracarich, Jr.
Production Editor:	Jamie Snavely
Cover Design:	Lisa Tosheff

Published in the United States of America by
Information Science Reference (an imprint of IGI Global)
701 E. Chocolate Avenue
Hershey PA 17033
Tel: 717-533-8845
Fax: 717-533-8661
E-mail: cust@igi-global.com
Web site: http://www.igi-global.com

Copyright © 2011 by IGI Global. All rights reserved. No part of this publication may be reproduced, stored or distributed in any form or by any means, electronic or mechanical, including photocopying, without written permission from the publisher. Product or company names used in this set are for identification purposes only. Inclusion of the names of the products or companies does not indicate a claim of ownership by IGI Global of the trademark or registered trademark.

Library of Congress Cataloging-in-Publication Data

International exploration of technology equity and the digital divide :
critical, historical and social perspectives / Patricia Randolph Leigh,
editor.
      p. cm.
  Includes bibliographical references and index.
  Summary: "This book explores and presents research that centers on the
historical, political, sociological, and economic factors that engender global
inequities"--Provided by publisher.
  ISBN 978-1-61520-793-0 (hardcover) -- ISBN 978-1-61520-794-7 (ebook) 1.
Technological innovations--Social aspects. 2. Information society. 3.
Information technology--Social aspects. I. Leigh, Patricia Randolph.
  HM851.I58 2010
  303.48'33091724--dc22
                              2010024438

British Cataloguing in Publication Data
A Cataloguing in Publication record for this book is available from the British Library.

All work contributed to this book is new, previously-unpublished material. The views expressed in this book are those of the authors, but not necessarily of the publisher.

Table of Contents

Detailed Table of Contents

Chapter 1
Caste, Class, and IT in India 1
Elizabeth Langran, Fairfield University, USA

While India has transformed itself into a global leader in IT, its major cities have become enclaves for growth that only benefits a middle- and upper-class minority. Despite the increases in employment, this prosperity has not "trickled down" to the millions of impoverished in a highly fragmented, stratified society. The caste system and the history of India's middle class have engendered political decisions that have not benefited the poor's use of technology. It is important to examine Indian culture in order to more fully understand reasons behind the duality of a modern, consumerist IT India and a terribly impoverished India. While there are some initiatives underway to address the digital divide, it is challenging to create replicable models in a complex and diverse country that changes rapidly from one location to another. Providing access to technology is not eliminating the existing social, cultural, political and economic factors that have created the inequities.

Chapter 2
India's Dalits Search for a Democratic Opening in the Digital Divide 20
P. Thirumal, University of Hyderabad, India
Gary Michael Tartakov, Iowa State University, USA

This chapter seeks to recognize and read the marginal presence on the Internet of India's most oppressed and stigmatized community, the Dalits, simultaneously as acts of resistance and acts of constituting both self and community. It is an exploratory exercise in writing the social history of technological media in general, and the Internet in particular. The medium of Internet, unlike most associated with script and print that preceded it, offers emerging users access without the censorship of either established authorities or canons of authoritative formal regulation. More significantly, it provides the Dalit community, heretofore excluded from all but the most marginal voice in civil society, with an entrance into the national discourse. And quite as important, it furnishes them with their first meaningful media platform for a nationwide internal discourse, which they have previously been denied.

Sunnie Lee Watson, Ball State University, USA
Thalia Mulvihill, Ball State University, USA

This chapter aims to explore the historical, sociological, and economic factors that engender inequities related to digital technologies in the East Asian educational context. By employing critical social theory perspectives, the chapter discusses and argues for the notion of "Technology as a Public Good" by examining the Chinese, Japanese and Korean societies' digital divide. This chapter examines how East Asian societies are exhibiting similar yet different problems in providing equitable access to information communication technologies to the less advantaged due to previously existing social structures, and discusses the urgency of addressing these issues. Based on the analysis of the digital divide in the East Asian context, this chapter also proposes and argues for the notion of "technology as a public good" in public and educational policies for information communication technologies. Finally, the chapter invites policymakers, researchers and educators to explore a more active policy approach regarding the digital divide solution, and provides specific future research recommendations for ICT policies and policy implementation in digital divide solutions.

Janinka Greenwood, University of Canterbury, New Zealand
Lynne Harara Te Aika, University of Canterbury, New Zealand
Niki Davis, University of Canterbury, New Zealand

Māori people have a history of adaptation of new technologies. In recent decades Māori innovators have taken and adapted digital technologies for a range of purposes that can be broadly defined as educational. In this chapter the authors examine three cases where groups have utilised, and 'colonised', a range of particular technologies in order to build capacity for their tribal groups and wider community. In this way they use technologies as tools to overcome some of the financial, social and political deprivation caused by historic and continuing colonisation. The authors initially locate their exploration in a discussion of the historical context of colonisation, Māori movement towards self-determination, and in a discussion of Māori values and approaches to knowledge. They then present the three cases, beginning with one from a formal tertiary education programme (a Māori one), then examining a tribal initiative for language revitalisation and finally looking at a national use of digital media through Māori television.

Cynthia J. Alexander, Acadia University, Canada

Inuit in the Eastern Arctic of Canada reclaimed their homeland on 1 April 1999 when the newest territory in Canada, Nunavut, was created. Inuit are using new media technologies to preserve and promote their language, traditional knowledge, and ways of being. In this chapter, the reader is offered an exploration of the challenges northerners face in the digital era, including affordable, reliable access to the Internet. However, the author shows how the resilience that characterizes Inuit culture extends to their innovative adoption of new media technologies. The author offers insight into one web development project, a partnered initiative with Inuit, which enables Inuit youth to learn from their Elders, and for users around the world to learn from Inuit via an interactive online adventure. The case study of The Nanisiniq Inuit Qaujimajatuqangit, or IQ Adventure, provides an interesting example of how harnessing the power of new media can support indigenous peoples' decolonization efforts.

Just as refugees fleeing to escape Zimbabwe have struggled to cross the crocodile-hungry waters of the Limpopo, so are Zimbabweans battling to find ways to traverse the abyss of a digital divide affecting their country. In 2008-09, Zimbabwe was rated third worst in the world for its national information communications technology (ICT) capability by the World Economic Forum, being ranked at 132/134 nations on the global ICT 'networked readiness index'. Digital divide issues, including severe deficits in access to new technologies facing this small Sub-Saharan country, are therefore acute. In terms of global power relations involving ICT capability, Zimbabwe has little influence in any world ranking of nations. A history of oppression, economic collapse, mismanagement, poverty, disease, corruption, discrimination, public sector breakdown and population loss has rendered the country almost powerless in ICT terms. Applying a critical social theory methodology and drawing on Freirian conceptions of critical pedagogy to promote emancipation through equal access to e-learning, this chapter is written in two parts. In the first place, it analyses grim national statistics relating to the digital divide in Zimbabwe; in the second part, the chapter applies this information in a practical fictional setting to imagine life through the eyes of an average Zimbabwean male farm worker called Themba, recounting through narrative an example of the impact on one person's life that could result from, firstly, a complete lack of ICT resources in a rural situation and, secondly, new opportunities to become engaged with ICT training and facilities.

This chapter explores why there is a need for scholars to not only systematically couple discussions about technology use along with technology access, but ground their inquiries in a theory of critical multiculturalism as they seek to fully understand ways for minimizing the digital divide. In order to help explain why using this critical framework is important, this discussion is set against the historical backdrop of the country of Brazil whose past in many ways parallels the United States with regard to its history of oppression and servitude of people based upon their racial heritage. Moreover, this work provides a brief discussion of Paulo Freire's work with African Brazilians and how he helped them to develop critical understandings about how hegemonic structures limited the extent to which they were able to experience their own humanity. This chapter draws from the historical experiences of African Brazilians as a way to deconstruct how issues of technology and educational inequalities are examined in the U.S. The author of this chapter claims that if U.S. educators are to help prepare students to become productive and reflective decision-makers, they must first acquire tools for understanding their own social realities and learn ways for re-creating them to reflect the ideals of democracy and social justice. Furthermore, the author made calls for educational scholars develop a new language that captures the spectrum of questions at the center of the digital divide debate concerning access and use, but also foregrounds issues of liberation, agency and social change.

In this chapter, the author examines the history of the colonization of Brazil through the transatlantic Black slave trade and the effects this history had upon digital equity experienced by Black Brazilians in the information age. This examination is conducted using the philosophical lenses of critical theory and critical race theory (CRT). Coming from these perspectives, the author join other scholars in the belief that racism does, in fact, exist in Brazilian societies and join with those who aim to dispel 'the myth of racial democracy' and the myth of racial harmony in a country with roots in a race-based system of slavery and peonage. The author contends that issues of digital equity and equality of opportunity can only be effectively addressed if one has a deep understanding of the factors that led to inequities that preceded the information age. With this in mind, the author shares various culturally-based, grass-roots efforts along with government initiatives she observed during the first stages of an ethnographic investigation of digital equity in this segment of the African Diaspora.

In this chapter, the authors explore the history of the Gullah people of the Sea Islands of South Carolina. In examining the history of oppression and isolation of Black Americans of Gullah descent, the authors look at how a history of racism and inequity set the stage for the digital inequities experienced by Gullah communities since the onset of the information age. They find that despite the Gullahs' tenacious struggles for education and literary during enslavement, many were left behind in this age of digital technology. The authors examine the effects that the isolated and closed Gullah communities, which were forced conditions during slavery, had upon many Gullahs' reluctance and resistance to engagement in information communication technologies (ICTs) centuries later. They contend that this continued, though self-imposed, isolation inadvertently contributed to the loss of Gullah land and a form of gentrification that severely compromises Gullah traditions and values.

A great part of the rhetoric accompanying the rapid diffusion of information and communication technologies (ICTs) in Western societies in recent decades has put the spotlight on their potential for generating economic growth and development in the socio-political arena. Yet mechanisms that generate disparities among citizens do not go away with the advent of electronic citizenship, as asymmetric access to economic and political resources limit access to new technologies. This contribution is divided in three sections. In the first part, the concept of "digital divide" is analysed by considering its first formulation in the US political debate during the Nineties, as well as the more recent efforts to consider the multidimensional nature of such category. In the second section, significant quantitative measures of digital disparities between countries is provided. Finally, it shows how developing countries adopting proprietary software are becoming dependent on the power of providers of ICT goods and services, which are mainly concentrated in the United States.

Foreword

A POWERFUL EXPERIENCE: MEETING THE MARGINALIZED PARTICIPANTS IN THE INFORMATION AGE

It is not uncommon for me to engage in conversation with those who speak of the Digital Divide in the past tense and assume that the abundance of computers and technology in our homes, schools and workplaces has decreased the need for attention to issues of digital access and equity. This naïve argument that the Digital Divide is now disappearing due to increasingly more accessible, cheaper digital resources usually extends to the developing nations of our global arena as well. When I participate in these conversations, I remain deeply concerned about the inequitable affordances provided by technology in both developed and developing countries for learners in our school systems and for adult citizens in our cultures and I am worried that the recent proliferation of technology will move focus away from this issue. Reading this book has helped me address this concern and has provided me with a much deeper understanding of the specific issues that make the Digital Divide an important ongoing issue and of the complexities of providing truly equitable opportunity through technology. I am confidant that this book will provide similar positive experiences for other readers.

International Explorations of Technology Equity and the Digital Divide: Critical, Historical and Social Perspectives provides a riveting account of the deep historical, political and cultural issues underpinning the potential opportunities and challenges provided through digital technologies. The reader will have the opportunity to experience and learn from the often-unexamined inequities in the global arena and will emerge from this experience with a much deeper understanding of the current state of the Digital Divide and the complexities that surround digital inequities. Appropriately, the book does not provide solutions for this complex issue, but it does provide important knowledge and tools for understanding and addressing the Digital Divide.

The first two chapters in the book provide an ideal backdrop for later chapters through a careful examination of the caste system in India and a thoughtful explanation of the inequities connected with the caste system in India for potential use of technology. Taken together, these chapters provide an ideal introduction to the book since these chapters describe both the immense challenge of making technology available to underserved populations (Chapter 1) and the potential for using technology to create a "democratic opening" for marginalized populations such as the Dalits in India. The challenges and opportunities for making digital technologies accessible and powerful for people in India provide a useful beginning frame for similar challenges and opportunities in New Zealand, Brazil, Africa, Canada, China, Japan and South Korea, and the United States.

Readers will become immersed in captivating stories and descriptions of marginalized peoples from a variety of countries, including the United States. The depth and detail in these descriptions makes a strong point about the uniqueness and value of each of these cultures and the need to address issues of technology from a perspective of knowledge and appreciation for each of these cultures.

Although one of the basic premises of this book is that uses of technology are unique for different cultures, the situations in each of the cultures described in the book provide knowledge and insight for those interested in understanding and addressing Digital Divide issues. The idea that technology has the potential to help activate the silenced voices of marginalized people in the global arena runs through several chapters in this book. Authors point to the capabilities of technology in communicating visual and auditory messages from cultures for which visuals and the spoken word are two of the major ways stories are handed down and communicated. Facilitating the use of technology to help preserve and strengthen different cultures is an exciting approach to addressing the Digital Divide. Through using technology in this way, technology can become a tool to reinvigorate people and cultures that have been oppressed. Ultimately providing tools to strengthen and celebrate different cultures and different people has the potential to truly use technology to address the deep historic issues that have created the conditions that feed the Digital Divide.

This book effectively addresses the naïve assumption that more technology is the key to addressing issues of digital inequity. The rich descriptions of indigenous people and their cultures makes it very clear that different populations will use technology in very different ways and that part of a solution to the Digital Divide issue involves understanding and addressing the complex contexts of different cultures. Although it remains tempting to assume that cheaper and more accessible technology will solve our digital equity challenges, experiences in schools in the United States have demonstrated that just providing technology for children in low SES schools does not solve the Digital Divide issue. The issues for children in these schools involve how the technology is used and how well prepared the teachers are to use technology to address the individual needs of the students. We have compelling evidence from these experiences on the ineffectiveness of merely making the technology available. The cases described in this book provide similar insight on the need to adapt the access and use of technology to the individual needs and goals of marginalized groups.

Several authors in this book demonstrate an ability to make the cultures of indigenous people come alive for the reader. Descriptions of individual personalities within cultures help the reader begin to experience and celebrate the diversity of cultures often marginalized in today's world. The rich, in-depth stories of the Inuits, Maori, Gullahs and Black Brazilians use specific cases to make clear the important point that different cultures and traditions will use technology in very different ways. The potential of technology to help the Gullahs address their land issues, the Maori to create community and the Inuits to share their culture provide specific examples of the diverse ways technology might serve diverse cultures.

The informative articles in this book also provide a strong opportunity for scholars to learn from the experiences of countries that have been relatively successful in providing ICT access for a majority of their populations. The chapter on East Asian countries provides useful and transferable insights into specific national policies that have worked for Japan and South Korea, two countries who are among the international leaders in providing access for citizens.

It is clear to me that this book has deeply affected my understanding of the complexity of the global Digital Divide issue. Examining issues of digital equity through the lenses of other cultures provides help in understanding and analyzing our digital equity issues in the United States and these same lenses provide the same affordance for other countries in the global arena. I am confidant that the next time I am engaged in a discussion on the current state of the Digital Divide I will be able to speak much more specifically and knowledgably about the importance of this issue and the need to understand and address causes of the Digital Divide using individual contexts. One of my insights from the book is that appropriate access to technology may help groups of people begin to address the inequities and power imbalances that created the conditions for the Digital Divide in the first place. Reading this book will not only reaffirm the importance of this issue, but also provide intelligent, contextualized approaches for addressing digital inequities.

I can promise readers that they will find a "good read" in these chapters, as well as a possibly life-changing experience with marginalized participants in the global information age.

Ann D. Thompson

***Ann D. Thompson**, Ph.D., University Professor at Iowa State University, has devoted her career to designing, studying and implementing effective uses of technology in education. She is the Founding Director of the Center for Technology in Learning and Teaching at Iowa State University and has been involved in preservice teacher preparation and educational technology research and development for more than 25 years. She served as Chair of the Department of Curriculum and Instruction for seven years and the department received the AACTE Award for Best Practice in Technology Integration in 2000. She is currently pursuing initiatives in technology pedagogical content knowledge, technology professional development, and technology in teacher education. She is a past president of the Society for Information Technology and Teacher Education (SITE), co-editor of The Journal of Digital Learning in Teacher Education, and has led grant supported projects totaling more than $5 million. She was elected to the Iowa Academy of Education in 2001. She is the author of more than 50 refereed journal articles, 3 books, and numerous book chapters.*

Preface

In this edited volume, authors explore the historical, political, sociological, and economic factors that engender global inequities related to digital technologies. The contributing authors present the phenomenon of digital equity from various critical social theory perspectives. These critical perspectives, together with the histories of domination and oppression on different continents, provide contexts for understanding the fertile grounds made available for the international growth and expansion of digital inequity.

The digital divide is an international, global phenomenon that negatively affects groups around the world. Consequently, the objective and mission of this book is to explore and present research that centers on the historical, political, sociological, and economic factors that contribute to global inequities. Acquiring such insights and knowledge is an important step towards rectifying socially ingrained inequities and a necessary step in working towards global justice in meaningful ways.

All chapters are written from a perspective and philosophical lens derived from a critical social theory. For the purposes of this volume critical social theory is defined as one that examines power relationships and addresses issues of oppression and domination. Typically, such a theory addresses issues of racism, classism, sexism, and/or other forms of discriminatory practices, behaviors, and policies aimed at specific social identity groups that have been historically underserved. Chapters that describe how a particular critical social theory explains the digital divide and its historical development are written with an international focus. Chapters written to look at specific and typically underserved groups examine how their histories and related social and political factors, which were present before the information age, have affected their digital access and use today.

The target audience for this book is composed of academic scholars and educators and their students. Such a resource will aid those researching, teaching, and studying in the area of digital equity or in the broader contexts of social and global justice. Moreover, the book provides valuable insights for professionals and researchers interested in examining issues of technology equity from various critical social theories.

ORGANIZATION

In chapter 1, "Caste, Class, and IT in India" Elizabeth Langran provides a history of the caste system in India, which served to render large populations in critical need of food, health, housing, and basic literacy even long after its theoretical and political dismantlement. She contends that although India is now seen as a global leader in information technology, it is the middle-class and affluent elite "digetari" who are benefiting from the outsourcing of technology-related jobs from developed countries while the impoverished and historically oppressed millions are left even further behind in this digital age.

P. Thirumal and Gary Michael Tartakov pick up on Langran's theme concerning the effects of India's caste system but take a closer look at those whose were pejoratively referred to as the 'Untouchables'. In chapter 2, "India's Dalits Search for a Democratic Opening in the Digital Divide", Thirumal and Tartakov specifically examine how digital technology can be used to promote equity through democratic dialogue. While doing so, these authors give a historical background of the socially ostracized, oppressed, and marginalized group, who shed the 'Untouchables' label to become self-named Dalits. This history reveals how they were silenced through a lack of voice in non-digital print media but, nevertheless, offers hope for democratic engagement and participation in the open and uncensored digital medium of the Internet.

Chapter 3, "Exploring the Notion of 'Technology as Public Good': Emerging Characteristics and Trends of the Digital Divide in East Asian Education", also evokes notions of human rights and democratic participation. Sunnie Lee Watson and Thalia Mulvihill state, "…technology must be considered a public good rather than merely an article of trade or commodity in our societies." From a critical social theory perspective, they contend that all human beings have a right to participate in developing their communities and participating in their societies. Open and equitable access and uses of technology to all groups would then render new technologies as public rather than private goods. Further, these authors use this notion of public good to explore, from historical contexts, the state of digital divides in China, Japan, and Korea and the possibilities for embracing this notion of public good while creating solutions for existing problems.

Janinka Greenwood, Lynne Harara Te Aika, and Niki Davis provide insights into the history of New Zealand colonization and the culture of the indigenous Maori. In chapter 4, "Creating Virtual Marae: An Examination of How Digital Technologies have been Adopted by Maori in Aotearoa New Zealand", the authors describe technology initiatives and projects emanating from Maori and aimed in achieving self-determination specifically in terms of language revitalization and education. As such, they contend that technology adoption by Maori serves to lessen the disparities and divides brought on by a history of colonization and further serves to simultaneously repair some of the resulting damages.

Similarly, Cynthia J. Alexander explores the history of various forms of oppression experienced by the Inuit of Canada, the effects of that history on their current situation, and how digital media can be used to decolonize. In chapter 5, "From Igloos to iPods: Inuit Qaujimajatuqangit and the Internet in Canada", Alexander specifically states, "New media technologies *can* be designed to serve postcolonial objectives", and goes on to specifically describe a web design created for the purpose of revitalizing, passing on, and maintaining indigenous cultures with a focus on language, traditions, and values.

In chapter 6, "The Digital Abyss in Zimbabwe", Jill Jameson provides a political, economical, and technological history of the African country, Zimbabwe, within the context of the continent's history. She describes the complex challenges faced by people who find themselves at the absolute bottom of the digital divide, as reflected in the statistics she offers. Zimbabwe, as portrayed by Jameson, is representative of a developing country that some believe is in need of technology infusion and others believe would suffer further at the hands of such infusion. A unique feature of this chapter is that while it offers historical backdrops and supporting graphs and tables, the author presents a fictional narrative in which a character, Themba, navigates through it all.

Chapter 7, "Paulo Freire's Liberatory Pedagogy: Rethinking Issues of Technology Access and Use in Education" is first situated in the historical context of colonial Brazil and is specifically focused upon the past experiences of African Brazilians. In this chapter, James C. McShay uses critical multiculturalism and Paulo Freire's liberatory pedagogy to address issues of digital equity in K-12 schooling for these present-day Brazilians. He makes distinctions between liberal multiculturalism and critical multicultural and further contends that the former does not adequately address issues at the core of the digital divide. McShay looks to the Freireian approach as a model critical multiculturalist perspective, which aims to use the infusion and use of technologies for purposes of liberation and emancipation.

Patricia Randolph Leigh builds upon McShay's work by also examining the history of the colonization of Brazil through the enslavement of Black Africans. In chapter 8, "Digital Equity and Black Brazilians: Honoring History and Culture", she explores the history of African Brazilians and the impact this history had upon the maintenance of African Brazilian culture and their ability to engage what some claim as (whereas others dispute the existence of) a democratic and egalitarian Brazilian society. Leigh describes grassroots and government initiatives related to digital inclusion projects in the Black communities and favelas (slums) of Salvador, Bahia.

Chapter 9, "Digital Equity in a Traditional Culture: Gullah Communities in South Carolina", provides a window into an otherwise isolated group of African Americans in the U.S. Patricia Randolph Leigh, J. Herman Blake, and Emily L. Moore describe the colonization of the Sea Islands of South Carolina with enslaved Africans who would come to be known as Gullah. They maintain that Gullah, unlike other oppressed groups who suffered significant loss of culture, were able to maintain much of the culture of their homelands through their relative isolation on the islands. They further contend that this isolation subsequently prevented their inclusion in digital age networks and communities, which continues to have far reaching negative effects on the future of Gullah culture in this age and beyond. Francesco Amoretti and Fortunato Musella conclude this volume with chapter 10, "Governing Digital Divides: Power Structures and ICT Strategies in a Global Perspective". This chapter reminds us again that electronic citizenship, which is seen by many as the outcome of increased access to ICTs, does not guarantee the elimination of disparities between individuals or between countries. As other authors within this volume have stated, Amoretti and Musella point out that the factors and mechanisms that led to these disparities are still present in the information age and, in many cases, the gaps that were present prior to the advent of twentieth and twenty-first century technologies are exacerbated between developed and developing countries. The persistence or even widening of these gaps is purportedly caused by the technology diffusion policies that are driven by a Western capitalistic model.

Patricia Randolph Leigh
Iowa State University, USA

Acknowledgment

This edited volume represents the concentrated efforts of many people whose work is reflected in these pages. I am honored to be among the authors who brought their experience, perspectives, and scholarship to this project and who responded to my comments as well as to the comments and suggestions of the reviewers. Each chapter fits well within the theme of this book yet offers the reader varying critical, historical, and social perspectives into issues concerning technology and digital equity. For their commitment and dedicated work, I would like to thank the following authors: *Cynthia J. Alexander, Francesco Amoretti, J. Herman Blake, Niki Davis, Janinka Greenwood, Lynne Harara Te Aika, Jill Jameson, Elizabeth Langran, James C. McShay, Emily L. Moore, Thalia Mulvihill, Fortunato Musella, Gary Michael Tartakov, P. Thirumal,* and *Sunnie Lee Watson.*

The quality of this volume is also due to the rigorous review carried out by its Editorial Advisory Board and other dedicated reviewers. The critical and detailed assessments of each chapter lifted the burden of editing such a volume. The Board and all reviewers were chosen for their expertise in their respective areas and I would like to thank the following for their time, attention, and dedication to this work: *Beverlyn Lundy Allen, Christine Clark, Karen Donaldson Dade, Jeong-Hee Kim, Nana Osei Kofi, Rema Nilakanta, Raymond M. Rose, Bonnie Bracey Sutton, Carlie Tartakov, Kay Ann Taylor, Sherry K. Watt,* and *Peter K. Yu.*

I would also like to thank *Ann D. Thompson* for providing the forward to this book and for her leadership as the director of the *Center for Technology in Learning and Teaching (CTLT)* in the College of Human Sciences at Iowa State University. I am indebted to the faculty and staff of CTLT for their support, throughout the years, of my social justice scholarship as it relates to technology equity. I am particularly grateful for their support of my research in Brazil that is represented in chapter eight of this volume.

As editor I am taking advantage of this space to also thank other individuals who contributed to the preliminary investigative research carried out in Salvador, Brazil, which made the writing of chapter eight possible. Thanks to the following individuals for helping me in ways too numerous to mention here: *Cibele Brito, Rubia Carvalho, Lazaro Cunha, Gustavo Farias, Maisa Flores, Silvio Humberto, Estaban Moreno, Conor O'Sullivan,* and *Paula Santos.*

Finally, special thanks goes to *IGI Global* for this opportunity and to the support staff, from acquisitions to publications, who made the completion of this project possible. I am particularly indebted to *Christine Bufton,* IGI Editorial Communications Coordinator and my primary contact throughout the process, for her expertise, input, patience, encouragement, professionalism, and flexibility.

Heartfelt thanks to all,

Pat Leigh
Editor

Chapter 1
Caste, Class, and IT in India

Elizabeth Langran
Fairfield University, USA

ABSTRACT

While India has transformed itself into a global leader in IT, its major cities have become enclaves for growth that only benefits a middle- and upper-class minority. Despite the increases in employment, this prosperity has not "trickled down" to the millions of impoverished in a highly fragmented, stratified society. The caste system and the history of India's middle class have engendered political decisions that have not benefited the poor's use of technology. It is important to examine Indian culture in order to more fully understand reasons behind the duality of a modern, consumerist IT India and a terribly impoverished India. While there are some initiatives underway to address the digital divide, it is challenging to create replicable models in a complex and diverse country that changes rapidly from one location to another. Providing access to technology is not eliminating the existing social, cultural, political and economic factors that have created the inequities.

INTRODUCTION

India's highly skilled middle class, abundance of relatively inexpensive labor, and liberalization of the domestic economy have made it an attractive destination for outsourcing call centers and software development. The industry employs 2 million Indians and is expected to grow over the next few years, renewing hopes for India to evolve into a major economic power. The ability of countries like India to compete in information capitalism has led authors, most notably Thomas Friedman (2005) in his bestseller, *The World is Flat*, to posit that technology is leveling the global competitive playing field. Neo-liberals view this trend in globalization as a tide that lifts all boats, as free market economics bring prosperity to previously untapped markets.

Yet there are estimated to be only 47 million Internet users in India – less than 5% of the population; the average monthly family income of these Internet users is 3.2 times the national average (JuxtConsult, 2009). Most rural Indians have never made a phone call. While English is one of the official languages of India with about 90 million speakers, there are millions more who speak one of the hundreds of other Indian languages that do not have a presence on the English-dominated

DOI: 10.4018/978-1-61520-793-0.ch001

Copyright © 2011, IGI Global. Copying or distributing in print or electronic forms without written permission of IGI Global is prohibited.

Internet, and nearly 450 million illiterate adults (UNICEF, 2008). Although India's economy is growing, there's a marked disparity in the distribution of wealth and millions are being left behind.

Several groups have cited the potential of the Internet for e-governance, local economic regeneration, and mobilizing rural resources for community development (Fukuda-Parr & Birdsall, 2001). This cyber-libertarian approach asserts that access to the Internet will create a more democratic and equitable society. In India, these efforts have most often taken the form of information center kiosks. Studies of these initiatives have indicated that the projects are not delivering what they originally envisioned (Keniston, 2004; Govindan, 2006; Sreekumar, 2006; Thomas, 2006; Tiene, 2002). Providing access to technology is not eliminating the existing social, cultural, political and economic factors that create the inequities.

This chapter will highlight the complexity of technology integration in India. The issues of the digital divide go beyond mere connectivity; giving the poor more access to computers will hardly transform such a highly stratified society. The disproportionate access to technology is a symptom, not just a cause, of inequities. While India has transformed itself into a global leader in IT, this prosperity has not "trickled down" to the millions of impoverished due to a number of historical, political, sociological and economic factors. A strong case would need to be made to justify the expense of technology investment for the rural poor in a country that suffers greatly from critical needs in food, health, housing, and basic literacy, and is home to one-third of the world's poor. Using social stratification theory, this chapter provides a unique way to understand the digital divide by examining in particular how the caste system and the history of India's middle class have engendered political decisions that have not benefited the poor's access to and use of technology, and what efforts are being made to address the digital divide.

BACKGROUND

In 1991 Dr. Manmohan Singh, then Finance Minister and later Prime Minister of India, engineered a series of economic reforms to liberalize the economy by removing some of the state controls. At the same time, changes in technology were transforming international business practices: transportation became cheaper and faster, and lower telecommunication costs made connectivity easier. With these changes, businesses began to segment the production process, outsourcing pieces as a cost-saving measure. Once one company in a particular market begins this process, competition drives others to do the same, and India became a particularly popular outsourcing destination as companies experienced increased computer systems debugging needs leading up to the year 2000 (Y2K) problem (Chandrasekhar, 2006). These outsourcing companies are not producing for local markets, but rather are intending their products to be exported back to the home country or a third country. While this model results in local employment, profits do not go back into the local economy. Despite the increases in employment that are being brought to India by the outsourcing movement, the cities of Bangalore, Delhi, Mumbai, Hyderabad and Chennai have become enclaves for the IT growth that is only benefiting a middle-class minority. Instability in some regions of India (such as the Northeast) has discouraged companies from wanting to invest there, as well as other regions that have unreliable power and low bandwidth. There are few links from the IT sector to the rest of the economy, which is still largely agrarian and has few sectors that make significant use of technology, and in a country that has a great need to put many people to work there is always the fear of computers replacing people. Unreliable power, low bandwidth, slow hardware, low purchasing power, and low literacy outside of the metropolitan cities ensure that technology diffusion will face difficulties.

Cyber-libertarians argue that access to technology would create a more democratic and equitable society accompanied by rapid economic transformation, and thereby eliminate the need to make any radical social transformation (Sreekumar, 2006). However, 70% of India is rural, and almost half of Indian rural households do not have electricity; those who do face constant disruptions. While three out of four urban Indians have access to a telephone, only 13.21% do in rural areas; out of the 5,97,655 recognized villages in India, there were nearly 70,698 villages at the end of September 2008 which had not yet received their first telephone line (Singh, 2008). Most rural households do not have access to sanitation and clean water; 40% of villages are unconnected to the road network. In order to provide full access in power to these areas, 552.43 billion rupees (more than 11 billion U.S. dollars) would be needed, and 926.90 billion rupees (nearly 20 billion U.S. dollars) for full access in telecommunications (NCAER, 2007). Only 2% of urban India pays taxes and most of India's top companies do not pay taxes, so this revenue would be difficult to raise. India's population growth rate will likely outgrow its economic growth, leading to more pressure on infrastructure and social programs on top of already-limited resources. Many children already do not attend school and more will continue to leave school to support their families as they take on the role of providing "social security" for their parents. Estimates by human rights groups have put the number of child laborers in India at 17 million. With high rates of illiteracy, particularly among the rural poor, many Indians may not even be aware of the need for information that can be found on the Internet, let alone how to use or access it, and there is an absence of culturally and locally relevant content in local languages (Chandra, 2002). Neo-liberals believe that those who are poor are so because of their failure to help themselves, but in order to understand the problems contributing to the digital divide, we must examine the social and cultural context that keeps Indian society stratified.

SOCIAL STRATIFICATION DIMENSIONS: CASTE, CLASS, AND POWER

Social stratification theory examines the hierarchies of domination and subordination, superiority and inferiority that exist within a given social system. Sociologist Max Weber's (1946) concept of social stratification uses three main categories to examine social stratification: status (social power), class (class power), and party (political power). The possession of power comes from an individual's ability to control social resources (land, capital, social respect, knowledge). An example of *social power* is the authority to have a person obey another because that other is seen as a social superior. *Class power* refers to the unequal access among classes to material resources. *Political power* is the access organized groups have to political structures and includes their ability to influence laws. We have come to know that there is significant interaction among the three domains of stratification, and most Indians recognize that political and economic factors are highly connected with the attribution of status; it becomes difficult to study one stratification system without the recognition that it interacts strongly with another; socio-economic status is a way of viewing these dimensions together (Gupta, 1991; Weber, 1946; Fuller, 1991). The function of social stratification is to provide mechanisms for social control, and thus adherence to order and conformity; mobility as well as illegitimate use of power is kept at a minimum in a highly stratified society.

Caste provides one way of stratifying status. Unlike classes, status groups are communities that have expectations for a specific socially visible "style of life" with restrictions on social interactions, including marriage. When status groups become stratified and guide a certain distribution of economic power, legal privilege follows. The status group then closes into a caste grouping. Status distinctions include rituals and a monopoly on materials, including wearing of special attire, eating special dishes, etc., and a disqualification against performing physical labor. According to Weber (1946), caste goes beyond ethnic segregation as it places groups into a vertical system of superordination and subordination. Acceptance of the system creates a social and economic interdependence.

A major difference between caste and class is that when a person changes to a new job, that does not change that person's caste, but it may change his or her class. Weber's theory of class fits well in capitalist, market-oriented societies where economic rewards determine one's class, as classes are stratified by their consumption of goods (Gupta, 1991). In India however, people are not separated simply into classes, but into castes as well. Where there is status stratification, there can be no genuine free market competition. Status groups hinder the free development of the market when status groups withhold or monopolize goods from free exchange. Weber (1946) stated that technological and economic transformations threaten status stratification, and indeed we did see social democratic reforms of the industrial age all over the world. However, what we are seeing now are technological and economic transformations that have the potential to create greater stratification, an economic divide between participants of the knowledge-based economy who are rewarded by globalization and those who participate in physical labor economy. The neo-liberalism of the new knowledge-based economy is a threat to the social democratic reforms of the industrial age all over the world. Knowledge is proprietary and rights are claimed by large companies, who have resources for innovation and protection under the Trade-Related Aspects of Intellectual Property Rights (TRIPS) of the World Trade Organization (Shiva, 2005). Rich countries can function well within this circle of technology innovation, creating more jobs and economic opportunities, thereby accumulating more wealth that can be reinvested in resources for innovation, thus creating an even bigger divide (Parayil, 2006).

Caste Hierarchies

The caste system, broadly defined, creates separation and hierarchy among Hindus through rituals, dress, marriage, and other visible means of differentiation, and provides the basic social structure of Hindu society (Hindus making up approximately 83% of the population). Other religions, such as Buddhism and Sikhism, were formed in a reaction against the caste system; however, differences in castes are acknowledged among Muslims and Christians as well. Earliest records of the caste system date back to an account by a foreigner visiting India in the 3rd century BC, referring to the use of ritual and religion to create a division of labor. With the connection between vocation and religion, continuing one's family trade became a sacred duty. There are thousands of *jati* (communities, also sometimes referred to as sub-castes) that often define geographical and occupational groups. An overly-broad classification of the caste system uses *varna* categories: Brahmins are traditionally priests and teachers, Kshatriyas are warriors, Vaisha are merchants and Shudras are laborers, with others (tribes and *Dalits*, formerly known as Untouchables) remaining outside of this varna system and considered the bottom of the social structure. This pan-Indian system does not take into account the local social orders that are based on jati, regulating marriage, occupational rigidity, village politics, and social taboos (Gupta, 1991). Thus the status social stratification is largely based on local-specific factors.

With hierarchy so accepted a feature of this society and dating back thousands of years, India is known as one of the most highly stratified societies in modern history. Those born poor or rich are said to be destined to stay that way in accordance with their karma from a previous life. If they fulfill their prescribed role in society, the poor will escape their situation in their next life, with no need for human intervention, though individuals are free to seize opportunities to escape their current lot. It is questionable whether the existing upper- and middle- class members or those who manage to break into the middle class will demonstrate any remarkable concern for those left behind; as Paulo Freire noted, the oppressed will tend to replace and copy their own tormentors (Varma, 2007). Hindu ideology utilizes the concepts of dharma and karma to justify the caste system. Dharma refers to the duty to fulfill one's role in life (determined by one's caste), and karma is the reward or punishment determined by how well one's dharma has been fulfilled. Thus good dharma and good karma lead to rebirth into a higher class, and visa versa. Lower castes are supposed to look to a brighter future in another life. Dharma and karma are more likely to be used by higher castes as a rationalization for their relative position in society. Hindu ideology also relies heavily on the concepts of purity and pollution (Fuller, 1991). According to this belief, pollution can be conveyed by touch, even indirectly, i.e. the "impure" are barred from drawing water from village wells for fear of polluting the water. Having different castes drink from the same tap led to controversy when the first public water system was introduced in the city of Calcutta. Even today it has been problematic to set up a community computer center in rural areas, where higher castes do not allow access to lower caste members for fear of pollution.

Caste Interaction with Class and Power

With new opportunities to acquire power, relationships between caste, class, and power have become more complex today, (Béteille, 1991). Caste and class hierarchy correlate significantly (Gough, 1991; Shah, 1991). The highest caste members, the Brahmins, have traditionally been able to combine their high ritual status with political and economic power (Mukherjee, 1991). They were the first to take advantage of the economic opportunities from an English education, so they gradually moved into more urban areas; this meant members of other castes took over ownership of their land, becoming the new local "dominant caste" (Srinivas, 1991a; Béteille, 1991). Until World War I, Brahmins monopolized education, widening the gap between themselves and other castes; with education came more economic opportunities from urban jobs—clerical, executive, managerial, professional—which they monopolized, using caste ties to recruit more Brahmins. Being educated, the Brahmins were able to become closer to the British, entering the prestigious and powerful Indian Civil Service and government bureaucracies, dominating law, medicine, and education, and even the leadership of the Congress party for a time (Béteille, 1991). Thus they were able to successfully combine the power dimensions of status, class, and political party to reinforce their position in the social stratification hierarchy.

In a reaction to the political power that the Brahmins had amassed, non-Brahmins began to agitate and become politically active, and succeeded in having seats reserved for them in government and higher education since the 1920s; by the mid-fifties non-Brahmins had considerable control of Congress, the State Legislature, and the cabinet, as they do today. The exclusivity of Brahmin status bred hostile reactions from non-Brahmins, but the ensuing anti-Brahmin movement also contributed to a feeling of unity among Brahmins, further widening the gaps between these groups (Béteille, 1991). By keeping others at a distance, Brahmin superiority is now an obstacle, as the skill of organizing people is important now in today's politics. Non-Brahmins now have an advantage over the old elite Brahmins who are not close to the people, look at politics as something dirty, and are unable to get out the vote by engaging the masses (Béteille, 1991).

It is generally accepted that Brahmins are at the top of the status power dimension hierarchy, and tribes and Dalits are at the bottom, but within the middle of the status stratification system, the hierarchy can become more vague and there is even some caste mobility possible. Before British rule, there was some caste mobility through warfare (Srinivas, 1991b). Today, the acceptance of rites, beliefs, ideas and values from sacred Hindu texts all influence a caste's diet, occupation, acceptance of food and water from other castes, and how they are regarded by Brahmins, thus determining a caste's position in the social status hierarchy. Caste mobility has always been slow, and to acquire land and move up the hierarchy in class takes one or two generations (Béteille, 1991). A caste that is locally dominant, strong in numbers, and has some wealth can move up the status dimension hierarchy, but this is still tied to the economic dimension hierarchy. While it is illegal to discriminate on the basis of caste, recent studies show caste plays a role in capital formation and in developing a marketing network (Varma, 2007). Sheer numbers of voters alone are not enough to benefit certain groups of people, as there must also be the ability to organize; thus people who already have social and economic standing in a community are at an advantage. In villages, most who have political power have land and some other source of income, allowing them to entertain other public officials in their homes and network with others who have power. Politicians might use government funds to their own advantage, and give contracts to those who can provide reciprocal benefits (Béteille, 1991). Thus there are still strong connections between the political and economic power dimensions. Having political and economic power enables groups of people to control access to resources, whether it is to information and communication technologies (ICT) or education.

Today, membership in the legislature and thus political power still depends largely on caste and class. The Backward Classes Movement of the 20th

Century came from the recognition that political power was needed to secure participation in education and to share in economic opportunities. If a caste was labeled "backward," members could receive preference in education, scholarships, and jobs in administration (Srinivas, 1991b). Dalits and the Denotified Tribes (mostly nomadic or seminomadic tribes) make up the majority of the 34% of the population making less than $1 per day. To improve their situation, the government sets aside public sector jobs, parliamentary seats and university places for them. Despite this "reservations" model that has extended political power beyond the highest castes, many who should benefit do not; Dalit leader B.R. Ambedkar (1944) criticized the Justice Party who, despite their anti-Brahmin platform, "…had been holding on to the spirit of [Brahminism] as being the ideal they ought to reach…During the twenty years the Party had been in office, it forgot the 90 percent of the Non-Brahmins living in the villages" (as cited in Jaffrelot, 2005, p. 82). Additionally, in order to have political power one must have economic power. Certainly the landless laborers are not benefiting from the current system, as they have little power and sometimes cannot work within the system to even obtain what benefits they are entitled to have. The government may try to help Dalits, but often the local dominant caste will hinder government efforts, since the dominant caste is still in need of people to perform labor cheaply and often to do degrading tasks. Still today there is resentment to sharing wells and temples with Dalits (let alone ICT resources), a resentment that at times has manifested itself in violence against Dalits.

This system of reservations has been criticized (largely by elite who never had to compete for these jobs before) for using criteria other than merit to fill positions, and in fact currently many faculty positions remain unfilled due to the claimed lack of "qualified" applicants to take these reserved seats. Without more investment in primary education to prepare young people to qualify for such positions, the reservations policy will be only of limited benefit to those for whom it was designed. The reservation system treats tribes as if they are all the same, even though they may be distinctly different, with some benefiting more than others (Shah, 1991). While several of the castes have benefited from this to grow in economic and political power, it is still largely only in rural areas, and middle class status cannot be sustained by income from agriculture alone. On the other hand, there are few routes to non-agricultural employment available, including into the IT world (Varma, 2007). Even within agriculture, some tribes who have qualified for land under government programs are new to agriculture and have limited knowledge of cultivation. Educated tribals aspire for public sector jobs and want quick promotion, but they are hampered by traditions and socialization, and face resistance from the middle and high castes (Shah, 1991).

It is essential for organizations interested in improving access to education and technology to have an understanding of the systemic barriers imposed by the caste system. Non-governmental organizations (NGOs) must be aware of local power dynamics and work through the traditional village headman. This unfortunately limits the participation of the local people in projects to only the village elites, reinforcing rather than bridging the divide. In rural areas in particular, caste will still largely determine a person's power and access to resources. More affluent classes and castes may hinder developmental progress because they want to retain their greater access to sources of power. Lower caste members may be barred from using a village computer because of caste, such as in the Information Village Research Project, a project of the M.S. Swaminathan Research Foundation in Pondicherry which had to close when Dalits were not allowed to enter the information kiosks (Sreekumar, 2006).

Caste is also a barrier to literacy. In Nambisan's (2000) examination of Bihar, among the poorest states in India, he recounted the problem encountered by an upper-caste headmaster: "He had had the same problems in his village which other Brahmins have when they teach lower-caste children. 'The Brahmins of my village once threatened to deny me food and water [the local form of excommunication] because I interacted freely with the Musahar [low caste] children'" (p. 134). There exists in India one of the world's largest pools of trained workers yet also the world's largest number of out-of-school children (Varma, 2007).

The Woman Jati

In a country that worships "mother India" and thousands of prominent female goddesses and has had a number of prominent female politicians, there also exists a long history of subjugation to the patriarch, with higher female infant mortality, child marriage, lower female literacy and school attendance, and considerable loss of status for widows (Smith, 2007; Clark & Sekher, 2007). The "workplace" in India was conceptualized during the country's industrial development, where a worker was defined as a male, factory-based employee, and women's economic activity was not recognized as "work" but rather an extension of their role as wife and mother in a gendered division of labor. Today, 96% of working women are engaged with the "informal economy"—they work as vendors, home-based outworkers, and daily laborers, outside of the protection of labor legislation (Broadbent & Ford, 2007). While there have been some attempts to organize and protect women workers, most notably by the Self Employed Women's Association (SEWA), the Working Women's Forum in Chennai, and the Annapurna Mahila Mandal in Mumbai, their efforts have been hampered by the cultural constraints of women's independence and the difficulty of organizing women in the informal sector who are working in homes, streets, and fields, in addition to being busy with household duties in a double-burden of labor.

Some organizations, both NGOs and corporations, are offering free technology classes designed to improve the career opportunities of women. Microsoft, for example, funded the non-profit Datamation to offer technology classes to women in Kanpur, a region known for the production of leather goods and embroidery. Datamation's goals are to enable women to design their embroideries on-screen using drag-and drop software and gain access to markets to work directly with customers instead of the middlemen who have traditionally exploited them. Many parents did not want their daughters to go out in public and attend these classes. *The Financial Times* (Merchant, 2006) reported that even though some parents are allowing their daughters to take these classes—more than 500 women from poor Muslim families and lower-caste Dalit communities have participated—they are not allowing them to seek employment, especially for jobs that would require moving to a city or interacting directly with buyers, designers, and shopkeepers.

Within the IT sector in India, women often still do the least skilled work for the least pay, and are found primarily in jobs with repetitive and routine tasks such as data entry, transcription, telephone sales, and consumer services. Few women are programmers, designers, technicians, or engineers—those who produce and make decisions about the technology. Globalization allows for the cheap labor to be performed away from the main office, and many women working in these end-user, lower-skilled IT jobs find it difficult to move into better paying jobs such as design and management (Pande, 2005). Additionally, participation in the new knowledge-based economy requires education and fluency in English, thus those who are working within the IT sector are nearly all from the upper-middle class. Clark and Sekher's (2007) study of women working in call centers noted that often these young women were highly educated, yet doing low-level work. With the cheap availability of educated labor, there is little incentive to invest in a primary education system to benefit the poor.

Caste and Fragmentation

The caste system fragments groups and contributes to stratification. Castes are their own small social worlds, as different groups repel each other and stay isolated through marriage and social interactions, thereby keeping groups from uniting. Allegiances are to self, then family, and then caste, contributing to a lack of concern for the community as a whole. Weber (1946) states that "communal actions" and "societal actions" that emerge from mass actions depend largely on general cultural conditions. Castes became the pressure groups for India's new democracy, and internal conflicts have become problematic among non-Brahmins, as competition among impoverished groups for short-term resources becomes more important than long-term cooperation (Bandopadhyay & Von Eschen, 1991; Béteille, 1991; Srivinas, 1991b). While elite families typically are more organized, have informal ties with other families, and are better able to mobilize, the poor do not have time, resources, or space for social networking and lending assistance to others. They must depend on the rich for favors so they fight with one-another for these favors, and are prevented by ideas of caste pollution from interacting, thus stratification leads to fragmentation and visa versa (Bandopadhyay & Von Eschen, 1991). Additionally, fragmentation impairs technology integration—diffusion studies show that adoption happens through friendships, and personal networks; when resources are low, cooperation is required so groups that are hindered by the fragmented nature of the caste system are not able to diffuse knowledge of new technologies.

Bandopadhyay and Von Eschen's (1991) study of rural West Bengal found rural society to be highly stratified in the dimensions of class, status, and power, with close interactions among all three and that social mobility over time is low. They found that a few large families dominated the best-paying non-agricultural jobs such as doctors, lawyers, large shop owners, and head schoolmaster. Their economic positions were strengthened by supplementing farm income with these jobs. At least 25% of the villagers were landless, unemployed for at least half the year, malnourished, and living in insufficient shelters. Even for some of those who do own land, many owe debt, sometimes going back over a generation, to a local wealthy family and live essentially in a state of semi-servility. These researchers found disproportionate access to the new higher-paying jobs, with inheritance also contributing to the rigid stratification with low mobility. Higher income gives the added advantages of ability to bribe and to travel to government offices for information and applications for government assistance. Most government officials are educated and are urban, so it is harder for them to empathize with the poor (Bandopadhyay & Von Eschen, 1991).

The caste system is still an important factor in rural areas to determine vocation, marriage, and social standing. It became difficult in emerging metropolises to be able to retain a physical separation among the castes, unlike in rural areas, where separate parts of the village would be reserved for a particular caste. While years ago the installation of a water system in Calcutta lead to fears of pollution among high-caste Brahmins who did not want to drink from the same source as untouchables, urban caste members have had to learn to share amenities and space (Mukherjee, 1991). In a village, family backgrounds are known and there is a long tradition regulating *jati* interactions. In cities where people come from many distant areas, these relationships and traditions are unknown so broader classifications such as the *varna* system are more common. New urbanites may use reference behavior, emulating higher-caste middle class lifestyles. More wealthy urbanites have occupations unrelated to traditional caste occupations and live mostly in class, not caste hierarchies. Even with these changes in occupations, the majority of Brahmins still do

not do manual work, and marriages are commonly within the same caste; Indians may move out of the country to be outside the caste system and to pursue economic opportunities, but even Indians overseas often seek same-caste marriages and largely come from India's middle and upper classes. Lower-caste politicians have risen to power, such as K.R. Narayanan who became the first Dalit president in 1997, and B. R. Ambedkar, architect of the Indian Constitution and leader of the Buddhist revival movement. However, there is still a wide cultural and economic gap between the leaders and the other caste members. Politicians must keep links to their own caste to stay in power, but professional elites do not, and can leave their fellow caste-members behind (Srivinas, 1991b).

Class Hierarchies

India, home to one-third of the world's poor, is economically stratified; most of the country is rural and poor, urban slums exist right next to wealthy areas, and the top 20 cities make up less than 10% of the population but generate more than 30% of income (National Council of Applied Economic Research, 2007). Along with the IT boom in India came an increasing number of consumers with greater purchasing power; however, economists' estimates of the size of the Indian middle class vary widely, from 50 million (just 5% of the population) to 250 million (Ablett,, Baijal, Beinhocker, Bose, Farrell, Gersch, Greenberg, Gupta, & Gupta, 2008). If you can have three balanced meals a day, send your children to school, or access health care you are considered privileged in India, a country where even water and electricity cannot be taken for granted (Varma, 2007). According to research from the McKinsey Global Institute, 2% of the population earns more than 1 million rupees a year and almost 25% of India's income, while 25% live on less than $1 a day (other economists put this figure of less than $1 a day at 34%). In 2005, 54% of the population were considered "deprived," earning less than 90,000 rupees, or $1,970 a year, and 41% of the population considered "aspirers," earning 90,000 to 200,000 rupees a year; these are considered by most economists to be below the middle class (Ablett et al., 2008). No matter which economists are to be believed, there is ample evidence that there are significant disparities in income distribution. It is important to examine the Indian attitudes toward individualism, education, acquisition of wealth, and concern for the poor in order to more fully understand some of the reasons behind the duality of a modern, consumerist IT India and a terribly impoverished India.

Individualism and pursuit of material gain are valued Indian traits. In Hinduism, one of the four purusharthas, or goals of life, is Artha, the pursuit of material prosperity. Hinduism is not institutionalized as are many other religions, having no organized church where people gather together in community; rather the focus is on an individual and the consequences of his or her own karma, within a moral relativism that may contribute to a sense of self over community (Varma, 2007).

While India is a democracy, the individual's participation in bettering the situation of the poor has been extremely limited. Most development strategies have generally been "top-down," absolving the individual from having to take any action. The poor have been around forever, so they have become part of the "accepted landscape" (Varma, 2007). Even the term "digital divide" has been subject to quite a large variety of definitions in the Indian press. *The Times of India* has published numerous articles using the term "digital divide," yet many of them discuss the generational divide between parents and their children rather than a gap between rich and poor, or the fact that the digital divide is disappearing or nonexistent. Additionally, the representatives in the early years of the Indian parliament were largely in the middle and upper classes, and also directed the businesses, controlled the media, and owned most of the land. Government policies tended to favor industry at the expense of agriculture, reflecting a bias for technology—it was through superior technology

that the British were able to dominate and unite India, after all. So even though 80% of the people (most of the poor) were in agriculture, agriculture and community development only received 15% of the funding allotted by the first three five-year economic plans following independence (Varma, 2007). Again, the class and political power dimensions are closely linked. There is a belief that what is good for the middle class is good for the country, but in reality policies that benefit the middle class would take decades to trickle down to most Indians.

Digerati

Cities like Bangalore, Mumbai, Chennai, and Hyderabad are home to an affluent elite class of young cosmopolitans with disposable income working in call centers and software companies; these "digerati" with an attraction to Western lifestyles are far removed from the illiterate who make up nearly 40% of the population (some estimates place this rate even higher). An Indian call center operator working for an overseas-based company receives an average income of between 10,000 and 12,000 rupees a month, or $225 - $270, several thousand rupees higher than what is earned at the average Indian company (but a fraction of the cost of what the overseas company would pay back home). While the middle class has been growing since Independence in 1947, it remains upper caste in character, and following the economic reforms of 1991, consumerism became more acceptable and even promoted as the symbol of a new India. India's new prosperity has not spread to rural villagers and urban poor. The euphoria over India's emergence as a global leader in IT has, at times, made it less concerned for the underprivileged. The government, historically controlled by the middle class, invests heavily in higher education at the expense of primary education, contributing to a "scarcity value" for those who hold a higher education degree and ensuring the continued prosperity of this middle class. In 2008, six new Indian

Institutes of Technology (IITs) were opened, with a planned three more due to open by 2010, more than doubling the number if IITs to sixteen in just a few years while creating a serious issue of quality control. There is a concern for a growing skills gap, that the quick growth and resulting quality issues will mean that students will sit in class to pass exams and obtain their degree, but will not have sufficient skills to keep India globally competitive.

During colonial rule, the British neglected primary education but supported higher education to develop a class of people who could work in administrative posts, creating a native elite whom they could use in their rule of India. English became a language of social exclusion (Varma, 2007). There was a fear among the Indian elite that the spread of English education among the lower classes would hurt the interests of the elite as laborers would demand higher wages, better jobs, and more equality (Mukherjee, 1991). In response to the Indian middle class needs, institutions of higher education were expanded, particularly in the early years following Independence when the middle class largely had control of government. A degree from an IIT is one key way to enter into employment and join the ranks of the digerati. The entrance examinations for the IITs are highly competitive; in 2007, nearly 300,000 students sat for the exam, and 2,000 were selected—less than 1%. Parents who can afford it send their children to private training institutes to prepare for the exams. Those who cannot afford the private training rely on the public schools for preparation. The quality of primary education in India varies widely, from schools equipped with computer labs to those that have no books and in some cases no teachers. The literacy rate nationwide is at 61%, but when broken down further the lopsidedness of this statistic becomes evident: in the impoverished state of Bihar, the literacy rate for females among Dalits is less than 5%. India has the highest number of out-of-school children in the world and with no social security system and high poverty rates, many

children are being sent out to work and are unable to go to school; human rights groups believe that India has an estimated 60 million child laborers. Thus even though there is government funding available for Dalit students, many are unable to qualify for the IIT spots due to inadequate primary education preparation. Until these larger issues are solved, the poor will largely be left out of the prosperity of India's IT boom. After the economic reforms of 1991, there were even more jobs, more computer schools, and more training institutions in IT for those who could get admitted to them. This attention to higher education at the expense of primary education is a result of the stratified economic and status hierarchies that exist; it also perpetuates them.

In addition to blocking the pursuit of higher education and better paying jobs in the IT industry, low literacy also affects the use of technology. In a study of information centers in two rural areas of Andhra Pradesh and Kerala that had been set up by the state governments and Hewlett Packard in 2002 to create an "i-community" and to impart IT literacy to at least one member of each household, Jayan Jose Thomas (2006) found that there was a correlation among computer usage, level of education, and level of assets. Additionally, the research indicated that the diffusion of technology was faster among literates as well—possibly because the awareness of the usefulness of and demand for the information provided by technology was higher among these literates.

Solutions and Recommendations

Some economists believe the IT industry will be able to pull the rest of India's economy toward change through both market forces and government interventions (D'Costa, 2006). However, a strong economy has close links with domestic sectors, and has a good balance of services and production for a diversified market—this is not the case in India at the moment. Despite what neo-liberals say, the state has an important role in economic development and labor productivity, and many other countries (Russia, Eastern Europe, China, the Philippines) could challenge India in the future. While government policies promote the export of software services, right now India does not develop its own software, which could produce good revenue. It also does not sell to its own domestic market and lacks significant hardware manufacturing, so not much revenue flows into the domestic economy. The Indian economy is dependent on external markets and foreign firms. In 2008, hardware exports from India totaled $500 million, consistent with the last 5 years. The other $40.4 *billion* in exports came from IT services and software and engineering services; services exports account for 98% of the total exports (NASSCOM, 2008). While there has been some increase in domestic hardware manufacturing (from $4.4 billion in 2004 to $11.5 billion in 2008), private industry and government policies need to work on further developing domestic market links. With the United States representing 61% of the export market and the United Kingdom 18%, there was great concern over the financial crisis in 2008 and the effects on India of the economic downturn for their major market. Between September 2008 and December 2009, India experienced an economic slowdown, with GDP growth slowing to 4.9 percent, but industry leaders continue to be optimistic about India's growth going in to the next decade despite spiraling inflation ("Inflation at 10-Year High," 2009; Ranganathan, 2009; "Worst Over," 2009).

If India does succeed in promoting domestic market links and greater market diversification, a larger pool of skilled workers is needed. With projected growth in the IT sector, it is in the best interests of government and private industry to be concerned about the potential skills gap resulting from degrading quality of the rapidly-expanding higher education sector. Direct employment in Indian IT-business process outsourcing was 2 million in 2008, an increase of 389,000, and currently accounts for 5.5% of the Indian GDP

(NASSCOM, 2008). This number may seem high, but in a country of more than one billion people, most are being left out. The per capita income of this largely agrarian country is among the lowest in world and income distribution is getting worse.

Ways of getting resources from the profitable IT sector to other areas are needed. This could be through job-generating investments, technological upgrading, research and development, increased domestic hardware production, worker re-training and infrastructure development (D'Costa, 2006). The government has historically favored higher education, creating a level of scarcity and structurally excluding those without a strong primary education. Therefore, investment in primary education is essential to allow greater numbers of people access to the knowledge-based economy; information literacy means nothing if basic literacy does not exist (Singh, 2002). A knowledge-based economy needs not just technology and software services operators, but also engineers, social scientists, and managers. Currently there is a weak user base in technology and low technology diffusion, which restricts collective learning and the ability to create revenue for the domestic market. It is important to create an awareness of technology at the grassroots level, thereby increasing the size and quality of the user base (Thomas, 2006).

One area the weak user base might be addressed is within schools.[1] Many of the schools affiliated with a "central" nationwide school system and following a common curriculum do have computer classes, though the concept of educational technology and using computers to teach other content areas beyond computer classes is very novel in India; aside from the computer science teacher, few teachers know how to integrate technology in their content areas. Schools affiliated with each state or tribal schools rarely have these resources. Occasionally school administrators use limited budget resources to acquire computer equipment but neglect the teacher's role, so without proper teacher training, the technology does not get adopted (Tiene, 2004). Some schools do not even have textbooks; currently there is an initiative to get just maps and basic educational tools to some of the tribal schools. Many tribal schools have been interested in acquiring a television. Television has been a very powerful influence in some of the more isolated areas of the country—it is seen in these areas as a window to the world, opening students' and teachers' eyes to new possibilities and creating a hunger for a different life. In some areas, computers might not be the answer, but another older technology already in use, such as the television or radio, might be adopted more easily because many people already know how to use these technologies (Tiene, 2004). A rethinking of the appropriate technologies to push in these areas may lead to strategies that rely more on cell phones, already in use in even the most remote area of the country, rather than computer kiosks.

Private industry, local governments, and the national government all have a responsibility to address the issues associated with digital inequities. Other strategies to address the weak user base should include an assessment of local needs and the promotion of the appropriate technology, which may be wireless technology. Local governments need to play a more proactive role in promoting and regulating service providers (NCAER, 2007). New economic policies can allow for competition at the local level, not just the state level, by promoting small, localized service providers. Indians are widely known for an entrepreneurship that would benefit from greater availability of microfinance. Without addressing these issues, India will not be able to stay on track with its economic growth goal of around 8% each year. While it has been able to meet that target over the last five years, India currently faces the same economic uncertainties as the rest of the world. With its growing population and an additional 14 million being added to the labor market each year (50% of India's population is under the age of 25), these issues urgently need to be addressed right now. If it does not, it is likely that the government will feel the repercussions from dissatisfied voters, particularly in light of

the increasing security concerns on the minds of Indian citizens. This would not be the first time discontented Indian voters threw out a ruling party due to the slow pace of development outside the cities. In the general elections of 2004, despite national economic growth, the ruling Bharatiya Janata Party was defeated ("How India Voted," 2004; "Poll Results," 2004).

Because the domestic market will not be developed until poverty and income gaps are addressed, and the private industry has a vested interest in addressing the skills gap to ensure a qualified workforce, private industry partnerships and initiatives that address education, health, and infrastructure are a mutually-beneficial arrangement. "e-Choupal" is an initiative of ITC Limited, one of India's largest exporters of agricultural commodities, to use the Internet to give farmers information and services to increase farm productivity, consult current market prices, and rely less on the middleman. Websites are in the local language, and currently there are 6,500 e-Choupals in 10 states and available to 4 million agriculturalists. At the International Institute for Information Technology in Andhra Pradesh, Motorola and Microsoft have endowed professorships, and Satyam Computer Services funded the Satyam School of Applied Information Systems. Keane of India (global IT services) established the Keane School of Excellence with a focus on object-oriented technologies, Oracle has a School of Advanced Software Technology, and there is a Motorola School of Communication Technology. Some IITs have a training center on their campus belonging to a private company, and the IIT faculty may even teach at the training center; this space, when not in use for training, ideally can be used by the community.

The reservation system has received criticism from both higher education and private industry. The founder of InfoSys is doing his own experiment by creating a Special Training Program to train 89 scheduled castes and tribes students (35 girls and 54 boys) to show that given the right training, they would be able to get software engineering jobs in major software companies through the same selection procedures as everyone else at campus recruitments, whereas reservations would "permanently damage the psyche" of those they target. The best and most committed teachers from Infosys and the Indian Institute of Information Technology Bangalore were used to give each candidate as much attention as wanted, and candidates were given a stipend and three meals a day. In 2007, 84 of 89 candidates were placed with well-known software companies (Murthy, 2008). When Prime Minister Singh visited the IIT in Guwahati in 2008, the director made an appeal to suspend reservations in faculty hiring so that they could fill more positions with "qualified" people; currently, especially with the rapid expansion of the IIT system, many faculty positions reserved for scheduled castes and tribes remain unfilled. It has become a matter of controversy as to whether there is truly a lack of unqualified applicants, or if this is an attempt to resist the hiring of Dalits and scheduled caste applicants while perpetuating discriminatory practices.

IITs are engaged in a number of initiatives to address the quality of education. One is the National Programme on Technology Enhanced Learning. This initiative seeks to introduce IIT faculty to educational technology, i.e. using multimedia and Web technology to enhance learning of basic science and engineering concepts. The Open Courseware initiative uses YouTube to host videos of experienced engineering faculty teaching a class. While this has the potential to make engineering education more accessible to anyone with sufficient Internet connectivity, its purpose is to provide models for new and less-experienced faculty.

eGovernance has the potential to empower marginal groups to participate in their own developmental efforts, but must be worth the effort and expense (Chandra, 2002). Media, especially cable TV, can also impact the involvement of marginal groups in the democratic process through the dissemination of information. It can also inspire rurals to seek entry into the middle class, as most television shows feature well-to-do Indians (Singh, 2002; Varma, 2007). With almost 110 million mobile phones in India (National Council of Applied Economic Research, 2007), this larger user base may be a better target of initiatives than computers. While currently there are numerous initiatives by non-governmental organizations, the private sector, and central and state governments, these groups are not coordinating with one another and many projects are simply pilots, rather than strategic initiatives. Projects that are not self-sustaining, have limited local participation, lack local resources, and are not linked to the local economy will end as soon as the funding leaves; it is essential to have an understanding of these local-specific factors, including local power dynamics, before investing in an initiative.

The government of India is trying to deal with such emergencies as floods and violence; education, health, water, electricity, and agriculture are social priorities, and it can be difficult to justify the expense associated with addressing the digital divide in a country where many people are too poor to buy necessities (Chandra, 2002). In a complex country that changes rapidly from one location to another due to many local-specific factors, it is challenging to create replicable models to address the digital divide. When technology is poorly implemented and sometimes saddles those who where supposed to be helped with extra expenses, even by a well-intentioned non-governmental organization, those who were burned become wary. Cyber-libertarians use technology as a solution without a clear idea of what the problem is. Mobile phones and Internet access can give market and price information, helping to cut out the middlemen. The Internet can help with governance, healthcare, education, distance learning, and connecting isolated communities to communicate their needs. But the belief that technology alone can change society is technological deterministic. With a lack of physical infrastructure and lack of relevant content in local languages, technology can make inequalities even greater between urban and rural, educated and illiterate, and haves and have-nots (Thomas, 2006). Parayil (2006) argues that the "digital divide, income dispersion, uneven development and asymmetries in intra-regional economic growth… after the onset of the global information revolution are not unrelated" (p. 198) and we are now dealing with the "digitisation of inequality."

FUTURE RESEARCH DIRECTIONS

Research that provides data on replicable models is difficult in a large and complex country that changes rapidly from one location to another. Local-specific factors, particularly class and status hierarchy dynamics, must be included in any evaluation of initiatives to address the digital divide. A research study may indicate that technology is able to give farmers information about better agricultural practices, but if there is landlessness, no irrigation, and no institutional credit, there is no need for this type of information (Thomas, 2006). If people do not have an education, there will be no demand for information on jobs and higher education. Future research should not only address the ability of initiatives to provide physical infrastructure, but also the demand for the information and technology. Poverty, income inequality and illiteracy are barriers in a knowledge-based industry for transforming the economy, and should be taken into account in any studies or initiatives.

Even where other countries do not have an explicit caste system, social stratification continues to exist. An examination of status and class hierarchies within a given social system has the potential to uncover some of the contributors to digital inequity beyond mere access to the technology itself. Stratification segments groups of people while creating barriers to information sharing, technology diffusion, and uniting to advocate for change. These barriers are not unique to India, and as other countries develop their potential as low-wage IT workers in the continuing globalization trend, they too will need to examine how their own history and culture has contributed to social stratification and the digital divide. Without doing so, there is a danger that IT policy will institutionalize inequalities, rather than address them.

CONCLUSION

Thomas Friedman's view of globalization as a leveling force of the economic playing field has prompted reactions that include the view that globalization is in fact a *recolonization*, led by corporations in partnership with powerful governments that dismantle trade protections, workers protections, and environmental protections (Shiva, 2005). Dalit leader Ambedkar (1930) once stated:

the depressed classes welcomed the British as their deliverers from age-long tyranny and oppression by the orthodox Hindus… When we compare our present position with the one which it was our lot to bear in Indian society of the pre-British days, we find that, instead of marching on, we are only marking time (as cited in Jaffrelot, 2005, p. 93).

Ambedkar placed blame on Britain's lack of understanding of India, failure to reform society, and decision to establish its authority through a network of administration dominated by upper castes. In the same vein, globalization is failing to reform the highly stratified Indian society, and continues to establish itself through a network dominated by the privileged. Proponents of globalization welcome technology as a "deliverer from age-long tyranny and oppression," but the poor worldwide may simply be "marking time" while corporations and powerful governments reap the benefits of globalization.

If India is to be prosperous, it must invest in its people. Development strategies that include communication technologies that give information on prices, markets, and opportunities are vital; however, the capability to *use* the information and the demand for information are also important. Problems of landlessness, lack of irrigation, lack of credit, and illiteracy hinder the growth of agricultural incomes in rural India, and affect the demand of information and technology (Thomas, 2006). Technology can be useful only if basic obstacles are overcome by major changes in land reform, credit, rural infrastructure, and primary education. Writer Vandana Shiva (2005) encourages us to look beyond the worldwide Web of IT and instead to the web of life, the food web, the web of communities and the web of local economies and local cultures being destroyed by globalization.

REFERENCES

Ablett, J., Baijal, A., Beinhocker, E., Bose, A., Farrell, D., Gersch, U., et al. (2007, May). *The "bird of gold": The rise of India's consumer market. McKinsey Global Institute.* Retrieved December 14, 2008, from http://www.mckinsey.com/mgi/publications/india_consumer_market/index.asp

Bandopadhyay, S., & Von Eschen, D. (1991). Agricultural failure: Caste, class and power in rural West Bengal. In Gupta, D. (Ed.), *Social stratification* (pp. 353–368). New Delhi: Oxford University Press.

Béteille, A. (1991). Caste, class, and power. In Gupta, D. (Ed.), *Social stratification* (pp. 339–352). New Delhi: Oxford University Press.

Broadbent, K., & Ford, M. (2007). *Women and labour organizing in Asia: Diversity, autonomy and activism. Retrieved from Ebook Library database.* Boca Raton, FL: Taylor & Francis.

Chandra, S. (2002, Sept.). Information in a networked world: The Indian perspective. *The International Information & Library Review, 34*(3), 235–246. doi:10.1006/iilr.2002.0202

Chandrasekhar, G. P. (2006). The political economy of IT-driven outsourcing. In Parayil, G. (Ed.), *Political economy and information capitalism in India: Digital divide and equity* (pp. 35–60). New York: Palgrave Macmillan.

Clark, A. W., & Sekher, T. V. (2007). Can career-minded young women reverse gender discrimination? A view from Bangalore's high-tech sector. *Gender, Technology and Development, 11*(3), 285–319. doi:10.1177/097185240701100301

D'Costa, A. (2006). ICTs and decoupled development: Theories, trajectories, and transitions. In Parayil, G. (Ed.), *Political economy and information capitalism in India: Digital divide and equity* (pp. 11–34). New York: Palgrave Macmillan.

Engler, M. (2008, May-June). The world is not flat: how Thomas Friedman gets it wrong about globalization. *Dollars and Sense, 276*(6), 20.

Friedman, T. (2005). *The world is flat: A brief history of the twenty-first century.* New York: Farrar, Straus and Giroux.

Fukuda-Parr, S., & Birdsall, N. (2001). *Human development report 2001: Making new technologies work for human development.* New York: Oxford University Press for the United Nations Development Programme. Retrieved August 11, 2008, from http://hdr.undp.org/en/reports/global/hdr2001/

Fuller, C. (1991). Kerala Christians and the caste system. In Gupta, D. (Ed.), *Social stratification* (pp. 195–212). New Delhi: Oxford University Press.

Gough, K. (1991). Class and economic structure in Thanjavur. In Gupta, D. (Ed.), *Social stratification* (pp. 276–287). New Delhi: Oxford University Press.

Govindan, P. (2006). Introduction: Information capitalism. In Parayil, G. (Ed.), *Political economy and information capitalism in India: Digital divide and equity* (pp. 1–10). New York: Palgrave Macmillan.

Gupta, D. (1991). Caste, class and conflict. In Gupta, D. (Ed.), *Social stratification* (pp. 303–306). New Delhi: Oxford University Press.

How India voted: Verdict 2004. (2004, May 20). *The Hindu*. Retrieved December 14, 2009, from www.hindu.com/elections2004/verdict2004/index.htm

India. (2009). The World Factbook [online]. Retrieved April 23, 2009, from U.S. Central Intelligence Agency: https://www.cia.gov/library/publications/the-world-factbook/geos/in.html

Inflation at 10-year high: Food prices to spill over to other areas. *The Economic Times*. (2009, December 11). Retrieved December 11, 2009 from http://economictimes.indiatimes.com/news/economy/indicators/Inflation-at-10-year-high-Food-prices-to-spill-over-to-other-areas/articleshow/5324792.cms

Jaffrelot, C. (2005). *Dr. Ambedkar and untouchability: fighting the Indian caste system*. New York: Columbia University Press.

JuxtConsult. (2009). *Internet usage behaviour and preferences of Indians: Snapshot 2009*. Retrieved April 16, 2009, from http://www.juxtconsult.com/download.asp

Keniston, K. (2004). Introduction: The four digital divides. In Keniston, K., & Kumar, D. (Eds.), *IT experience in India: Bridging the digital divide* (pp. 11–36). New Delhi: Sage Publications.

Kozma, R., McGhee, R., & Quellmalz, E. (2004). Closing the digital divide: Evaluation of the World Links program. *International Journal of Educational Development*, *24*(4), 361–381. doi:10.1016/j.ijedudev.2003.11.014

Merchant, K. (2006, May 1). The women behind closed digital doors: Technology development: Social conservatism in India is harming efforts to increase computer literacy - and wasting talent. *Financial Times (North American Edition)*, 8.

Mukherjee, S. (1991). The Bhadraloks of Bengal. In Gupta, D. (Ed.), *Social stratification* (pp. 176–182). New Delhi: Oxford University Press.

Murthy, N. R. N. (2008, Sept 25). Caste away: Great ideas, great minds – state accountability. *India Today*. Retrieved October 23, 2008, from http://indiatoday.digitaltoday.in/index.php?option=com_content&issueid=72&task=view&id=16150&acc=high

Nambisan, V. (2000). *Bihar is in the eye of the beholder*. New Delhi: Penguin Books India.

NASSCOM. (2008). *Indian IT-BPO industry factsheet: NASSCOM analysis*. New Delhi: National Association of Software and Service Companies. Retrieved December 12, 2008, from http://www.nasscom.in/Nasscom/templates/NormalPage.aspx?id=53615

National Council of Applied Economic Research. (2007). *India rural infrastructure report*. SAGE. Retrieved December 16, 2008, from http://books.google.com/books?id=xjpqUSei38kC&printsec=frontcover#PPP16,M1

Pande, R. (2005). Looking at information technology from a gender perspective: The call centers in India. *Asian Journal of Women's Studies*, *11*(1), 58–82.

Parayil, G. (2006). The political economy of informational development: A normative appraisal. In Parayil, G. (Ed.), *Political economy and information capitalism in India: Digital divide and equity* (pp. 196–217). New York: Palgrave Macmillan.

Poll results verdict against globalisation: CPI(ML) (2004, May 17). *The Times of India*. Retrieved December 14, 2009 from http://timesofindia.indiatimes.com/city/thirupuram/Poll-results-verdict-against-globalisation-CPIML/articleshow/681922.cms

Ranganathan, C. (2009, December 10). IT companies will tweak HR policies to remain competitive: Lakshmi Narayanan. *The Economic Times*. Retrieved December 10, 2009 from http://economictimes.indiatimes.com/opinion/interviews/IT-companies-will-tweak-HR-policies-to-remain-competitive-Lakshmi-Narayanan/articleshow/5323977.cms

Shah, G. (1991). Tribal identity and class differentiations: A case study of the Chaudhri tribe. In Gupta, D. (Ed.), *Social stratification* (pp. 288–302). New Delhi: Oxford University Press.

Shiva, V. (2005, May). The polarised world of globalization (A response to Friedman's flat earth hypothesis). *Z-Net*. Retrieved May 7, 2009, from http://www.zmag.org/zspace/commentaries/2299

Singh, J. (2002). From atoms to bits: Consequences of the emerging digital divide in India. *The International Information & Library Review*, *34*(2), 187–200. doi:10.1006/iilr.2002.0194

Singh, T. (2008, Dec 11). Only 13% of rural India has access to telephone. *The Times of India*. Retrieved December 20, 2008, from http://timesofindia.indiatimes.com/articleshow/3820808.cms

Smith, D. (2007). *Hinduism and modernity*. Chichester, England: John Wiley & Sons, Ltd.

Sreekumar, T. (2006). ICTs for the rural poor: Civil society and cyberlibertarian developmentalism in India. In Parayil, G. (Ed.), *Political economy and information capitalism in India: Digital divide and equity* (pp. 61–87). New York: Palgrave Macmillan.

Srinivas, M. N. (1991a). The dominant caste in Rampura. In Gupta, D. (Ed.), *Social stratification* (pp. 307–311). New Delhi: Oxford University Press.

Srinivas, M. N. (1991b). Mobility in the caste system. In Gupta, D. (Ed.), *Social stratification* (pp. 312–325). New Delhi: Oxford University Press.

Thomas, J. J. (2006). Informational development in rural areas: Some evidence from Andhra Pradesh and Kerala. In Parayil, G. (Ed.), *Political economy and information capitalism in India: Digital divide and equity* (pp. 109–132). New York: Palgrave Macmillan.

Tiene, D. (2002, Sept.). Addressing the global digital divide and its impact on educational opportunity. *Educational Media International*, *39*(3/4), 211–222.

Tiene, D. (2004). Bridging the digital divide in the schools of developing countries. *International Journal of Instructional Media*, *31*(1), 89–97.

UNICEF. (2008). India statistics. Retrieved October 23, 2008, from http://www.unicef.org/infobycountry/india_india_statistics.html

Varma, P. (2007). *The great Indian middle class* (2nd ed.). New Delhi: Penguin Books.

Weber, M. (1946). Class, status, party. In Gerth, H., & Wright Mills, C. (Eds.), *From Max Weber: Essays in Sociology* (pp. 114–124). New York: Oxford University Press.

Worst over for Indian economy, may grow at 6.5% in FY10: EIU. (December 10, 2009). *The Economic Times*. Retrieved December 10, 2009 from http://economictimes.indiatimes.com/news/economy/indicators/Worst-over-for-Indian-economy-may-grow-at-65-in-FY10-EIU/articleshow/5324299.cms

ADDITIONAL READING

Afele, J. S. (2003). *Digital bridges: Developing countries in the knowledge economy*. Hershey, PA: Idea Group Publishing.

Ambedkar, B. R., & Moon, V. (1979). *Dr. Babasaheb Ambedkar, writings and speeches*. Bombay: Education Dept., Govt. of Maharashtra.

An elephant, not a tiger: A special report on India. (2008, December 13). *The Economist.*

Basu, T. (Ed.). (2002). *Translating caste.* New Delhi: Katha.

Friedman, T. (2008). *Hot, flat, and crowded.* New York, NY: Farrar, Straus and Giroux.

Gerth, H., & Wright Mills, C. (Eds.). (1946). *From Max Weber: Essays in Sociology.* New York: Oxford University Press.

Guillen, M., & Suarez, S. (2005). Explaining the global digital divide: Economic, political and sociological drivers of cross-national Internet use. *Social Forces, 84*(2), 681–708. doi:10.1353/sof.2006.0015

Gupta, D. (Ed.). (1991). *Social Stratification.* New Delhi: Oxford University Press.

Gupta, D. (2000). *Interrogating caste: Understanding hierarchy and difference in Indian society.* New Delhi: Penguin.

Jondhale, S., & Beltz, J. (Eds.). (2004). *Reconstructing the world: B. R. Ambedkar and Buddhism in India.* New Delhi: Oxford University Press.

Keniston, K. (2002). *IT for the Common Man: Lessons from India. NIAS Special Publication SP7-02.* Bangalore: National Institute of Advanced Studies, Indian Institute of Science.

Keniston, K., & Kumar, D. (Eds.). (2004). *IT experience in India: Bridging the digital divide.* New Delhi: Sage Publications.

Narayan, U. (1997). *Dislocating cultures: Identities, traditions, and Third World feminism (Thinking gender).* New York: Routledge.

Parayil, G. (Ed.). (2006). *Political economy and information capitalism in India: Digital divide and equity.* New York: Palgrave Macmillan.

Saraswati, B. (Ed.). (1998). *The cultural dimension of education.* New Delhi: Indira Gandhi National Centre for the Arts & D.K. Printworld Pvt. Ltd.

Shiva, V. (2005). *Earth democracy: Justice, sustainability, and peace.* London: Zed Books Ltd.

ENDNOTE

[1] Both the central government of India and each state have control over education in India. Primary (10 years) and secondary (2 years) schools are established by each state and provide a free education, as established in the Constitution. The central government also maintains a "central" school system that was established to provide a common curriculum nationwide for children of government employees and other professions who are required to relocate frequently. Additionally, there are many private schools that vary in quality. Universities largely fall under state jurisdiction, while those specializing in business, medicine, science and technology (such as the IITs) are controlled by the central government. More information on India's school system can be found in Baidyanath Saraswati's (Ed.) (1998) collection of essays, *The Cultural Dimension of Education.*

Chapter 2
India's Dalits Search for a Democratic Opening in the Digital Divide

P. Thirumal
University of Hyderabad, India

Gary Michael Tartakov
Iowa State University, USA

ABSTRACT

This chapter seeks to recognize and read the marginal presence on the Internet of India's most oppressed and stigmatized community, the Dalits, simultaneously as acts of resistance and acts of constituting both self and community. It is an exploratory exercise in writing the social history of technological media in general, and the Internet in particular. The medium of Internet, unlike most associated with script and print that preceded it, offers emerging users access without the censorship of either established authorities or canons of authoritative formal regulation. More significantly, it provides the Dalit community, heretofore excluded from all but the most marginal voice in civil society, with an entrance into the national discourse. And quite as important, it furnishes them with their first meaningful media platform for a nationwide internal discourse, which they have previously been denied.

INTRODUCTION: A DEMOCRATIC OPENING IN THE DIGITAL DIVIDE

Most of the time when we think about the Digital Divide we are considering issues of class, the economic divisions of our societies and the social distance between those with access to information technology and those with little or no access. There are, however, other meaningful measures within the perspective of social power available. There exists in India and the United States, and indeed most, if not all, nations, a division that both trumps and parallels class, offering us a valuable refinement worth considering in our view of the

DOI: 10.4018/978-1-61520-793-0.ch002

Copyright © 2011, IGI Global. Copying or distributing in print or electronic forms without written permission of IGI Global is prohibited.

Digital Divide. In the United States we find this refinement when we bring the social construct of "race" into the picture.[1] In India we have a close variant of the same perspective in terms of "caste."[2] In both of these cases we go beyond the simple hierarchy of classes to embrace the case of an antithetical class, a population treated as an enemy of the system as a whole, we can think of as an anathematized, stigmatized or criminalized class, caste or "race" (Tartakov, 2009). In both cases we have segments of the population occupying not only the bottom of the social hierarchy but an identity that is treated as if it was established against the system, and so we may say *doubly divided* from access to all social amenities from essentials like food, housing and employment to more extraordinary opportunities such as information-technology.

Though we include these *most-discriminated-against* in our vision of the society and the Digital Divide among those at the bottom of the class system, they actually play a distinctive role that is somewhat different from the other excluded populations they tend to be lumped into, a role that is worthy our interest. Observing the situation of the anathematized classes not only reveals discrimination at its most extreme—and so the divide at its most extreme—but also models of resistance at their most crucial and so a situation to watch for suggestions about how those denied access are resisting that denial and how we may contribute to that resistance.

India's caste system is a traditional, religious class system that categorizes people into a hierarchy on the basis of their supposed ritual purity as determined by their birth into historically fixed marriage circles, dividing power along the borders of those circles. Though its traditional categories are not precisely equivalent to India's economic classes, they parallel them closely enough to offer useful insights into their social dimensions (Shah, Mander, Thorat, Deshpande, & Baviskar, 2006; Deshpande & Darity, 2003). In India we see access to digital technology largely confined

to the upper three of Hinduism's four great *varna* (macro-castes), the Brahmins and two other "twice born", so-called "Aryan" castes, that compose the national elite, along with a small upper layer of the fourth *varna*, the Shudra. The great majority of the fourth (lowest) *varna*—which comprises the great bulk of the Hindu population—has no access to digital technology. Outside and below the caste system, at bottom of the class system, are India's Dalits, (those defined by the government as Scheduled Castes), the Scheduled Tribes, and the religious minorities of Muslims and Christians, who are largely descendents of those groups. It is since the early 1970s that these people at the bottom of the social system have begun adopting designation of Dalit (*oppressed* or *ground down*), as opposed to the demeaning Brahmanical epithets like *achut* (untouchable) or governmental technical designations like Scheduled Castes. [3] The significance of this new title comes from it being a self-definition based on their experiences, that recognizes the source of their situation in their exploitation by the upper castes rather than a negative definition placed upon them by those seeking to exploit them. What is most important for us may be to realize that Dalits represent seventeen percent of the Indian population or nearly 200 million people. They are a minority but quite a substantial one. If we include Scheduled Tribes among the Dalits, as more and more do, we will be up to nearly 25 percent of the population, or over a quarter of a billion people. [4]

Dalits belong to a fifth category of castes or communities excluded from the traditional four-*varna* caste system.[5] When we look at their treatment it is strikingly comparable to the criminalized status of similarly stigmatized groups around the world: the Burakumin in Japan, the Roma (and earlier the Jews) of Europe, and Blacks in the United States (Thorat, 2009). They are segregated, denied equitable access to health care, employment, education, economic advancement and so digital technology.

Most of the time when we think about the Digital Divide we are considering issues of class from the perspective of the upper end of scale and read the *divide* in a passively neutral manner without associated agency, as if the social or economic distance between those with access to information technology and those with little or no access just happens to exist. In this essay we are going to take a more assertive perspective from the lower end of the scale, and recognize the division as a characteristic of the social and economic walls erected to separate those who have from those who have not. That is, not as a matter of *technological distance* but one of material barriers maintained to defend the class interests of those with access from those they prefer to deny access.[6]

Our subject here is the *democratic opening* in the Digital Divide, and in modern society, that information technology has made available even to Dalits. It is our contention here that the Internet has opened spaces in the walls of the caste system, and so the class system, that Dalits in India and similarly stigmatized classes in other countries, can use to the advantage of those seeking an end to the super-discrimination they presently endure. Through these openings a small but vigorous group of Dalits are using information technologies to transcend barriers of caste in ways not possible before, and thus to take advantage of democratic opportunities that can lead to breaking through caste and the ritual walls to share understandings and interests with each other and from those who have previously been beyond their reach.

Through the fissures that digital technology is opening in the dividing walls, information is seeping and traveling, communicating all sorts of ideas and tactics. The upper and controlling classes are allowed to see the actual state of the Dalits—from Dalit points of view—as they never could before, while some Dalits have been allowed to see and learn from the ruling classes heretofore beyond their view. More important, Dalits have found a means of communicating with each other beyond the control of others. Important for all of us, there has opened up the potential for a digital dialogue through which a democratic interclass exchange becomes possible, something not as possible before even in the written print communications media that was so important to the founding of the modern world (as depicted in Benedict Anderson's discussion of print capitalism) (Anderson, 1983). There is a way now for anyone with access to the information technology to not only read but to *view* Dalits in the words and pictures that flow through the Internet. There is a terrain for dialogue to take place, a terrain of meeting. There is a sort of civil stage upon which a discussion may take place, and what ever sort of civil society exists can be accessed by those kept out [of touch] by earlier class and caste rules and languages.

Though this democratic opening through the barriers of caste and class is only a minor phenomenon today, we see it as valuable already and potentially one of great significance. As such it may provide us with not only some hope for progress but with some guidance in moving ahead toward cogent development of the democratic potential that has drawn our shared interest in the Digital Divide.

Our objective here is to recognize the existence of a significant opportunity that Information Technology, and the Internet in particular, have opened up for even the most marginalized classes of modern society to participate in the national discourses of civil society and to develop a discourse among themselves. We begin by sketching the manner in which India's stigmatized classes have been restricted from access to mass media and continue to offer examples of how they have endeavored to overcome this restriction by use of the Internet. We include a particular notice of the Internet's potential to incorporate and exploit visual imagery.

BACKGROUND: DALITS AND THE DIGITAL DIVIDE

There are two significant issues that need to be addressed with regard to Dalits and the Digital Divide. The first is the lack of access to the Internet for this quarter of a billion people in India's population. The second, our theme here, is the significance —individual, social and transcendental (religious)— of the few thousand mostly, if not exclusively, male Dalits from this population who have acquired access to Internet.

In terms of their actual life priorities, the millions of illiterate and semi-literate Dalits, who make up the vast majority of the community, would consider access to the Internet to be one of the least important items on their agendas. For them there are far more pressing issues to attend to, like shelter, economic livelihood, basic education, and mitigation of the deeply oppressive material and social conditions of their everyday existence. Dalit women command the lowest literacy rates in India[7], Dalit men the highest rates of unemployment. The major political parties—the leftish Bahujan Samaj Party, the centrist Congress Party and the rightist Bharatiya Janata Party—have all attempted to attract their votes through emancipatory ideologies, promises and symbols. Non-governmental organizations have worked with Dalit communities to raise their standards of living and political awareness. Banned radical armed political parties have attempted to mobilize Dalit youth in the rural hinterlands. The community's great twentieth-century leader, Dr. B. R. Ambedkar, is constantly invoked by all of these institutions as an icon symbolizing a commitment to the dream of Dalit emancipation, across all political persuasions.

Over the centuries sections of every non-Brahmanical religious community in India have sought the conversion of Dalits from Brahmanical Hinduism. Christians have proselytized a large number of Dalits into their religious fold. There are many Muslims throughout India, who claim to be Dalits or the descendants of converted Dalits.[8] First-generation-educated Hindu Dalits rarely convert, but there is a growing trend today among the second generation of educated Dalits to adopt Buddhism as their faith, as it delivers them from the caste prohibitions and restrictions of Hinduism into which they have been born.[9] For some this amounts to less a spiritual conversion but more of a theological rejection of caste-mongering Hinduism.

Over the past two generations, a tiny Dalit middle class has developed. Composed almost exclusively of college-educated men with government jobs and only a modicum of economic development, it constitutes only a fraction of this vast socially disenfranchised urban and rural community with, as yet, little presence in the public sphere. While the community as a whole speaks local languages and focuses itself largely on the politics of daily survival, this small fraction, sometimes referred to antagonistically as its *creamy layer*, has taken up the English discourse of India's middle class. And they have used it to enter the ongoing national discourse on civil rights at the same time as they consciously proceed with the process of fashioning themselves. This class has already produced a nationally and internationally recognized Dalit literature of autobiography and poetry in several regional languages (Zelliot, 1992). And it has now embarked on the more directly political project of bringing its community's interests to the nation by way of the Internet, and most powerfully in English.

As they gather online to search among themselves for a way out of their historical predicament, they search to reignite the vision of Dr. B. R. Ambedkar, the community's most historically important intellectual and activist model and leader, who produced a compelling interpretation of civic activism and Buddhism with a growing reach, that has been dismissed as politically insignificant by the main stream caste media. In response, most Dalit collectives and online groups regard mainstream media and politics as unable to transcend their medieval, caste-Hindu roots and meet the goals of either the nation as-a-whole or modernity.

We argue here that Ambedkar's intellectual and activist descendants, the Dalit activists and intelligentsia now beginning to inhabit the Internet, pose valuable material, spiritual and aesthetic questions for themselves and for the world at large. Choosing not to escape into their new class opportunities, a good number regularly invoke the wellbeing of the less fortunate masses of their community. This miniscule but growing Dalit middle class, acting as interlocutors for their broader community, are attempting to provide a non-coercive form of bonding and leadership by which it may move ahead. And, as such, it offers a particularly valuable subject for our investigation and interest.

The Social History of India's Mass Media

What history we have of the mass media in India, since the nineteenth century introduction of the printing press, is the story of a particular structure of affect and privilege of the upper-caste "Hindu" elites. It is a history omitting any mention of either Dalit or lower-caste participation before the middle of the twentieth century, and showing the arrival of Dalits only haltingly after that. To make up for this omission, as some historians have begun only recently to do (for example, see Aloysius, 1997), will not lead to simple additions to what is already recognized, but will demand the subversion of that caste-dominated, historically constituted national history. It will require nothing less than creating a new discourse, that exteriorizes caste and includes recognition of the Dalits, who the Brahmin-dominated discourse considered unmentionable as well as "untouchable."

In 1945, two years before Independence from Great Britain, when modern India was about to be materialized, Dr. B. R. Ambedkar pointed to this incompleteness:

[The Untouchables] have no press and the Congress Press is closed to them. It is determined not to give them the slightest publicity. They cannot have their own press...for obvious reasons... The staff of the Associated Press of India, which is the main news distributing agency in India, is entirely drawn from Madras Brahmins — indeed the whole of the press in India is in their hands...who for well known reasons, are entirely pro-Congress and will not allow any news hostile to the Congress to get publicity. These are reasons beyond the control of the Untouchables. (Ambedkar, 1945, p. 200).

In that context, the situation of Dalits appears in the form of the absence of an Untouchable press and presence of hostility towards making available the voice of Untouchables or any serious consideration of untouchability in the public sphere of colonial India. Two years later in the creation of the modern Indian state, there was more than one kind of violence to be noted. Along with the bloodbath[10] of the partition of British India into *Hindu* India and *Muslim* Pakistan there was the less bloody, but hardly-less violent, constitution of a Hindu India incorporating a caste system epitomized by its unmentioned and unrecognized, "untouchable" Dalits. While the media of the day, in and outside of India, was filled with reporting on the carnage of Partition, they were both relatively silent on the problems of the Untouchables. The unarguable source of this silence was the main news distributing agency, the Associated Press,[11] staffed—as Ambedkar pointed-out—solely by the English-educated 'Madras Brahmins.'[12]

Dr. Ambedkar, along with Mohandas Gandhi and Mohammad Ali Jinnah, may be considered one of the founding narrators of the new Indian nations.[13] At the same moment that Gandhi was spreading the story of Hindu Civilization as an alternative to Western colonization, and Jinnah was creating Pakistan as a Muslim state,[14] Ambedkar, trapped with his fellow Dalits in Hindu India, had to settle for projecting a Dalit emancipation inside an officially secular India.[15]

Beginning in 1920, Ambedkar had launched a Dalit newspaper *Mooknayak* (Leader of the Silent), a periodical that pointed out the political neglect of the Untouchables. Prior to Ambedkar, Ayothee Dass—the most significant Untouchable leader of Madras Presidency—had started a newspaper "Oru Naiya Piasa" (Pandian, 2007). While in the North, Swami Achutanda had produced a newspaper in the early twentieth century. These were marginal attempts at producing and circulating media products to rectify the undemocratic nature of the making of the anti-colonial nation. Anti-colonial nationalism was based on processes of exclusion and the Untouchables were there constituted in the anathematized form of a potentially antagonistic other.

Did things change after Independence? Not really. While the Press remained under private initiative, India's audio and audiovisual media, including the All India Radio and Doordarshan (the national television network), came initially under the control of the Federal Government. There is interesting work on how the first Federal Minister for Information and Broadcasting, B. V. Keskar deliberately pursued a policy of Brahminizing the content of All India Radio (Lelyveld, 1994). Very little work has been done on the nature and social composition of these institutions. Doordarshan, at its height in the mid-eighties, telecast serial versions of the Brahmanical epics, the Ramayana and the Mahabharata,[16] trumpeting yet again a religious construction of the Indian nation as one encouragement among many of the arrival of the Hindu fundamentalism of the emerging Bharatiya Janata Party (Rajagopal, 2001). The Joshi Committee Report (1985)[17] suggested that both these institutions operated from a Delhi centric elite point of view. With regard to the Press, Robin Jeffrey (2003), who has made an extensive survey of the newspaper industry in India, writes

The fact that no Dalit men or women worked in even minor editorial jobs on Indian-language dailies meant that the reality of the lives of Dalits

was neglected. And the fact that no sizeable daily in India was owned or edited by Dalits meant that stories about them were unlikely to receive the constant, sympathetic coverage of stories about, for example, the urban, consuming middle class. (p. 178)

Jeffrey's observation is similar to Ambedkar's but it has come five decades later, and not from one of the occluded Dalits but from an essentially neutral, foreign observer. And, it comes at a time when capitalism has entered into the print industry in a big way. The regional language press grew at an astonishing pace over the period following independence. Those previously known by the Brahmin's condemnation as polluted "untouchables" had risen up and redefined themselves as Dalits (people who had been *oppressed* by the Brahmanical caste system). But little had changed in either the media's personnel or its attitude.

Lacking access to mainstream media over the decades, Dalits have developed their own severely limited media and practices with which to constitute their social identities. One example of this would be the circulation of Ambedkar icons across the length and breadth of the country in plethora of formats, with middle class, politicized Dalits buying Ambedkar-inscribed dairies, calendars, photographs, posters, badges, figurines and all manner of objects for use and display. The commodification of Ambedkar along with his iconization has given currency to a Dalit multiverse, that may give us clues about the discursive character of non-institutionalized forms of opinion formation.

It is not as if the Indian State was ignoring Dalits entirely. The State has declared Ambedkar's birthday on April 14th a national holiday, as he is recognized widely as the principal author and so "Father of the Constitution." For middle and lower middle class Dalits living in major cities like Mumbai, Chennai, and Hyderabad, and even smaller urban spaces, huge rallies and congregations are held to perpetuate the memory of Ambedkar and to celebrate the solidarity of the community. Most of the Dalits who participate, work in public sector enterprises and affirm a *Dalithood* in both an ontological fashion and in an epistemic fashion [a real and a theoretical fashion]. Books relating to Dalit issues by Dalit authors and others, calendars, badges, flags, pens, pen-stands, rings, and all manner of trinkets and instruments alluding to Ambedkar circulate freely on that day[18] and on other occasions and venues where Dalits gather in demonstrations of solidarity.[19] It is true, however, that this recognition has not extended to access to national print or broadcast media.

The single, quite modest, contradiction to this situation may be found in film. As some chroniclers of popular media have observed, the popular cinema has thrived on an undifferentiated audience suggesting the medium's accessibility to the lower caste communities. Indeed cinema may be the only medium where there have been no strict economic barriers to a thoroughly heterogeneous audience. In postcolonial India, cinema claims a place for mobilizing the masses politically. In South India especially, for those deprived of entitlements, cinema becomes the ground where they constitute and participate in popular governance. They become subjects of their own making. Though there has been little notice taken by Bollywood, a number of film scholars have noted an interest in issues of caste and even Dalits to be found in the South Indian cinema (Prasad, 2004). Of late, there have been a number of visual documentaries being commissioned by various development agencies to make ethnographic films on Dalits and Adivasis.[20] The middle class genre of documentary traces its history in the work of people like Anand Patwardhan and others.[21]

Internet and Dalit Cyberspace

India has over 80 million Internet-users and is ranked fourth globally.[22] For most Indians the Internet has yet to become banal and ordinary. Even in these days of global recession, the Internet remains suggestive of the mixed consequences of liberalization and technological development. It may not be wrong to say that for Indians the history of Internet fits smoothly into the history of liberalization and the entrenchment of upper castes and elites in the postcolonial state. The Satyam Industries scandal not withstanding, the Information Technology sector has been seen as synonymous with the merit and dynamism of India's upper caste elite,[23] in contrast to the supposed corruption and inefficiency of the federal and regional government bureaucracies and the public sector industries, subject to the nation's affirmative reservation policies.[24] Until the scandal, various levels of government guaranteed support and legitimacy for the growth of the IT sector.[25]

As of the first decade of the twenty-first century it is only a few thousand, mostly-urban male Dalits with minimal English and technological competence, who have discovered the Internet as a public and a private medium. For those few privileged Dalits it has become an unsurpassable novelty. For most Dalits not just the Internet, but the English language, urban culture and modern lifestyles remain restricted in access (Thirumal, 2008). But for these few educated Dalits, the English language and the medium of information technology, with their instantaneous access to the production and consumption of texts and images, are features of the cyber world they have been keen to appropriate and use it for individual, social and political purposes. Thus, members of a prototypically subaltern community have discovered that access to the Internet has opened access up to a vital political and social terrain unlike any ever available to them before.

Going beyond use of the net as they have found it, these few-but-energetic intellectual activists have created online sites addressing Dalit issues. There, besides a defensive articulating of human rights against exclusionary and discriminatory practices based on caste, a good number of websites and Internet groups run by Dalit collectives and Dalit organizations have sprung up to engage in a vigorous questioning of the normative structures of Indian modernity. There are websites demanding not merely procedural justice but substantive justice. Hence, we find Dalit websites such as the portal of National Campaign on Dalit Human Rights, [26] Dr. Ambedkar and His People,[27] and 'Scholars Without Borders'[28] supporting the development of affirmative action. These issues can be classified into two categories, agitating at the same time both for political recognition and redistributive justice.[29]

Although Dalits may not find a substitute in the Internet for mainstream media like newspapers, radio, television[30], and cinema, the Internet does constitute a locus for widely disseminated communication with significant ideological independence and first class production values; it is indeed a spectacular, new and potentially powerful intellectual terrain. On the Internet people can express themselves beyond the control of caste censorship,[31] on topics ranging from social vision to the most intimately embodied pain, and for anyone looking on, their words appear as crisply polished on the pixilated pane of the computer monitor as anything published by the most established and well-financed newspaper or magazine. Their pictures are as clear as any magazine or documentary films. The Internet provides a legitimizing format that could not be afforded previously in the other media. And it can be read all over India, indeed throughout the English-speaking world.

Production-values on the Net offer an interesting phenomenon. On one hand, as has just been pointed-out, texts and graphics on the net are strikingly finished and professional looking on their surfaces. People whose handwriting, speaking voices and wardrobes might need class-bound enabling in other media can appear quite polished on the computer screen. Publications that would look sub-standard if published on inexpensive materials by modest technology can appear quite equal to any other on the monitor. At the same time, texts or illustrations, which other media might require to pass through editorial or

production gauntlets, are freed from the threat of exclusion by upper caste gatekeepers.

One crucial contrast to this parity of production values is found in the realm of English syntax. To make yourself understood on the Internet, as in daily speech, one is not confined to the elite's English or as we might put it in the Indian caste-bound context: an upper-caste editor's English. Semantic English is good enough for most purposes. On the net are wide realms where your language is judged as it is in the coffee shop or on the street. If you can be understood, you are 'good to go'. Higher style may get a more sympathetic viewing in some quarters, but irregular grammar, punctuation or style is of relatively little matter in many contexts. The semantics of getting out what you intend to be understood is what counts and the perfection of the syntax matters relatively less. Where proper English is hemmed in with proper breeding, Internet English can do pretty much what it needs and wants to, without worrying about who's "gonna" find it good enough for them. Unlike the elite in India, Dalits have rarely studied in the public schools where they could acquire enough proficiency in the finer niceties of the English language. But since the Internet displays a certain disregard for the etiquettes of literary and journalistic English, the use of a semantically comprehensible—albeit sometimes syntactically unfriendly—English opens up a symbolic universe populated with liberal notions of citizenship and democracy, without requiring their being dressed for a proper middle class dinner table. On the Dalit use of English in the Cyberspace, Thirumal and Melinda observe:

The distinction between semantic grace and syntactic grace in writing the English language presently eludes the literate Dalit. Fortunately, the genre of e-mail demands an etiquette, which makes this distinction fuzzy. It is this nature of the medium and the language that the Dalits have been able to exploit. For instance, it would be difficult for educated Dalits to write for a mainstream national newspaper or a recognized academic social science journal because these demand journalistic, academic and linguistic protocols that the Dalits are yet to acquire.[32]

The Internet for Dalits who can access it has become an important social medium. The cultural capital needed for networking in Cyberspace is equally receptive to Dalit users and consumers. It is true that Upper Castes use the media more frequently, and possibly for a wider range of purposes including commerce and sustaining immigrant emotional and family ties. The importance of Dalits' use of the Internet includes their ability to constitute a Cyberspace outside the realm of the Hindu cosmic order and the nation-state. Thus, for the Dalits, the Internet offers a terrain for exploitation of their community interests in social activism that is relatively casteless, nationally, and even internationally, extensive and so potentially useful in ways no previous medium has been before.

Communication and Caste

Communication in a democratic society presupposes a theory and practice of equality. Caste is a theory and practice of inequality. If successful communication is about mutual rational acknowledgement, it is important to note that neither mutual nor rational acknowledgement is possible in a caste-based society. For there to be democratic development in society, it becomes necessary to find locations and means in which true communication can flourish. One thing Dalits need for their development in the India, where the majority has condemned them through caste, is a place for thinking and acting that is outside the caste system and outside communications systems dominated by the caste system. If Indian democracy is to develop beyond the most minimal level, the rest of the nation needs this place as much as Dalits.

The Internet offers a space for the constitution of a Dalit self from the outside through law and from the inside through appropriation of Buddhism or another non-discriminatory religious or transcendental practice.[33] The Dalit self that is under construction today demands recognition from non-Dalit others through forms of sociality that accord dignity through the process of law, as envisioned in the Constitution, and is still in the process of constituting non-legal forms of social and religious relations. Speculative thesis on what ought to be the practices of the self, along with corresponding practices of personal association, may inform many of the Dalit Internet practices. Prevention of negative practices of association through the enforcement of right against harassment and atrocity inform a significant proportion of Dalit activism on the Internet. But resort to normative modernity may not be the sole anchoring point for all time to come.

Here we are identifying a small section of the stunted community of middle class Dalits who are seeking to break through the regime of exclusion and isolation the world of caste has enforced upon them, by means of their access to the world of digital communications. In their reworking of the regime of exclusion, they appropriate the language of modernity and command known as the Law, through their haltingly acquired powerful medium of English. As both the language of law and the use of English provide self-definition from outside, the emerging emphasis on Buddhism for some appears to summon a discipline from within. This complex social use of Internet may be considered as an attempt to fabricate a utopia around practices derived from law on one hand and religion on the other to re-construct the outer and the inner self of the Dalit.

This new Dalit-being is constituting itself in part around these new texts in this new virtual terrain. Is there a logic that the Dalit institutes towards this textuality of word and image?[34] Does the self of the Dalit stand relinquished the moment he or she gets seduced by the word, text or visual image? For the Dalit, the self that mediates between the body and the consciousness requires a communication situation that disrobes the invisible structures that support a world of naturalized inequality within which they live most of their material lives.

The spoken word, in India's many regional vernaculars, continues to be the major medium of interaction for the majority of Dalits. But, among these few middle class Dalits, most of whose economic status is based on their English medium education, English is the common language and the Internet is the medium of communication. The Internet extols the possibilities of transcending 'caste' and recasting an encumbered 'historical individual subject' through a strategic maneuvering of the institutions of law, English language and the religion of Buddhism. In other words, among other things, the Cyberspace is deployed to produce an unencumbered Dalit vis-à-vis the unmarked citizen—a Dalit freed from the chains of caste, which is something different than an undifferentiated citizen.

What the access to Cyberspace has offered the Dalit, who must spend her or his normal working day in the material reality of caste dominated social and intellectual interaction, is access to a digital terrain in which they may develop their personal and social skills of language and thought, argumentation and exploration, invention and self-development. With other Dalits they may explore who they are and who they want to become. With those beyond the world of Dalits they can explore their actual personal experiences and the reconstitution of the world they share in ways never before possible. In Cyberspace they can explore what it means to be a Dalit and what it means not to be a Dalit.

The Entitlements of a Stable Addresses

To be a recognized citizen subject in a modern nation state it is necessary to possess a fixed (legally recognized) address. Such citizens are legal claimants and belong to civil society. The sites of a home, office, club, worship place and other kinds of habitation involve the invocation of a fixed address. A fixed address is needed to procure ration cards for groceries, gas connections, bank loans, employment and education, voter card for electoral purposes, passports for travel abroad and so on. In short, it is needed for entry into the active zone of civil society. In third world societies like India, a major chunk of the population does not possess a stable address and so they do not fully—or so *truly*—belong to the civil society. Thus one element of their most basic politics consists of transforming their existing tenuous addresses into concrete addresses, because that inscription in a place produces entitlements necessary for citizen subject status.

Possessing the entitlements that depend upon an address has currency not only in the material realm, but also in the cultural politics of recognition. Those in a population with only tenuous addresses—such as those living in the segregated Dalit "colony" of a rural village[35] or an illegal urban slum—maintain more the identities of their traditional communities than those of the citizen subjects of the modern state. They are defined by their fellow-citizens—*and even themselves*—more as members of the communities they come from than as individuals or citizens. India's Dalits fall largely into this situation, considered Dalits before they are recognized as individuals or citizens. Thus every Dalit manifesting a personality on the Internet adds an individualized citizen to the civil world that has previously treated Dalits as a faceless depersonalized category.

While writing and print culture inscribe territoriality as a fundamental requirement of the nation-state, the new media have the potential to add cyber location to physical location, or street address, as a site for exercising democracy. An e-mail campaign is fought in virtual space, and cannot be regulated by the usual conventions of the democratic nation-state. The Dalit elite—through access to the English language, the Internet and human rights discourses—is mobilizing a population that is severely disenfranchised in terms of access to legal addresses and their entitlements (Thirumal, 2008). While most of the Dalit population is constantly in search of jobs within restricted rural geographies or engaged in migration to metropolitan centers, the activist middle class is likely to possess both postal and cyber addresses.

Cyber-savvy Dalits now have the Internet for campaigning against cases of atrocities like those at Khairlanji or Navatan (Teltumbde, 2008; Rawat, 2008). These Dalits are able to make the boundaries between civil society and political society porous. As campaigners they use the coercive power of law to lay claim for dignity and freedom even as they realize that it is a law that has previously been manipulated to deny their claims to basic human rights (Narula, 2008, pp. 25-28).[36]

Establishing stable addresses for the mass of impoverished Dalits remains a defining task at this historical juncture, along with such basic essentials as stable livelihoods, medical care, weather-proof housing, adequate schools and gender equity. As the struggle for these essential human rights proceeds, it is our contention that there is a role for the nascent Dalit middle class to play in Cyberspace. So far, it appears that middle class Dalit mobilization has focused more on a politics of recognition than on redistributive justice, but both are growing. Acquiring the access to a cyber address offers the opportunity of an intervention at the level of civil debate and popular politics that has not been available in the past, and with it the opportunity not only to be heard and seen, but to require a response.

It is our belief that the use of the Internet by the small but growing Dalit elite offers a significant step toward the citizenship the entire community is searching for, both in terms of a place from which to address the nation as citizens and as a location in which their legitimacy as citizens can be established.

Visual Imagery

Though we don't think of it automatically when we are considering Information Technology, visual imagery has been a part of popular (mechanically reproduced) media from the beginning. Printing of pictures is as old as printing of words. Both forms were born with the earliest coins and continued with the woodblock printing that was among the earliest forms of mass reproduction. Throughout the nineteenth century pictures appeared along with writing, in the form of drawings and paintings reduced to line cuts, halftone engravings and lithographs. By the later part of the century, we see photographs, with their striking air of particularly authentic-seeming documentation. What we have not always considered, as we may here, is how such visual images convey a somewhat different quality of information, or evidence, than written descriptions, simulating authenticity and intuitive availability in ways that verbal descriptions cannot approach.[37] It is one thing to read a description of a historical figure or a personal relative, and quite a different thing to see a drawing or painting of them. Then there is the leap to the photograph. By the middle of the twentieth century cinema documentaries and newsreels added an even greater suggestion of untampered and deeply penetrating simulation of reality.

It is true of course that photographs, even moving pictures, can be staged, or edited, or labeled to suggest things that are not, or were not, as they appear or have been portrayed to be.[38] And it is equally true that knowledge of this, if not as well or widely understood, has been understood by many. But that hardly matters. Before the spread of digital technology at the end of the twentieth century, when the reverse awareness—that photographs are as vulnerable to manipulation as writing—most people assumed automatically, if not pragmatically, that photographs were indeed proof of the realities they portrayed. Even with full knowledge that particular images have been altered, it is difficult not to believe them to portray reality to some extent.

Mohandas Gandhi, London-trained barrister though he was, conducted his Congress Party politics in a peasant dhoti and had himself photographed that way, beside the *charka*, spinning wheel, that became the symbol of national independence, while his Dalit critic, Dr. B. R. Ambedkar, was recorded in the international business suit of a Bombay lawyer. These were pictures of national leadership the likes of which the people of India had never witnessed before. They did not depict fabled beings from another world, but men who in surprisingly intimate ways resembled themselves. It's easy to say "a baniya in a dhoti"[39] or "a Dalit in a business suit", but seeing them is something else altogether, something far more personally compelling. Something one may identify with in tangible terms that far exceed anything on the written page. Additionally, these images are important in different ways depending on the community with which one identified.[40]

Visual images play a role in our thinking somewhat differently than verbal images in two major ways. First, they have a presence, and convey a verisimilitude and apparent psychological depth that verbal records cannot match. Verbal records can tell us what someone thought they saw; we respond to visual images intuitively in terms of what we ourselves think and feel. Second, unlike the concrete rationality that forms the basis for verbal communication, pictorial images float a bit more rationally and emotionally free from their contexts. When sentences express thoughts that conflict with our previous knowledge we are immediately alerted to the contradictions, but when pictures do, we tend to ignore the contradiction while absorbing the new data.

The major historical facts concerning Dalits and visual imagery are hardly different than the reality discussed above for textual print media. Dalits were largely invisible. Everyone who has seen India's traditional art knows what the gods and Brahmins look like, along with the forest creatures of the epics and even the fantastic beings of artistic imagination, like lionine kirtimukha and half-bird half-reptilian makara. But they don't likely have such a clear idea of what the Chandala (the Dalits of the ancient tradition) looked like. Nor do modern newspaper readers and film goers have much idea of what Dalits look like outside of the exceptional example of Dr. Ambedkar or the stereotype of the victimized slum dweller. The modern media portrays India as casteless, with Brahmin priests. This is pretty much the same approach as found for African Americans in the United States before the civil rights advances of the later twentieth century. In both places the only time the stigmatized classes reached the newspapers was as victims of an atrocity[41] too dramatic to ignore, or as particular individuals who accomplished something altogether unique. Otherwise both "Black" and "Untouchable" meant *invisible*.

But what happens when Dalit graphic designers and photographers escape the control of caste-bound editors? They are able to show and share imagery that the caste-based media has shunned. They are able to share imagery with each other like the archives of Ambedkar's history and pictures of Dalit history and daily life, to which they have had little or no access previously. And they are able to share imagery of their current struggles with those outside the community who would otherwise be less able to understand their experience.

As the widespread publication of Ambedkar's written work in the 1990s spawned a vast and growing literature on the subject of Dalit experience,[42] opening up a wave of self-awareness and activism, so the pictures of Ambedkar archived and available on CD-ROM and on the web have augmented that wave, providing both imagery for its illustration and for the further consideration of what it can means to be a Dalit intellectual and activist. Along with his collected written works we can find Ambedkar's life displayed in the 304 photos of him and associates at the Ambedkar.org website[43], or on the CD-ROM.

Figure 1. Dr. Ambedkar perusing papers during the making of India's Constitution

Those images are now available and being used to make posters and book covers that not only stand out on book shop displays, but on the tables and on the walls of Dalits who previously had no imagery of their own community to display and think about, or share with each other. An example of such is image 68 from the site (Figure 1).

If this is the first image the average international reader of this essay has ever seen of an "untouchable", we may suppose it is not what they envisioned previously. If it is an image printed out at an Internet café or purchased at a fair, we can think about what it may mean, attached to the wall above a young Dalit's bed, or in the room where she or he sleeps.

A second use of visual media is for communication between Dalits and the rest of their world. It is virtually impossible for the discussions of atrocities to have a meaning in words alone that they can in words and pictures. It quite as true, and quite as important to acknowledge, that it is just as unlikely for pictures of an atrocity, or anything else to have the meaning they deserve without accompanying texts. Pictures are not only worth the proverbial thousand words. To be understood by anyone not already intimately familiar with the situations they record, they require texts explaining them, identifying the figures and explaining the social relations recorded in them.

The caste atrocity that took place at Khairlanji in Maharashtra on September 29, 2006 is only one instance of a constantly recurring phenomenon of multiple lynchings one reads of in India's newspapers weekly, and occurring daily around the nation. What this means is sometimes explicitly detailed in the news articles, though more often than not it may seem to be there more for the sensationalism than the reality.[44] But so common are such occurrences and so outside most people's normal realm of experience, photographs have the ability to not only give their unique individuality, but to bring them to life. For example, when exploring certain web sites[45], one will find images akin to the one depicted in Figure 2, which is not at all the most disturbing the may be found.

One does not need to know the famous American case of the Emmitt Till photographs published in *Jet* magazine, (July 23, 1964), one of only a few national publications controlled by the African American community in the 1950s,

Figure 2. Mutilated bodies of son and mother, Khairlanji. (© 2006, Manuski Advocacy Center. Used with permission.)

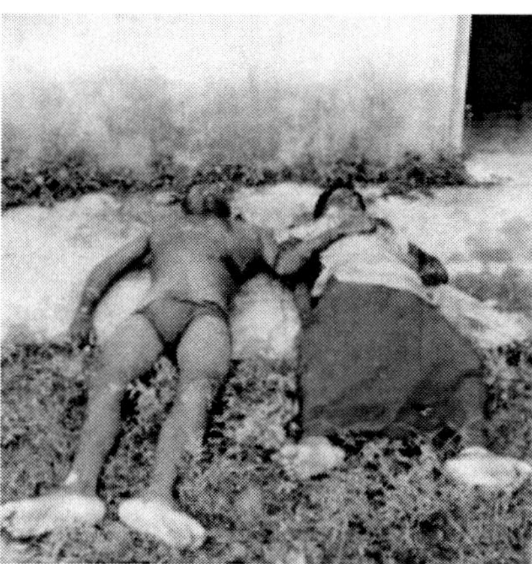

to know what a powerful tool the photograph of a taboo-subject can be for a dominant community's education and development. The essential issue here is "community-ownership." Such pictures aren't as likely to come from or be broadcast from those who do not want to them seen.

The following is the fourth among the 470 responses to "Khairlanji news pictures" found on one website. On October 24, 2006 the writer said:

The dasterdly act indeed was only reported in some Marathi daily but the brahmanical media never gave any space to this incident. The way many react to Jassica and Matto case why the very same people do not take up the issues of this nature where the atrocities are committed in living day lights. Though all the political parties organized a protest but it remained a mere ritual. The ruling congress and NCP have once again shown their colours and the perpetuator of this crime should be killed in front of every one and let the so-called Hindus witness such events.

Such blogs as the one this response comes from, exist due to the interests of the Dalit community and those concerned with it. [46]

Even a simple ID photo (Figure 3) brings the personality and the humanity of the victim to us, as writing her name, "Priyanka Bhotamange", can only hint. It is for this reason that we want to remind ourselves—as too often it may be forgotten—that the power of the Internet goes beyond the printed word to the visual image, which is to say a good ways beyond the abstraction of words to the concrete features of a particular woman we cannot see without identifying with as a fellow human being.

In relation to pictures, even more than with print, the technical quality and availability of visual imagery on the net has a significant ability to legitimize and equalize the rhetorical contest of ideologies the Democratic Opening creates. As semantic speech tends to rival syntactic speech and nullify the hegemonic community's advantages of class tradition in language, Dalit graphic and photographic imagery can hold its own on the net in technical terms, as it never could in the other media, where economic power over-rules all else and Dalit-related imagery has been invisible. In the world of the bourgeois market place the economic power of the dominant community can out-produce most anything the subaltern community can muster by means of higher production values based purely on financial investment. But on the net, basic production values suffer little by comparison with the most expensive.

FUTURE RESEARCH DIRECTIONS

Looking ahead from the standpoint of the discussion here we see four main directions for further investigations of the Dalit interest in this democratic opening into the world of information technology. As far as the world of mass information media is concerned, there continues to be a profound difficulty in accessing print media for Dalits. For this reason we are concerned to understand better how the accessing digital media

Figure 3. Priyanka Bhotmange (© 2006, Manuski Advocacy Center. Used with permission.)

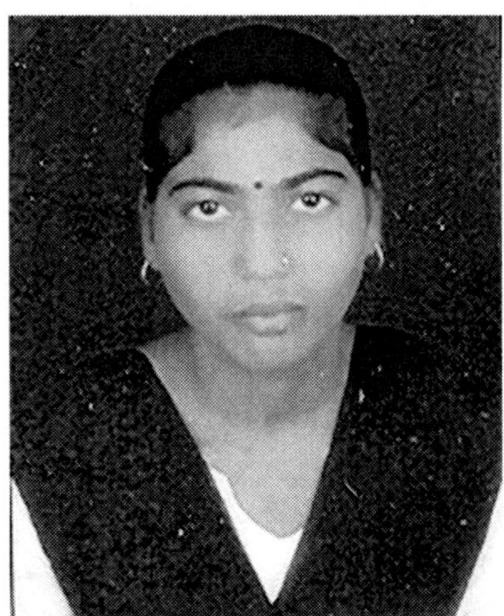

affects Dalits' access to analog media, and vice versa: how Dalits accessing of print, audio and visual media will effect their use of digital media.

From a practical social perspective, it will be important to ascertain what difference access to literacy and print media have actually made to the lives of Dalits and their community and what is the spillover of this on Dalit use of digital media will be.

Further, as the miniscule Dalit middle class we have been discussing becomes the interlocutor between the State and the rest of the disenfranchised Dalit community, we will want to know if this role leads to their monopolizing the material and symbolic resources available to all Dalits, as conservative critics have warned? That is, whether or not there is a problem in developing a cadre of college educated Dalits.

And finally, reaching beyond the political realm we have discussed here, what are the non-Ambedkarite strategies for mobilizing Dalits? Are there such strategies available for Dalit activists on the Net in places like Bihar, Karnataka and Eastern India, where an Ambedkarite approach has found only limited footing? In this case, where such a strategy has not evolved previously, will the evolving internal Dalit discussion, that access to the Net has opened up, lead to a radical re-interpreting caste rather than the approach taken by Ambedkar's followers, of annihilating Caste?

CONCLUSION

The digital divide between the impoverished *have-nots* and the digitally empowered *haves*, can be found in India as in every nation around the globe. As an anathematized class, India's Dalits represent a most-impoverished segment among her have-nots, a social status that can also be found in most if not every nation around our digitally wired modern world. The point of this chapter has been to call attention to the reality of the democratic opening made by Dalits in India's Digital Divide, through which activists of this otherwise subordinated and walled-off minority community have made use of Information Technology in significant ways, to communicate with each other and to reach out to the upper castes that have until now been largely if not entirely beyond their sphere of communication.[47]

In India today, a microscopic section of the nation's vast subordinated most-exploited minority is using the Internet as a means to both develop its community's internal potential and to alert the nation's digitally-connected classes to its interests and the nation's as well. It is our contention that this "Democratic Opening in the Digital Divide" can be exploited in India and every other nation around the world where the digital divide exists. This is not to say that we have found an answer to the deep and significant social problems that the Digital Divide both reveals and exacerbates through its existence. But it does demonstrate that like some other technologies that divide the exploited classes from the exploiting classes around the world, it is not the technology itself but our use of it that counts.

The uses that India's Dalits are making of the Internet suggest ways in which members of the exploited classes and their democracy-seeking allies may use the Internet to further the causes of democracy and equality of opportunity in every nation where a significant subordinated minority is struggling to be recognized and understood.

Depending on the context of its use and who is using it, technology has as great a potential to ameliorate historically produced societal inequities as it does to exacerbate them. India's Brahmanical caste system is a historically produced institution that has questioned itself only slightly across the years through its literary, academic, journalistic and entertainment media. This essay considers a growing, if still rather limited struggle by a small number of Dalit activists, to face the wider Indian society with an uncensored critique produced by the previously voiceless upon whose backs they have for too long progressed without hearing their words or seeing their faces.

REFERENCES

Aloysius, G. (1997). *Nationalism without a nation in India*. New Delhi: Oxford University Press.

Ambedkar, B. R. (1945). What Congress and Gandhi Have Done to the Untouchables. Bombay, Thacker & Co. Reprinted in Vasant Moon, (Ed.), *Dr. Babasaheb Ambedkar Writings and Speeches*, 9. Bombay, Government of Maharashtra, 1990.

Anderson, B. (1983). *Imagined communities: Reflections on the origin and spread of nationalism*. London: Verso.

Deshpande, A., & Darity, W. (2003). Boundaries of clan and color: An introduction. In Deshpande, A., & Darity, W. (Eds.), *Boundaries of clan and color: Transitional comparisons of inter-group disparity* (pp. 1–13). London: Routledge.

Ganguly, D. (2005). *Caste and Dalit Life Worlds: Postcolonial Perspectives*. New Delhi: Orient Longman.

Jeffrey, R. (2003). *India's newspaper revolution: Capitalism, politics and the Indian-language press*. New Delhi: Oxford University Press.

Lelyveld, D. (1994). Upon the subdominant: Administering music on All-India Radio. *Social Text, 39*, 111–127. doi:10.2307/466366

Narula, S. (2008). Equal by law, unequal by caste: The "untouchable" condition in critical race perspective. *Wisconsin International Law Journal, 26*, 255.

Pandian, M. S. S. (2007). *Brahmin Non Brahmin: Geneologies of the Tamil political present*. New Delhi: Permanent Black.

Prasad, M. (2004). Reigning stars: The political career of south Indian cinema. In Fischer, L., & Landy, M. (Eds.), *Stars: The film reader* (pp. 97–114). New York: Routledge.

Rajagopal, A. (2001). *Politics after television: Hindu nationalism and the reshaping of the public in India*. Cambridge: Cambridge University Press. doi:10.1017/CBO9780511489051

Rawat, V. B. (2009). *Fire incident in Deoria, Mushhar families hurt: Please help*. Retrieved August 11, 2009, from http://awareness-2009. blogspot.com/2009/03/fire-incident-in-deoria-mushhar.html

Shah, G., Mander, H., Thorat, S., Deshpande, S., & Baviskar, A. (Eds.). (2006). *Untouchability in rural India*. New Delhi: Sage.

Tartakov, G. M. (2009). Why compare Dalits and African Americans? They are neither unique nor alone. In Natrajan, B., & Greenough, P. (Eds.), *Against stigma: Studies in caste, race and justice since Durban* (pp. 95–140). Hyderabad: Orient Black Swan.

Teltumbde, A. (2008). *Khairlanji: A strange and bitter crop*. New Delhi: Navayana.

Thirumal, P. (2008). Situating the new media: Reformulating the Dalit question. In Gajjala, R., & Gajjala, V. (Eds.), *South Asian Technospaces* (pp. 97–122). New York: Peter Lang.

Thorat, S. (2009). Caste, race and United Nations' perspective on discrimination: Coping with challenges from Asia and Africa. In Natrajan, B., & Greenough, P. (Eds.), *Against Stigma: Studies in Caste, Race and Justice since Durban* (pp. 141–167). Hyderabad: Orient Longman.

Weisskopf, T. (2004). *Affirmative action in the United States and India; A comparative perspective*. London: Routledge.

Zelliot, E. (1992). *From Untouchable to Dalit: Essays on the Ambedkar Movement*. New Delhi: Manohar.

ADDITIONAL READING

Anand, S. (2005). Covering caste: Visible Dalit, invisible Brahman. In Rajan, N. (Ed.), *Practising journalism: Values, constraints, implications* (pp. 172–197). Thousand Oaks, CA: Sage.

Aquil, R., & Chatterjee, P. (Eds.). (2009). *History in the vernacular*. New Delhi: Permanent Black.

Axford, B., & Huggins, R. (Eds.). (2001). *New media and politics*. Thousand Oaks, CA: Sage.

Bayly, S. (2002). *Caste, society and politics in India from the eighteenth century to the modern age*. Cambridge: Cambridge University Press.

Bhargav, R., & Reifeld, H. (Eds.). (2005). *Civil society, public sphere and citizenship dialogues and perceptions*. Thousand Oaks, CA: Sage.

Blackburn, S. H. (2003). *Print, folklore and nationalism in colonial south India*. New Delhi: Permanent Black.

Bourdieu, P. (2005). *Language and symbolic power*. Cambridge, MA: Harvard University Press.

Chatterjee, P. (2004). *The politics of the governed: Reflections on popular politics in most of the world*. New Delhi: Permanent Black.

Cordner, C. (2008). Foucault, ethical self-concern and the other. *Philosophia, 36*, 593–609. doi:10.1007/s11406-008-9138-4

Febrve, L., & Martin, H. J. (2006). *The coming of the book; The impact of printing 1450-1800*. London: Verso.

Fraser, R. (2008). *Book history through postcolonial eyes, Rewriting the script*. London: Routledge.

Jaffrelot, C. (2005). *Dr Ambedkar and untouchability*. New Delhi: Permanent Black.

Naregal, V. (2001). *Language politics, elites, and the public sphere*. New Delhi: Permanent Black.

Natrajan, B., & Greenough, P. (Eds.). (2009). *Against stigma: Studies in caste, race and justice since Durban*. Hyderabad: Orient Black Swan.

Preston, P. (2001). *Reshaping communications, technology, information and social change*. Thousand Oaks, CA: Sage.

Rogers, E. M., & Singhal, A. (2001). *India's Communication revolution: From bullock carts to cybermarts*. Thousand Oaks, CA: Sage.

Thirumal, P., & Smriti, K. P. (2004). Print and digital imaginations: The post colonial Shudra. *Journal of Karnataka Studies, 1*, 133–158.

Thorat, S. (2009). *Dalits in India: Search for a common destiny*. Thousand Oaks, CA: Sage.

Vasudevan, R. (Ed.). (2001). *Making meaning in Indian cinema*. New Delhi: Oxford University Press.

ENDNOTES

[1]   In writing of "race" here, we are referring to the social construct recognized in everyday

speech, and do not suggest that race exists as a biological reality.

2   By caste, we are referring to the Brahmanical categories of *varna* and *jati*.

3   This was a conscious ideological assertion parallel to, and possibly inspired by, the decision of African Americans in the U. S. to take Black as a self-chosen title in contrast to designations put upon them by dominant Whites.

4   With most Christians and a growing number of Muslims now beginning to refer to themselves as Dalits, these numbers are increasing.

5   In traditional texts it is normally called the *chaturvarna*, the "four castes."

6   And for this reason, we will write of an *opening* in the divide rather than a bridge over it.

7   As per the National Literacy Mission data (1991), the literacy rate for the SC women is 23.76 per cent, against the national average of 52.21 per cent. There is an elaborate table appended to this online document. See http://www.nlm.nic.in/tables/f_scst.htm

8   Sky Baba, a Telugu poet is an important emerging ideologue of a Muslim Dalit identity. At Patna there has emerged a political party calling itself the All India Backward Muslim Morcha. In New Delhi, on November 18, 2009 Asia News headlined an article, "Thousands of Christian and Muslim Dalits march against discrimination." See http://www.asianews.it/index.php?l=en&art=16894&size=A

9   The set of disabilities that mark the Dalit condition are an explicit element of Brahmanical Hinduism and explicitly rejected by Buddhism.

10   The Partition of British India into India and Pakistan is considered one of the most violent events in twentieth century post-colonial history.

11   India's Associated Press was re-christened the Press Trust of India in 1947.

12   See Pandian (2007) for more on Madras. On the intolerant attitude of the Madras Brahmin towards the controversial "Non-Brahmin Manifesto" of 1916 issued by the nationalists in Madras, the Brahmin newspaper, *The Hindu*, quipped, "We do not wish to open our correspondence column to a discussion on this subject, as it cannot but lead to acrimonious controversy and as it would promote the invidious object of some of those who are engineering the movement" *The Hindu*, 20 December 1916, quoted in S. Saraswathi, *Minorities in Madras State: Group Interest in Modern Politics* (Delhi: Impex India, 1974), p. 42. Unlike the strict silence maintained against Untouchables in national dailies during the colonial period and afterwards, Pandian points out: "Soon *The Hindu* had to open its columns to discuss and criticize claims made on the basis of non-Brahmin identity" (Pandian, 2007, p. 3).

13   As the leader of India's Scheculed Castes, he represented quite as many as Jinnah, the leader of British India's Muslim community.

14   Though the caste system is essentially a Brahmanical Hindu institution, the fact that most of India's other religious traditions have developed with a Hindu- dominated world and most of their populations are descended from converts from Hinduism has meant that they have inherited the Brahmanical tradition of castes and untouchability. One of the contradictory outcomes of this situation can be seen in the Muslim-Pakistani declaration of untouchable scavengers, a special class who were not allowed to migrate to India during partition, and thus untouchability remains in Pakistan to this day.

15   Indeed he has written this projection into the new nation's very constitution, of which he was the primary author.

16 Ramayana and Mahabharata are two epics of the Indic civilization. Often, they have been used to construct a sacral community and the Indian nationalists, including Gandhi, deployed the epic Ramayana for imagining the modern nation.

17 See MIB (Ministry of Information and Broadcasting) (1985), *An Indian Personality For Television: Report Of The Working Group On Software For Doordarshan.*

18 Visitors witness celebrations of Ambedkar Jayanti (birthday) on Tank Bund Road, Hyderabad. Men, women and children of lower and middle class Dalits throng the public space on this day. They touch the feet and garland Ambedkar's formidable statue that stands in the middle of an otherwise busy public street.

19 Among the most prominent of these are the celebrations of the Navayana Buddhists, besides the anniversary of Ambedkar's birth, are the anniversary of Ambedkar's Paranirvana (passing away), Buddha Jayanti and Diksha Divas, the inauguration of the great conversion. V. S. Naipaul begins and ends his *India, A Million Mutinies from Now* with a [misdated] description of the day celebrating Ambedkar's Paranirvana in Mumbai.

20 The official nomenclature for Adivasis is referred to as Scheduled Tribes. Their percentage is 7.5% of the total population of the country. *Adivasi* means the original inhabitants.

21 For a look at the censorship of Patwardhan's approach to Dalits, see: http://www.patwardhan.com/writings/press/082402.htm

22 See http://trak.in/tags/business/2008/12/03/with-81-million-Internet-users-india-is-ranked-4th-globally/ dated 16/04/09. Its ranking is based on Internet users and not on connections.

23 The recent scam of a huge IT corporate sector Satyam based in Hyderabad shook the morale of the IT sector as a whole when the chairman of India's fourth-largest IT company acknowledged his culpability in hiding news that he had inflated the amount of cash on the balance sheet by nearly $1 billion.

24 In India, programs comparable to affirmative action policies in the United States are referred to as reservation policies, are established in the national constitution. See Weisskopf (2004) for a thorough comparison of the two.

25 Actually, the mainstream media were very apologetic about the scam because it hurt the middle class sentiments and the IT sector was among the largest advertising revenue earner for the industry. But the arrest of Mr Ramalingaraju, the former owner of IT giant, Satyam, brought down the prestige of the entire industry.

26 See the portal of National Campaign on Dalit Human Rights: http://www.ncdhr.org.in/

27 Dr. Ambedkar and His People http://www.ambedkar.org/

28 Scholars without Borders: http://www.scholarswithoutborders.in/item_show.php?code_no=DST038&add=y&ID=undefined&calcStr=

29 There is a shared world of asymmetrical recognition among caste groups in India, especially among the top-caste Brahmins and the lowest-caste Dalits. Even among Dalits themselves, there are subsections that look down upon other groups within the broad category known as Dalits. Most violence happens between the immediate caste groups above the Dalits, commonly referred to as "the backward classes," the bottom end of the Sudra *varna.*

30 There is a movement to constitute a Dalit news channel underway in late 2009.

31 If we are to be distressed over the censoring of IT in China, we must be equally pleased by the lack of such censoring of political

speech as currently available in India and elsewhere.

32   Conference Paper Unpublished: Thirumal, P. and Robbins, Melinda (2008, April) *Dality Identity Formation and Media Use in India,* International Communication Association, Montréal, Canada.

33   This is to say, traditions unlike Brahmanical Hinduism that do not incorporate caste hierarchy and discrimination in their essence.

34   Ganguly (2005) notes: "The pain of the Dalit is palpable and embodied. I cannot presume to reduce it to a text or even a series of texts, or even to discourse pure and simple" (p. 10).

35   The *cheri* that stands on the south of the Tamil village proper, the *ur*. In interior villages of Andhra Pradesh, the postmen reads out the names of Dalits under a tree and they are supposed to collect the letters from that place, rather than the letters being delivered to their homes.

36   The SC/ST Atrocity Act which is meant to provide legal protection against all sorts of harassment is the most dysfunctional law. Hardly anybody is convicted through this legal measure.

37   Indeed they convey the information they do to different parts of the brain.

38   Verbal information of course bears the same potential.

39   *Baniya* is Hindi for someone of the merchant castes. A *dhoti* is the north Indian style of length of cloth wrapped around the waist and below, pulling it between the legs and tucking it in to make leggings.

40   In preparing this essay I visited the ambedkar. org website, which I will discuss below and there ran into a recording of Ambedkar's voice in a parliamentary address. (See http://www.ambedkar.org/) And, needless to say, the same sort of verisimilitudenous experience ensued. It's not a matter of distrusting written communication or a personal quirk; analog records of sensuous reality have a presence —and apparent verisimilitude— that abstracted verbal recreations cannot match.

41   "Atrocity" is the technical term in government documents of violent attacks upon Scheduled Castes or Scheduled Tribes.

42   The state of Maharashtra published the works of Ambedkar in several volumes over the last two decades. The widespread circulation of this literature has resulted in the proliferation of a Dalit literary culture in and outside Maharashtra.

43   See http://www.ambedkar.org/images/movement1/index.html

44   For a fine detailed study the event, see Teltumbde (2009). For a daily record of news items that include rapes and other atrocities see Dalit Media Watch: http://dgroups.org/Community.aspx?c=51ecc6ec-9dc9-4aff-b399-8130698414ab

45   See http://atrocitynews.files.wordpress.com/2006/10/khairlanji.pdf

46   See <http://images.google.com/imgres?imgurl=http://atrocitynews.files.wordpress.com/2006/10/05bhnp_74_priyanka_bhotmange.thumbnail.jpg&imgrefurl=http://atrocityne>

47   We are not saying the democratic opening is a benefit of the divide, but that it is a benefit of information technology that the digital divide does not altogether hamper.

Chapter 3
Exploring the Notion of 'Technology as a Public Good':
Emerging Characteristics and Trends of the Digital Divide in East Asian Education

Sunnie Lee Watson
Ball State University, USA

Thalia Mulvihill
Ball State University, USA

ABSTRACT

This chapter aims to explore the historical, sociological, and economic factors that engender inequities related to digital technologies in the East Asian educational context. By employing critical social theory perspectives, the chapter discusses and argues for the notion of "Technology as a Public Good" by examining the Chinese, Japanese and Korean societies' digital divide. This chapter examines how East Asian societies are exhibiting similar yet different problems in providing equitable access to information communication technologies to the less advantaged due to previously existing social structures, and discusses the urgency of addressing these issues. Based on the analysis of the digital divide in the East Asian context, this chapter also proposes and argues for the notion of "technology as a public good" in public and educational policies for information communication technologies. Finally, the chapter invites policymakers, researchers and educators to explore a more active policy approach regarding the digital divide solution, and provides specific future research recommendations for ICT policies and policy implementation in digital divide solutions.

DOI: 10.4018/978-1-61520-793-0.ch003

Copyright © 2011, IGI Global. Copying or distributing in print or electronic forms without written permission of IGI Global is prohibited.

INTRODUCTION

The term digital divide, often referred to as "information gap" or "information inequality" has generated a great amount of policy and academic discussion. While it is important to note that the precise definition of digital divide varies by the context in which it is being used or the group of people discussing it, the meaning of digital divide includes the discrepancies in physical access to information communications technologies as well as the inequalities in resources and skills needed to effectively use digital information or participate in the digital society (Korean Ministry of Information and Communication, 2001; Seo, 2001; Cho, 2001). In other words, it is the unequal attainment of information and communications technology by some members of the society and the unequal acquisition of related skills.

When considering the digital divide, various researchers and policymakers often discuss a variety of contexts, including socioeconomic status, gender, race, age, region, or geography. These various discussions provide valuable insight in understanding the current state of digital inequity around the world. In this chapter, the authors aim to explore the historical, sociological, and economic factors that engender inequities related to digital technologies in the East Asian educational context. By employing critical social theory perspectives, the chapter discusses and argues for the notion of "Technology as a Public Good" by examining the Chinese, Japanese and Korean societies' digital divides in education. The examination and analyses of these three societies' digital divides, and the three countries' different approaches to digital divide solutions will provide further understanding of the international expansion of digital inequity worldwide and the facilitation of bringing social justice through digital equity.

We begin by conceptualizing social inequity and the worldwide digital divide. We then examine the digital divide in three East Asian countries, China, Japan and South Korea, with special attention to the historical and sociological characteristics of inequity. We conclude by presenting a series of questions and challenges regarding the notion of "technology as a public good" through a critical social theory perspective to technology policymakers, researchers and educators who work towards bringing technology to serve a better role in society.

WORLDWIDE DIGITAL DIVIDE

Digital Divide

The rapid distribution of the ICT (Information Communications Technology) across the population has led many to hypothesize about the potential effects of the new media on society at large. While many optimistic educators and information technologists have advocated for the potential promises of information technology to reduce inequalities in society, emphasizing the "leapfrogging" characteristic that will enable the disadvantaged to catch up (Negroponte, 1998), many others warn that the rapid and uneven spread of technology across the population will lead to increasing inequalities, advancing the situations of those who are already in privileged positions while disallowing opportunities for development to the underprivileged (Hargittai, 2003).

Over the past decade, researchers and policy makers have paid considerable attention to what parts of the population have access to ICT and what sort of effects these trends have on the society. Findings clearly state that ICT is not fulfilling its promise for positive impact; rather, it is leading to new divides and increasing inequalities in countries and communities (DiMaggio, Hargittai, Celeste & Shafer, 2004; Warschauer, 2002; 2004). Inequalities related to access and use of technology are now clearly a significant global public policy issue. In this chapter, we will explore the trends and characteristics of the digital divide in the East Asian context, which centers on the historical, sociological, and economic factors that engender inequities. South Korea, Japan and China are exhibiting similar yet different problems in providing equitable access to information communication technologies to the less advantaged due to previously existing social structures. Based on an analysis of the digital divide in the East Asian context, this chapter discusses the notion of "technology as a public good" in public policies for information communication technologies.

Worldwide Digital Divide

Although the most common understanding of the notion of digital divide has been centered on regional and/or other geographic differences, such as the North-South divide or references such as 'Western countries and the others', the issue of digital divide is not such a simple matter and encompasses far greater differences than just those suggested by geography alone.

Whether the emphasis is on computing, communication in general or Internet-based communication in particular, some form of digital divide exists within every region, between the affluent minority and the rest of the population. The digital divide is now undoubtedly becoming a worldwide crisis and is recognized as a palpable example of how political, economic and educational power is in the process of being renegotiated, commodified and regrettably more deeply entrenched within the hands of the few. Its negative impact on society in general and education in particular has already been recognized in many information-technology advanced countries (Tiene, 2002). The Global Information Infrastructure Commission (2001) has recognized this worldwide concern and identified seven critical obstacles to addressing the digital divide through a worldwide survey. The seven obstacles were: infrastructure (19%), poverty (17%), bureaucracy (14%), protectionism (13%), language (12%), corruption (11%) and culture (11%). There are obviously a number of different factors that are influencing the inequalities in ICT access worldwide.

Access to ICT means much more than just additional information or entertainment. Research studies consistently show higher pay rates among business employees, higher academic achievement in reading and math of students, and increased political participation and civic engagement of citizens resulting from the ICT use in everyday lives (DiMaggio et al., 2004). Literature show that people acknowledge that information, resources and educational opportunities people find online ultimately transfer to power in their lives (Cartier, Castells & Qiu, 2005; Zhang & Perris, 2004; Choudrie, Papazafeiropoulou, & Lee, 2003). Thus, many information technologists and online educators have hoped that because of the Internet's ease of use and ability to overcome distance, online knowledge, information and opportunities would have helped societies to overcome inequalities in society.

However, advancement of ICT is enlarging, rather than thinning, the gaps between the privileged and the less or least advantaged. Societies that have suffered from the primacy of particular subgroups in economic activities, population, and cultural opportunities are now experiencing more extreme divisions due to the digital divide (Randall, Reichgelt, & Price, 2003; Tiene, 2002).

East Asian countries are not an exception in this theme of digital inequality. While Asia adds up to over 60% of the world's population, the region's proportion of Internet users is only about 27%. Overall, Asia's Internet penetration percentage adds up to only 12% (World Internet Statistics, 2007). China makes up a big part of this digital inequality, and the digital divide that exists within Korea and Japan has also been a consistent problem the two governments have been wrestling with.

HISTORICAL AND SOCIOLOGICAL PERSPECTIVES: EAST ASIAN EDUCATION AND DIGITAL DIVIDE

The East Asian region's history of societal inequalities is directly reflected through the digital divide. The current East Asian region includes the nations China, North and South Korea, Japan, and Mongolia. East Asia is the most populous part of the world, with around 1.4 billion people, which is more than a quarter of the world's total. More than 1.2 billion people live in China alone, and population densities in East Asia are also among the highest in the world (Hawkins & Su, 2003).

Following World War II, the East Asian region was among the poorest in the world. Because of the high levels of illiteracy, wars and civil conflicts, the region's future was not viewed with optimism (Hawkins & Su, 2003). However, from 1965 to 1990, East Asia's economy grew faster than all of the other regions of the world. Currently, Japan, South Korea and China are very much recognized for their economic growth and development (OECD, 2006; IMF, 2006). Distribution of income and human welfare such as education also improved considerably in this region as well (Morris, 1996).

The dramatic economic development of East Asia is directly tied to the influence of Confucianism in the region. Because of the strong influence of Confucianism, East Asians have an extensive history of putting a strong emphasis on education. East Asians still view high-quality education as the most valuable way of gaining socio-economic success (Kim & Santiago, 2005; Kikkawa, 2004). The South Korean and Japanese governments have also acknowledged the importance of human resources (Harbison & Myers, 1964) and emphasized the role of high-level manpower in the countries' national economic development. This has resulted in a focus on the explosive expansion of secondary, tertiary, and vocational education in both countries. China's recent development and improvement in compulsory education has been radical as well, with the explosive increase in national public school enrollment in the recent years (Hawkins & Su, 2003).

However, East Asian countries are now facing new challenges to equity and equality in education, including unequal access to ICT. The Chinese, Japanese and South Korean governments' approaches to ICT policy have differed, and thus have been showing different results of bridging the digital divide. Then where *do* these countries currently stand regarding digital divide issues relative to education? For example, "Who has the most, consistent and easiest access to ICT resources?" "Who are the ones that gain the most educationally, economically, and politically from the technology resources?" "What kind of ICT / educational policies are in practice regarding digital equity?" And "What are the policy implications for technological equity?"

While we cannot answer these issues fully within the scope of this discussion, we believe that it is crucial to enliven the conversation among educators, researchers and policymakers, and to enable us to more effectively respond to building a sustainable argument for the notion that technology must be considered a public good rather than merely an article of trade or commodity in our societies.

The following sections discuss China, Japan and South Korea's status with respect to the digital divide with a focus on educational opportunities. The sections also discuss the policies in each country that address bridging the digital divides within their societies.

Digital Divide in China

Economist Hu AnGang stated that China currently faces three big digital divides: (a) between China and the world, (b) between the different regions of China, and (c) between the urban cities and rural areas of China (Hu AnGang in Yu, Wang & Che, 2005).

According to World Internet Statistics (2007), only around 10 percent of China's population currently has access to information technology. However, China's development in ICT access has been explosive during recent years, and the great population of China offers immense room for growth and development in ICT. China is also recognized worldwide to have enormous potential as an online market. Even though only 10 percent of China's population has ICT access, China's number of ICT users ranks second in the world and is growing rapidly. There has been a dramatic increase in ICT users in China, from 620,000 in 1997 to 90 million in 2004, and to about 162 million in 2007 (CNNIC, 2004; World Internet Statistics, 2007).

The recent political and economic changes have had a powerful impact on the development of educational opportunities in China; many reports show how China has made significant advances in educational equity and equality since the 1960s (Rong & Shi, 2001). The Chinese educational system now aims for nine-years of compulsory education, and the enrollment rate for primary education and secondary education has been over 100% and 70% since 1997 (Hawkins & Su, 2003). Regarding equity in educational opportunities, however, issues such as gender, geographies, and socioeconomic factors are critical obstacles. China faces major challenges in breaking the effects of the vicious cycles of these factors, jump starting the development and expansion of educational opportunities, and ultimately decreasing the gap between subgroups in educational resources, methods and systems (Yu, Wang & Che, 2005).

The digital divide in China reflects these historical inequalities in educational opportunities. Socio-economic status is one of the most important indicators to information technology access in China. The CNNIC (2004) survey reveals that 65 percent of China's Internet users earn an annual income of more than 6000 Yuan. However, only 15 percent of Internet users earn less 6000 Yuan annually. The level of education is closely interrelated to income, which evidently facilitates the acquisition of ICT and a more ICT equipped work environment. Compared with the general population of China, ICT users are much better educated. The digital divide in terms of education is remarkable as China National Statistics in 2003 show that 57% of Internet users have at least a college level education and 31% of users have a high school degree (China National Statistics Bureau in Chen and Wellman, 2003).

Huge regional and geographical discrepancies also continue to exist (Hawkins & Su, 2003) in all areas in China. As a result of the different levels of socio-economic development between the richer east coast region and the other poorer regions, the regional distribution of Internet access is very uneven. Most rural areas have not yet been networked, and where rural areas do have access, connection speeds are poorer than in the metropolitan cities. CNNIC (2003) showed that peasants, who account for approximately 80% of the Chinese population, make up only 1% of all Internet users in the country. Even telephone subscription of farm households in rural China also proves that farm households with higher annual incomes are more likely to subscribe to a telephone than are low-income farm households (Wensheng, 2001). In general, people living in the peripheral areas such as in the underdeveloped rural areas and inland provinces are experiencing serious inequality. There is clearly an enormous digital divide between urban and rural regions. Illiteracy is also a problem that contributes to the digital divide between the rural and urban areas.

Another main factor in inequality in the Chinese digital divide is gender. The predominance of males in Internet use is reflected in the gendered diffusion rate in the general population of China. While 5.3 percent of the overall male population is on the Internet accessing online resources and information, only 3.9 percent of the female population are; although, the percentage of female users among the overall Internet users has steadily climbed from 13% in 1997 to 41% in 2003 (Chen

& Wellman, 2003). Overall, however, the gender-based digital divide appears to be much smaller than education and income-based divides. CNNIC (2004) reports that 39 percent of Chinese Internet users in July of 2001, were women. Compared to the figure in July of 2000, there was an 8 percent increase in women Internet users. This implies that women are catching up in Internet access.

Disability or age differences also play key factors in unequal ICT resources (Rong & Shi, 2001). Disabled people in China have much less access to online resources. The generational digital divide dynamic also appears; younger Chinese form the majority of Internet users in the country (Zhu & Wang 2005; Chen & Wellman, 2003). Persons 50 or older make up only 3.7 percent of all Internet users, while more than 60% of users were younger than 35.

China's ICT and Digital Divide Policies

The gaps of inequalities in China are more extreme than other countries. However, the Chinese government does understand the importance of ICT development and have made ICT development a national policy and development strategy. Both in the 2001-2005 and 2006-2010 Chinese government Five-Year Plans, ICT policy have been the overall strategic measures for China's modernization construction (UNESCO, 2003). China is working to take advantage of ICT development to bring industrialization of China and remodel traditional industries in the course of promoting informatisation. In addition, the United Nations Development Programme (UNDP) in China is running a 2.5 million dollar project to bring Internet access to the rural areas of China (UNESCO, 2003).

Related to the issue of digital divide, China has notably made great progress in the use of ICT for education as a result of the national ICT strategies included in the several recent five-year plans. It is clear that the Chinese government is committed to the belief that modernization of education by applying information technology is essential in order to produce students who can be competitive in the information era. The Chinese government is putting a great effort into the "Educational Informationisation" project of introducing and integrating ICT application into the educational system.

The major plans implemented for ICT in education were China Education and Research Network (CERNET) bandwidth increase, administration informationisation, increasing the number of ICT specialists in schools, distance education projects for Communist Party members training and primary/secondary schools in the countryside. Constructing infrastructure, encouraging computer education, and supporting professional development for teachers, are some long-term goals that Informationisation project have set as ICT strategies as well (UNESCO, 2003).

Despite these significant efforts, however, the Chinese government's approach to ICT development has not contributed much to bridging the digital divide in China. The lack of a clear and shared vision and understanding of what informationisation is, the demand for people who are skilled in ICT in the education sector, the imbalance of development of infrastructure, facilities and investment in rural village information structures and schools, and the immaturity of the educational ICT industry are all major constraints that must be dealt with for successful ICT use in education.

While there is much to be done, the Chinese government's solution for digital divide is heavily based on the enthusiasm for trickle-down economics, with an optimistic outlook that somehow through the ICT development in the corporate sector in eastern urban regions, the disadvantaged population in western rural regions will eventually catch up. The capitalist model of the Chinese government has resulted in prioritizing and investing ICT development in only the cutting-edge areas of computer sciences and engineering, leaving little room for investment in ICT access policy or development for rural areas and disadvantaged population. Currently, China's internal forces, such as the government's current practice of small government and the global, external forces such as globalization are playing significant roles in cementing the free market structure of access to ICT resources, and ICT is still serving a role of expanding the gap between the privileged and least advantaged.

Digital Divide in Japan

Compared to other advanced industrialized countries, the diffusion of the Internet, especially the PC-based Internet, started relatively late in Japan. However, the Ministry of Public Management, Home Affairs, Posts and Telecommunications of Japan (MPHPT, 2003) reports that from 1996 to 2002, the percentage of Japanese households owning PCs increased 2.6 times. The number of Internet users in Japan jumped from 12 million in 1997 to 17 million in 1998, 27 million in 1999, 47 million in 2000, 56 million in 2001, 69.4 million in 2002 and 86.3 million in 2007 (World Internet Statistics, 2007).

A survey by World Internet Project Japan (Chen & Wellman, 2003) identifies some of the most popular uses of the Internet. The most important use of the Internet was accessing online resources; they used the Internet to search sites, transportation or travel course information, maps and weather forecasts. Secondly, e-learning or online teaching has fast become a preferred teaching mode in Japan in the areas of training, licensing, and university teaching. Furthermore, existing higher education institutions, corporate ventures, and virtual universities are adopting e-learning and distance education in Japan (Nakayama & Santiago, 2004; Uchida, 2004).

One very interesting aspect of Japan's Internet access compared to the other East Asian countries is that there is a difference between PC access and Internet access because of their unique trends of using mobile phones. Known to be a world leader in the use and manufacturing of mobile phones, the high rate of Japan's mobile phone use may possibly explain the gap between the high rate of PC use and the relatively low rate of PC-based Internet use. While most Chinese and Koreans who are connected to the Internet use PCs, the Japanese connect to the Internet in multiple ways. Among the Japanese who connect to the Internet, 85 percent access the Internet through PCs, 63 percent through mobile phones, and five percent through other technology devices. The share of users accessing the Internet through mobile phones soared during 2001 and 2002, and 83 percent of Japanese mobile phone users (62.5 million) were mobile Internet subscribers (MPHPT, 2003).

Similar to China, socioeconomic status is crucial in ICT access in Japan (Kikkawa, 2004). Japanese with high school or lower education have much less access to online resources and continue to be left behind, while those who already have a college education have much more access to the benefits of ICT education and resources (World Internet Project Japan in Chen & Wellman, 2003). The 2003-2004 report of MPHPT shows that around 50% of Japanese households with an income of eight million Yen or more had access to Internet resources, while only about one quarter of these households with an income less than two million Yen had access to ICT. The likelihood of Japanese having access to the Internet is greater when the level of education and household income is higher (Chen & Wellman, 2003).

These inequality trends are reflected by how different groups use the Internet as well. Miyata, Boase, Wellman & Ikeda's (2004) study found that the use of the Internet varied by age and gender, drawing data from a random sample survey in the Yamanashi region. Compared to the users connecting to the Internet via PCs, those who connect via mobile phones tend to come from a different socioeconomic background. Those who have access to the Internet through PCs are more likely to be male, older and better educated, while those who have access to mobile Internet are more likely to be female, younger and less educated (World Internet Project Japan in Chen & Wellman, 2003).

Geographic location is also an important factor in the digital divide in Japan. The percentage of people connecting to Internet resources via PCs decreases with the size of the city. Major cities have much higher Internet diffusion rates than smaller cities, followed by smaller towns and villages. However, the explosion of mobile Internet use during the last couple of years may be decreasing the gap of the digital divide. The MPHPT report (2003) shows that the digital divide, in terms of, city size dropped two percent in the year of 2002.

Finally, life stage and gender are factors in digital inequality. Japan has a substantial generational age divide in Internet use. The AMD Global Consumer Advisory Board's report on the digital divide shows how Japanese who are in their twenties are 30 times more likely to be connected to the Internet than Japanese in their seventies. Gender also has an impact; compared to 68 percent of Japanese men using online resources, only 56 percent of Japanese women were Internet users (Chen & Wellman, 2003).

Japan's ICT and Digital Divide Policies

While the reasons why these inequalities exist are similar to China and other countries, the inequalities of ICT access in Japan are not as extreme. Rather, the ICT access rate in Japan is one of the highest in the world (World Internet Statistics, 2007). The reason for the advanced ICT access situation in Japan can be found in the Japanese government's national ICT strategic plans. Japan has been aggressively pursuing national ICT development through the e-Japan (2001-2005) and u-Japan (2006-2010) plans and has been successful in accomplishing impressive economic advancement through ICT.

While the first push for ICT started in 1994, it was in 2001 that the IT Strategic Headquarters for the Promotion of an Advanced Information and Telecommunications Network Society was established with the goal to make Japan the world's most advanced IT nation by 2005. In January 2001, the e-Japan strategy was mapped out by the ICT Strategy Headquarters, which was set up within the Cabinet. The promotion of the use of ICT was publicized by an e-Japan Strategy to develop an energetic, worry free, exciting and more convenient society (Naito & Hausman, 2005). In the strategy mapped out for e-Japan (2001-2005), there are references to the bridging of the digital divide in Japan, calling for actions of the government to study measures to spread the use of the high-speed Internet in depopulated areas, remote islands and other areas experiencing disadvantageous conditions (UNESCO, 2003).

The policies call for closing of the digital divide stemming from geographical factors and problems arising from age and disability. Policies included specific initiatives to (1) support elderly, people with disabilities, and foreign nationals to be able to live securely regardless of physical or linguistic barriers, (2) equal access to information, (3) smooth communication and the installation of optical fibers, and (4) elimination of all areas where broadband service remains unavailable by the year 2010 (UNESCO, 2003).

Grants and subsidies to support private sector research and development, establishment of local infrastructure, provision of telecommunications-broadcasting services for the people with disabilities, competitive environment for ICT businesses, and promotion of support centers are some major investments that the policies prioritize. The Ministry of Internal Affairs and Communications, local governments, and telecom operators are also working in partnership toward the elimination of broadband-zero areas based on the Next-Generation Broadband Strategy 2010 (Naito & Hausman, 2005; UNESCO, 2003).

The fact that these policies have led to Japan's impressive accomplishments in ICT access is well acknowledged and identified worldwide. Along with South Korea, Japan has been providing the cheapest and easiest ICT access to the vast majority of citizens in recent years. However, much like the rapid economic development of Japan, the development in ICT has been explained with an overly simplified version of human capital theory (Morris, 1996), which results in too much of a generalized and undifferentiated call for investments in policies for the digital divide.

Digital Divide in South Korea

The diffusion of broadband availability in Korea has been incomparable, drawing worldwide attention (OECD, 2001). The number of Internet users in Korea was over 26 million, sixth in the world as of April 2003 (National Computerization Agency and Ministry of Information and Communications, 2003), and South Korea had over 34 million Internet users in 2007, according to World Internet Statistics Online (2007). According to an OECD report, South Korea already had nearly 14 connections per 100 inhabitants in 2001, followed by Canada with 6.2 and Sweden with 4.5 (Lee, O'Keefe, & Yun, 2003).

This advancement in infrastructure has often been explained from perspectives such as the density of population, competition in the information-technology market and government policies. With the burgeoning digital technology industry, having one of the largest online gaming industries worldwide, and the status as a primary region of Asia's pop entertainment industry with the emergence of Korean pop-culture Hallyu, the Korean economy and culture now appears to be inseparable with the online environment (Brender, 2004; Kim, 2001). In addition, Korea's major corporations, leading universities, lifelong learning institutions that operate through the Ministry of Education, and even K-12 schools are showing immense momentum for building cyberspace for online education and education resources (Rho, 2002).

However, while the South Korean government is striving to be at the cutting edge in the competitive information technology movement through the build up of information infrastructure in many sectors of its society and through the construction of advanced ICT environments (OECD, 2001), there are an abundance of examples that show the notable inequality in ICT access (Morse, 2004; Randall, Reichgelt, & Price, 2003; Tiene, 2002).

Historically, most of the equity or equality problems in Korean society come from the family's wealth or socioeconomic status (SES). The most influential aspect of society that determines privilege and discrimination in ICT access is socioeconomic status. Students enrolled in high-tuition universities or private schools are already able to afford an expensive education and are the ones receiving alternative or additional online classes from these schools' cyber universities and e-campuses. Similarly, already privileged K-12 students who are technology savvy through their wealthy parents' support for private tutoring are the ones using online educational resource banks. Life-long learning or adult learning is not an exception; use of online databases or information banks is central to selected occupations or socioeconomic groups.

Closely tied to the influence of socioeconomic status, geography is another key factor in unequal access to ICT resources. What region a person resides in makes a big difference; it is an indicator of which subgroup you belong to even when you live in the same city or rural area. The differences of distribution of PCs, access to the Internet and the World Wide Web are significant in different regions. The digital divide between the capital city, Seoul, and the rest of the country, between the urban and the rural areas is considerable.

Huh and Kim's (2003) study on regional Internet information flows in South Korea calls for explicit spatial policy measures of the central and local governments to tackle the problem of geographical imbalance. While Seoul prominently leads the nation with nearly half of the total ICT access and information flows (48.7%), the rural areas have almost no ICT access. An astonishingly large number of Internet hyperlink domains (85.4%) are registered only within Seoul. All together, Seoul is the single most prominent consumer and producer of Internet resources in Korea. The Internet, which is often advertised as the technology that will reduce the regional hierarchy in Korea, is paradoxically extremely concentrated in Seoul to form an "enclave" (Huh & Kim, 2003).

Occupations, disability, and gender also make significant differences in ICT access. A government survey with a sample of 30,000 families show that more than 85% of teachers and nearly 80% of public service personnel are able to use the Internet, while hardly any low income farmers (2.2%) have access to online environments (Rho, 2002). The lower income occupation groups such as the self- employed (25.3%), the disabled (36.5%) and housewives (31.8%) are other groups that do not have much online access. A Korea Network Information Center survey also shows that the proportion of males using the Internet is higher than that of females, 58.7 versus 44.6 percent (Korea Network Information Center, 2002a; 2002b).

South Korea's Digital Divide Policies

The Korean digital divide is due to factors such as geographies, socioeconomic status, age and gender as discussed above, similar to Japan and China. While the divide has been a consistent issue in South Korean society, the Korean government has had several comprehensive ICT development plans that have been implemented to achieve drastic development in the ICT industry. Similar to the Japanese government's approach to the ICT development plans of Japan, the Korean government's aggressive implementation of these plans is why the rates of ICT access and availability have been much higher than the worldwide average.

South Korea has made major strides in ICT over the last four decades. After the Korean War, in 1960, Korea had a telephone penetration of 0.36 per 100 inhabitants, barely one tenth of the then world average. Even after the rapid economic development between the 1960s and 1990s, South Korea had less than one Internet user per 100 inhabitants in 1995 (International Telecommunication Union, 2003). And yet in 1999, Korea's Internet user rate surpassed the developed nation average, and by the end of 2002, Korea had the world's fifth largest Internet market with 26 million users. The number of Internet users in South Korea as of 2007 was over 34 million (World Internet Statistics, 2007). The diffusion of broadband availability in Korea has also been incomparable, drawing worldwide attention (OECD, 2001). As a result, many international scholars and reports on ICT have referred to the Korean case as a miracle.

One interesting thing about Korea's ICT policies is the explicit attention to the digital divide. While Japan implemented the digital divide policies as a part of the overarching national ICT plans, the Korean government set a separate policy that functioned parallel to the national 1999-2002 CyberKorea 21 or 2003 e-Government Korea policies. In March of 2001, the *2001 Master Plan for Digital Divide Solution* was announced by the government and lead by the Ministry of Information and Communication. The policy set goals to achieve six main objectives: (1) ensure high speed broadband Internet through providing Asymmetric Digital Subscriber Lines (ADSL) and establish at least one or more free Internet centers in every single village, town, and city in the entire nation, (2) provide free computers and five years of high speed Internet in 50,000 homes of children-led families (families of children on welfare with no parent or foster parent direction), people on welfare or with disabilities, and social service centers, (3) ensure access to people with disabilities, and content for e-learning was to be developed and distributed to those of certain demographic groups that were traditionally excluded from information technology such as people with disabilities, the elderly, or farmers/fishermen, (4) provide 550 public libraries that still did not have computers and Internet access with digital resource centers for facilitating all citizens' Internet use and facilitate the opening of at least 14,000 Personal Computer Cafes (PC-bang) nationwide, at a price lower than one U.S. dollar an hour, (5) provide low-income families and people with disabilities the ADSL service at a 50% discount through government support, and (6) based on the successful program implementations through 1988-1999 on digital literacy for children, housewives, people with disabilities and low-income families, expand the programs for 10 million citizens, including the elderly, small business owners and farmers/fishermen in less developed regions.

This five-year policy plan was supported with 2.3 trillion won (approximately 2.3 billion U.S. dollars) by the government. Additionally, tax cut incentives, subsidies, direct underwriting, loans and other types of financial support for construction of new high capacity computer hardware and private high speed broadband Internet companies were supported in order to facilitate the equitable distributions of computer hardware and high speed Internet services (Choi, 2003).

In 2006, a newly updated digital divide policy was announced with more specific, detailed goals for addressing the issues of the digital divide (Ministry of Information and Communications, 2006). The 2006 plan is currently in implementation, and we will soon see what the Korean government and society make out of this phase of the digital divide solution policy implementation.

CRITICAL SOCIAL THEORY AND THE NOTION OF TECHNOLOGY AS A PUBLIC GOOD

The three countries experience societal inequalities manifesting into similar forms of the digital divide. Most inequalities are persistently apparent throughout the years of Internet use of these three countries. However, it is clear that the East Asian countries' different approaches to digital inequity is producing different outcomes in ICT access for the disadvantaged. China's inequalities have been decreasing; however, the gap is extreme and has a long way to go to improve the state of discrepancies. On the other hand, the digital inequalities in Japan and Korea are not as severe. The two countries are among those that provide the highest ICT access rates in the world. However, the inequalities that exist have been persistent in their factors and characteristics despite somewhat aggressive digital divide government solutions.

So how should governments, particularly in the light of education, approach ICT development and access? We might advance this discussion by considering how both the Japanese and Korean cases show how aggressive digital divide policies have made significant changes in digital inequity as well as significant contributions to the economic development. The Korean case in particular, despite not having the ideal situation for ICT development, accomplished a remarkable development in bridging the digital divide through the government's clear vision and policies for ICT as a public good. While the reasons behind South Korea achieving an ICT access rate of over 70% of the general public can be explained with various factors, international ICT researchers refer to the government's strong commitment to digital divide policies as the most important factor in Korea's success in having one of the highest rates of ICT access in the world (Han, 2003; ITU, 2003; Picot & Wernick, 2007). Studies on domestic and international policies for digital divide have consistently shown that the Korean government ensured a high level of confidence and assurance for the general public and private companies by establishing and communicating a clear vision and strategy for Information Communications Technology as a "public good".

Advancing an argument for technology as a public good within a society requires familiarity with a critical social theory approach, certain democratic principles, and the recognition of the multivocal nature of the concept of social justice. In general, critical social theories posit critiques that look at improving the human condition, but in particular they can be quite diverse based on other expressed or implied values about a notion of the "public good." Critical social theory approaches attempt to move beyond critique to some form of social action that will ameliorate the undesired tensions, such as the negative implications of prolonged and sustained forms of "digital divide."

What critical social theory perspective offers us within the analysis of "digital divide" is an explanation of the forms of power and control asserted through what may appear to be a benign, inconsequential, question of access to technology as a more neglectful, and perhaps intentional, form of decentering or erasing full participation of human beings in the creation of their social lives. A critical social theory perspective is based on the essential belief that all human beings have the right to fully participate in the construction of their communities including access to knowledge production and consumption arenas. Therefore, when technology becomes one of the dominant modes of creating and consuming knowledge, any conditions which exist that prevent, delay, or dismiss the need for it to be universal is suspect and unacceptable.

Radical-democratic principles are also good examples of why technology should be considered as a public good within a society. According to Cohen and Fung (2004), radical-democratic principles "value participation in public decision-making and deliberative democracy...[whereby]... [c]itizens should have greater direct roles in public choices or at least engage more deeply with substantive political issues and be assured that officials will be responsive to their concerns and judgments.... Instead of a politics of power and interest, radical democrats favor a more deliberative democracy in which citizens address public problems by reasoning together about how best to solve them..." (pp. 23 – 24).

Furthermore, Light and Luckin (2008) claim that "Social justice is an interventionist standpoint, in that it seeks to reorganise society's resources and structures to create a fairer social order." Furthermore, they advocate for a "user-centered design (UCD)" approach that values more inclusive and egalitarian involvement across all dimensions of societal users of technology as well as "technology-enhanced learning (TEL)" where "technology can be used to recognise and address everyone's differences, including the needs and desires of minority groups...[including] the way in which it can be used to enable more people to communicate, socialise, join in debates and play a greater role in society"(Light & Luckin, 2008, p. 5).

It is also helpful to reexamine the underlying principles in the Universal Declaration of Human Rights (UDHR) Article 26, which states, "Everyone has the right to education. Education shall be free, at least in the elementary and fundamental stages. Elementary education shall be compulsory. Technical and professional education shall be made generally available and higher education shall be equally accessible to all on the basis of merit" (UDHR, 1948). Embedded within the UDHR is the notion that all human beings have the right to fully participate in negotiating what ought to be determined as the "public good" for their communities and that the skills and sensibilities required for such creation must be obtained by way of an educative process. The UDHR identifies technical education as part of these universal human rights.

ICT is now proving to be one of the vital tools of life in the currently emerging global society, and the concerns of the digital divide are particularly important because of its potential for fostering a more inclusive and equitable society rather than the negative influence that it is currently proving to be serving. Despite the divides in ICT across the world, many believe that technology can be used to promote social inclusion rather than widening the gaps and maintaining inequalities in societies.

Currently, educators and policy makers worldwide focus more on equity and equality issues in traditional, institutionalized policies and overlook the problem of digital inequity. To this date, policy makers and ICT educators consider the issue of the digital divide with an overly simplistic notion, and most government policies merely focus on improving the physical numbers of computers and broadband available to the public. Even highly advanced infrastructure countries are facing challenging situations with digital divide and many countries have come to realize that advancement in infrastructure itself does not guarantee an improved ICT access for all. In order to truly address the challenges, it is most critical to address the wide range of physical, digital, human and social resources that are meaningful in ICT access (Warschauer, 2002; 2004).

The fundamental first steps to take are identifying the need for physical, digital, human and social resources that are necessary for ICT access in that country or region. Exploring the solutions in various contexts to meet these needs should follow these first steps. In this sense, government efforts to achieve a universal service for 'technology as a public good' are absolutely necessary. A national service of eliminating obstacles and barriers is needed, so that every member of society has the opportunity to use ICT for meaningful and effective participation in all aspects of society, from economy to culture, from political participation to community life.

FUTURE RESEARCH FOR TECHNOLOGY AS A PUBLIC GOOD

Research and evaluation on digital divide policies worldwide are mostly looked at from a macro-level perspective, primarily situated at a national level. While it is true that the Japanese and Korean governments' plans for bridging the digital divide were successful in bringing positive results, by taking on this macro-level perspective and dealing with the users as an aggregate population, evaluation and research for digital divide policies are also limited by this perspective. When data collection is aimed solely at obtaining and evaluating large-scale data, the analysis often misses insights that could be obtained from data about the day-to-day lived experiences of those disadvantaged from the divide. In addition, a review of literature reveals that there is an absolute lack of independent researchers conducting studies related to digital divide policies.

The adoption and diffusion of ICT within disadvantaged populations in China, Japan and South Korea raises many issues that need to be investigated for future digital divide policies. Yet, these issues are not being explored as they should be, and there is a critical need to explore these issues of digital divide with a more micro-level perspective, for a richer understanding of the contributing factors and how they interact (Picot & Wernick, 2007). Ultimately, both macro and micro-level studies should be pursued together in order to help policy makers to understand a theoretical framework that encompasses various factors within the policy.

It is vital to explore under what supportive conditions disadvantaged groups such as women, the poor, the elderly and those in less-developed countries can obtain access to ICT. There should also be more effort in trying to understand the impacts of digital divide policies in the diverse groups of beneficiaries' lives as well. What are the behavioral characteristics regarding their adoption and use of ICT? What does the adoption of ICT into their lives mean to them? What kinds of ICT programs were more successful and why were they successful in that context? What factors were most important for certain groups? What kind of impact did ICT access have in their small business revenue or family household? Did the revenue of the business or family household income rise by any measures? What kind of impact did ICT educational programs have on students? Are they more likely to set higher aspirations and goals that require higher technology skills in their jobs because of their exposure to ICT? Did they become more interested or feel inclined to participate in online discussions on social issues or political matters?

These are important questions that need to be considered in future research on digital divide policies not only in East Asia but also in the international community as a whole. In addition, attempting to understand a theoretical framework that encompasses many of these various factors within the policy planning and implementation process might be worthwhile. Future research should study the impact of disadvantaged user group behavior, personal demographic factors, pricing, educational factors and content on ICT access in combination with private sectors' competition and public good oriented governmental strategies in a more holistic, comprehensive way. Such research efforts should be aimed at both collecting and evaluating large scale, macro-level data and more micro-level studies that give us a richer understanding of these various factors and how they interact.

In addition to understanding how these disadvantaged user groups adopt ICT into their lives, research should also aim to understand how stakeholders can help each other in their efforts to adopt ICT. The focus here is to study how ICT researchers can engage in research work that can facilitate social change and contribute to more equitable and empowering use of ICT at the local, grassroots level. The important objective is to learn how practitioners and researchers can collaborate and contribute to closing the digital divide in ICT use for disadvantaged or marginalized populations.

CONCLUSION

In this chapter, we examined China, Japan and South Korea's digital divides with special attention to the historical and sociological characteristics of inequity. Clearly, the three countries that we examined have different approaches to the adoption and diffusion of ICT, and these differences raise many issues about digital divide policies worldwide that need to be investigated. Many policymakers, researchers and educators have hoped that information technology will help societies overcome their disparities, including educational opportunities (Huh & Kim, 2003; Attewell, 2001). However, the reality is that the newly but widely accepted ICT phenomenon is currently only assuming a permissive role and is being affected by the existing structure of assumptions and expectations of a given nation or region (Huh & Kim, 2003; Attewell, 2001).

For ICT to truly serve its promising role, it is critical that policy makers and educators aggressively act upon the emerging issues and patterns of the digital divide. In this sense, the notion of "technology as a public good" presents a timely challenge and the East Asian region's case provides an illuminating example to address the topic of digital divide. Future exploration by policymakers and researchers from many different areas of education on this topic are needed in order to identify what could be improved for disadvantaged populations in adopting ICT into their lives, to connect that to help shape policies in other countries with similar digital divides, and ultimately to facilitate bringing social justice through digital equity.

REFERENCES

Attewell, P. (2001). The First and Second Digital Divides. *Sociology of Education, 74*(3), 252–259. doi:10.2307/2673277

Brender, A. (2004). Asia's new high-tech tiger. *The Chronicle of Higher Education, 50*(4).

Cartier, C., Castells, M., & Qiu, J. L. (2005). The information have-less: Inequality, mobility and translocal networks in Chinese cities. *Studies in Comparative International Development, 40*(2), 9–34. doi:10.1007/BF02686292

Chen, W., & Wellman, B. (2003). *Charting and bridging digital divides: Comparing socio-economic, gender, life stage, and rural-urban Internet access and use in eight countries: GCAB*. The AMD Global Consumer Advisory Board.

Cho, J. (2001). 정보격차 해소 종합계획의 내용과 향후과제[The digital divide plan's content and future path]. The path to informationalization, 56.

Choi, D. (2003) 우리나라 정보격차의 특성 및 정보격차 해소를 위한 정책 과제[Korea's characteristics of digital divide and its solution]. Korean Information Culture Center Publication, 1-19.

Choudrie, J., Papazafeiropoulou, A., & Lee, H. (2003). A web of stakeholders and strategies: A broadband diffusion in South Korea. *Journal of Information Technology*, *18*, 281–290. doi:10.1080/0268396032000150816

CNNIC (China Internet Network Information Center). (2003) *Semiannual Survey on the Development of China's Internet*. Retrieved September 15, 2006 from http://www.ccnic.org.cn

CNNIC (China Internet Network Information Center). (2004) *Semiannual Survey on the Development of China's Internet*. Retrieved September 15, 2007 from http://www.ccnic.org.cn

Cohen, J. & Fung, A. (2004) Radical Democracy. *Swiss Journal of Political Science. 10*(4).

Cohen, J. & Fung, A. (2004) Radical Democracy. *Swiss Journal of Political Science, 10*(4).

DiMaggio, P., Hargittai, E., Celeste, C., & Shafer, S. (2004). Digital inequality: From unequal access to differentiated use. In Neckerman, K. (Ed.), *Social inequality* (pp. 355–400). New York: Russell Sage.

Global Information Infrastructure Commission. (2001). *Global Information Infrastructure Commission Survey*. Retrieved October 14, 2008 from http://www.giic.org/#survey

Han, G. (2003). Broadband Adoption in the United States and Korea: Business Driven Rational Model Versus Culture Sensitive Policy Model. *Trend sin Communication. 11*(1), 3-25.

Harbison, F. H., & Myers, C. A. (1964). *Education manpower and economic growth: Strategies of human resource development*. New York: McGraw Hill.

Hargittai, E. (2003). The Digital Divide and What To Do About It. In Jones, D. C. (Ed.), *New Economy Handbook*. San Diego, CA: Academic Press.

Hawkins, J. N., & Su, Z. (2003). Asian education. In Arnove, R. F., & Torres, C. A. (Eds.), *Comparative education* (pp. 338–356). Lanham, Maryland: Rowman & Littlefield Publisher, Inc.

Huh, W., & Kim, H. (2003). Information flows on the Internet of Korea. *Journal of Urban Technology*, *10*(1), 61–87. doi:10.1080/1063073032000086335

International Monetary Fund (IMF). (2006) *World Economic Outlook Database*. Retrieved September 26, 2006 from http://www.imf.org/external/pubs/ft/weo/2006/02/data/index.aspx

International Telecommunication Union (ITU). (2003). *Broadband Korea: Internet Case Study*. Retrieved February 10, 2002 from http://www.itu.int/net/home/index.aspx

Kikkawa, T. (2004). Effect of educational expansion on educational inequality in post-industrialized societies: A cross-cultural comparison of Japan and the United States of America. *International Journal of Japanese Sociology*, 13.

Kim, C., & Santiago, R. (2005). Construction of e-learning environments in Korea. *Educational Technology Research and Development*, *53*(4), 108–115. doi:10.1007/BF02504690

Kim, S. (2001). Korea's e-commerce: Present and future. *Asia-Pacific Review*, *8*(1), 75–85.

Korea Network Information Center. (2002a). *A Survey on the Number of Internet Users and Internet Behavior in Korea: Summary*. Seoul: Korea Network Information Center.

Korea Network Information Center. (2002b). *Internet statistics*. Retrieved February 14, 2002 from http://stat.nic.or.kr/sdata.html

Korean Ministry of Information and Communication. (2001). *2001 Master Plan for Closing the Digital Divide Solution*. Retrieved March 23, 2008 from www.mic.go.kr/eng/index

Korean Ministry of Information and Communication. (2006). *2006 Master Plan for Closing the Digital Divide Solution*. Retrieved March 23, 2008 from www.mic.go.kr/eng/index

Light, A., & Luckin, R. (2008). *Designing for social justice: people, technology, learning*. Retrieved December 15, 2008 from www.futurelab.org.uk/openingeducation

Ministry of Education. (2003). Masterplan 2 for IT in Education. Retrieved May 11, 2008 from http://www.moe.gov.sg/edumall/mp2/mp2.htm

Ministry of Public Management. Home Affairs, Posts and Telecommunications, Japan. (2002). *Information and Communications in Japan White Paper 2002: Stirring of the IT-prevalent Society*. Tokyo: Ministry of Public Management, Home Affairs, Posts and Telecommunications, Japan.

Ministry of Public Management. Home Affairs, Posts and Telecommunications, Japan. (2003). *Information and Communication in Japan White Paper 2003: Building a "New, Japan-inspired IT Society"*. Tokyo: Ministry of Public Management, Home Affairs, Posts and Telecommunications, Japan.

Miyata, K., Boase, J., Wellman, B., & Ikeda, K. (2004). The Mobile-izing Japanese: Connecting to the Internet by PC and Webphone in Yamanashi. In Ito, M. (Ed.), *Portable, Personal, Intimate: Mobile Phones in Japanese Life*. Cambridge, MA: MIT Press.

Morris, P. (1996). Asia's four little tigers: A comparison of the role of education in their development. *Comparative Education, 21*(1), 95–109. doi:10.1080/03050069628948

Morse, T. (2004). Ensuring equality of educational opportunity in the digital age. *Education and Urban Society, 36*(3). doi:10.1177/0013124504264103

Naito, S., & Hausman, B. (2005). *Information and communications technology in Japan. A general overview of the current Japanese initiatives and trends in the area of ICT*. Sweden: VINNOVA. Retrieved January 12, 2009 from http://www.vinnova.se/upload/EPiStorePDF/vr-05-04.pdf

Nakayama, M., & Santiago, R. (2004). Two categories of e-learning in Japan. *Educational Technology Research and Development, 52*(3), 91–111. doi:10.1007/BF02504680

National Computerization Agency and Ministry of Information and Communications. (2003). *2003 White Paper Internet Korea, 72*, NCA & MIC, Seoul.

Organization for Economic Co-operation and Development (OECD). (2001). *The development of broadband access in OECD countries*. Paris: OECD.

Organization for Economic Co-operation and Development (OECD). (2006). The development of broadband access in OECD countries. *OECD Composite Leading Indicators (CLIs) for OECD Countries and Major Non-Member Economies*. Retrieved September 14, 2007 from http://www.oecd.org/statisticsdata/0,2643,en_2649_37443_1_119656_1_1_37443,00.htm

Picot, A., & Wenicek, C. (2007). The role of government in broadband access. *Telecommunications Policy, 31,* 660–674. doi:10.1016/j.telpol.2007.08.002

Randall, C. H., Reichgelt, H., & Price, B. A. (2003). *Demography and IT/IS students: Is this digital divide widening?* Paper presented at the 17th Annual Conference of the International Academy for Information Management, Barcelona, Spain.

Rho, K. (2002). Uses of Internet in Korea. *Educational Technology Research and Development, 50*(1), 84–88. doi:10.1007/BF02504964

Rong, X. L., & Shi, T. (2001). Inequality in Chinese education. *Journal of Contemporary China, 10*(26), 107–124. doi:10.1080/10670560124330

Seo, Y. (2001). 디지털 시대의 평등사회 구현을 위한 정보격차 해소방안 [A digital divide solution for an equitable digital society]. Policy and Computers, 23(1).

Tiene, D. (2002). Addressing the global digital divide and its impact on educational opportunity. *Educational Media International, 39*(3-4), 211–222. doi:10.1080/09523980210166440

Uchida, H. (2004). Information Technology-Driven Education in Japan: Problems and Solutions. *Educational Technology Research and Development, 52*(3), 91–111. doi:10.1007/BF02504679

UNESCO. (2003). *Meta-survey on the Use of Technologies in Education in Asia and the Pacific 2003-2004.* Retrieved January 12, 2009 from www.iosn.net/education/Metasurvey

UNESCO. (2007). *ICT in education: Japan.* Retrieved May 12, 2008. http://www.unescobkk.org/index.php?id=1381

Universal Declaration of Human Rights. (1948) Retrieved January 19, 2009 from http://www.unhchr.ch/udhr/lang/eng.htm

Warschauer, M. (2002). Reconceptualizing the digital divide. *First Monday, 7*(7).

Warschauer, M. (2004). *Technology and Social Inclusion: Rethinking the Digital Divide.* Cambridge, MA: The MIT Press.

World Internet Statistics. (2007). Internet Usage and Statistics - The Big Picture. World Internet *Users and Population Stats.* Retrieved September 30, 2007 from http://www.Internetworldstats.com/stats.htm

Yu, S., Wang, M., & Che, H. (2005). An exposition of the crucial issues in china's educational informatization. *Educational Technology Research and Development, 53*(4), 88–101. doi:10.1007/BF02504688

Zhang, W., & Perris, K. (2004). Researching the efficacy of online learning: A collaborative effort amongst scholars in Asian open universities. *Open Learning, 19*(3), 247–264. doi:10.1080/0268051042000280110

Zhu, J., & Wang, E. (2005). Diffusion, use, and effect of the Internet in china. *Communications of the ACM, 48*(4). doi:10.1145/1053291.1053317

ADDITIONAL READING

Evers, H.-D., & Gerke, S. (2004). Closing the digital divide Southeast Asia's path towards a knowledge society: papers delivered at the Centre for East and South-East Asian Studies public lecture series "Focus Asia", 25-27 May, 2004. Working papers in contemporary Asian studies, 5. Sweden: Lund University, Centre for East and South-East Asian Studies. http://www.ace.lu.se/publications/workingpapers/evers%5Fgelke.pdf.

Groshek, J. (2009). The Democratic Effects of the Internet, 1994--2003. *International Communication Gazette., 71*(3), 115–136. doi:10.1177/1748048508100909

Rooksby, E., & Weckert, J. (2007). *Information technology and social justice*. Hershey, PA: Information Science Pub.

Stevenson, S. (2009). Digital Divide: A Discursive Move Away from the Real Inequities. *The Information Society*, *25*(1), 1–22. doi:10.1080/01972240802587539

Vrasidas, C., Zembylas, M., & Glass, G. V. (2009). *ICT for education, development, and social justice*. Charlotte, NC: Information Age Pub.

Chapter 4
Creating Virtual Marae:
An Examination of How Digital Technologies Have Been Adopted by Māori in Aotearoa New Zealand

Janinka Greenwood
University of Canterbury, New Zealand

Lynne Harata Te Aika
University of Canterbury, New Zealand

Niki Davis
University of Canterbury, New Zealand

ABSTRACT

Māori people have a history of adaptation of new technologies. In recent decades Māori innovators have taken and adapted digital technologies for a range of purposes that can be broadly defined as educational. In this chapter, the authors examine three cases where groups have utilised, and 'colonised', a range of particular technologies in order to build capacity for their tribal groups and wider community. In this way they use technologies as tools to overcome some of the financial, social and political deprivation caused by historic and continuing colonisation. The authors initially locate their exploration in a discussion of the historical context of colonisation, Māori movement towards self-determination, and in a discussion of Māori values and approaches to knowledge. They then present the three cases, beginning with one from a formal tertiary education programme (a Māori one), then examining a tribal initiative for language revitalisation and finally looking at a national use of digital media through Māori television.[1]

DOI: 10.4018/978-1-61520-793-0.ch004

Copyright © 2011, IGI Global. Copying or distributing in print or electronic forms without written permission of IGI Global is prohibited.

INTRODUCTION

A marae is a communal meeting ground of Māori in Aotearoa New Zealand. Long before the advent of westerners and western technology to New Zealand the marae with its whare whakairo, the carved meeting house, has served as an open access library of histories, genealogical connections, philosophical and social values, and a site for discussion about current events. Carvings, weavings and other decorative elements within the meeting house recorded the multiplicity of records, stories and debates that were important for the well-being and development of the people. In addition, in a society built on the value of the spoken word, the flow of oratory and debate developed a common space where knowledge (practical, philosophical and of the simple gossip variety) could be shared. In some ways the processes of storing and accessing knowledge were more like the contemporary technological web (weaving together discussion, documentary artefacts, video and music) the traditional libraries of western knowledge systems.

When the missionaries arrived, Māori embraced the new technology of print (as they had other new technologies such as steel tools for carving and the plough for farming). Their acquisition of literacy was extraordinarily rapid. In many regions during the first sixty years of contact the percentage of literate Māori outnumbered that of literate settlers (Reid, 1935). The processes of colonisation produced big changes in the conditions of the indigenous people of these mountainous volcanic islands that some have called 'devastating landslides', in particular the taking of Māori land and the consequent economic and political marginalisation of Māori people, seriously interrupting Māori adaptations of the newly encountered technologies, including writing and print. However, although the effects of colonisation remain, in recent decades Māori have been effective in asserting their right to self-determination in a range of spheres, including language revitalisation and education. As Māori have moved out of the *margins* (Spivak, 1996), where they had been relegated, to reclaim the *centre*, they have reached out to utilise modern technologies, including digital, to attain their goals of well-being (hauora) and self-determination (tino rangatiratanga).

In this chapter we are particularly interested in the ways Māori have taken and adapted digital technologies for a range of broadly defined educational purposes. We explore three cases, each of which illustrates Māori utilisation of different aspects of digital technologies to enhance education. To extend the topographical metaphor in the concept of 'digital divide' we offer the proposition that in terms of indigenous development within Aotearoa New Zealand, digital technology creates less of a divide than an opportunity to repair some of the devastating landslides created by colonisation.

In this context we call the words of the late Monte Ohia, a Māori leader in both mainstream political and traditional arenas:

If we turn our back on e-learning, we turn our back on the future. With the advent, or rather avalanche, of technology now and, increasingly, in the future, Māori need to be in the position of exploring and using it with confidence, as well as predicting what may be just around the corner. It will not be helpful to be either reactionaries to the new trends or, worse still, spectators.... Māori have enough entrepreneurial spirit and opportunism to be at the cutting edge of technological innovation and creativity, and lead in its engagement with Māori learners and resource people. This knowledge and skill will not jump out of the sky at us. In the best traditions of [the creation story of] Tawhaki, we have to retrieve it in cooperation with those who already have it. (Ohia, 2004).

CONTEXT AND BRIEF HISTORY

We appreciate that many of our readers will need some introduction to Māori culture and to the relationship in New Zealand between indigenous people and immigrants. At the same time we are aware of the dangers of offering an abbreviated account of history and of the current socio-cultural context. As a number of critical social theorists (McLaren, 1997; Giroux, 1988; Andreotti, 2008) and postcolonial theorists (Said, 1978; Spivak, 1996; Smith, 1999) remind us, thought is mediated by ideological assumptions and power relations that are themselves historically constituted and that can never be isolated from values. The histories and concepts we will summarise are rich, complex and contested fields of study. We encourage readers who want to know more to read, among others, Orange (2004), Walker (1990), Kawharu (1989), Smith (1999), Simon (1986), Bishop(2005), Greenwood & Te Aika (2009), Greenwood & Wilson (2006). Given these disclaimers we offer the following brief and perhaps subjective, contextualisation.

Māori are the indigenous people who populated New Zealand over a thousand years before the arrival of Europeans. They belong to a number of different tribal groups (iwi). Most Māori will identify themselves by their tribal affiliation to a varying extent, as well as identifying as Māori. For example, although the agenda of self-determination is a pan-Māori one, the structures for achieving it in education or in health are often tribal. There is a distinct language that is Māori, but each iwi has its own dialect variations, particularly in vocabulary and pronunciation. Similarly there is a cosmology and a value system that is distinctly Māori, although mythic stories vary from iwi to iwi. Two of the cases we examine come from an iwi context, though they both have implications for other iwi and have been developed in collaboration with other iwi. The third case comes from a pan-Māori context, but much of the content is shaped by particular iwi.

The formal act of British colonisation took place in 1840 with the signing of the Treaty of Waitangi. This had been proceeded by several decades of contact with European and American whalers and traders and of growing numbers of British settlers. The history of colonisation has many common elements with other countries in the former British Empire, such as Canada or parts of Africa. Such elements include loss of land, loss of language, and erosion of culture, economic deprivation, social and political marginalisation.

One significant difference is the Treaty of Waitangi. The treaty was signed in 1840 as a basis for colonisation and as a guarantee of protection of all the rights, including ownership rights, of Māori. The meaning of the Treaty, its status in law, the transgressions against it, the claims for redress, and the implications for a New Zealand constitution are important and contested fields of study. What we want to highlight here is the way the Treaty delineates the promise of partnership between Crown (now usually interpreted as Government) and Māori and the way that partnership defines New Zealand as a *bicultural* society. It should be noted that such definition is contested and that there are gaps between Treaty-based expectation and common practice. In addition, it is important not to create a dichotomy between biculturalism and multiculturalism. Clearly most New Zealanders, Māori and non-Māori, recognise the increasingly multicultural population, and seek to acknowledge the importance of resulting cultural diversity. Within a New Zealand context the term biculturalism expresses an over-arching framework, what could be called a quasi-constitutional structure in which both Māori and English are acknowledged as official languages and in which the struggles for sovereignty and self-determination are played out. At the risk of oversimplification, it might be said that biculturalism includes and embraces multiculturalism; what is contested is the centre into which multiple cultures will be included.

Claims to the right to self-determination have never been absent from a Māori agenda, but they gained particular force during the 1970s and 1980s of the previous century and continue today. In terms of education, Māori strongly contested the 'deficit' view (Simon, 1986; Bishop, 2005) that problems with Māori underachievement were caused by shortcomings within the home environment and could be resolved by remedial teaching. Instead they challenged the monocultural nature of the education system and sought both change to the system as a whole and greater autonomy for delivering educational opportunities that are congruent with Māori aspirations and values. The struggle for language recovery and revitalisation is reported more fully in our second case. Here it is important to note that cultural values and concepts cannot be separated from the language that gives them expression. Language revitalisation is not only about the socially accepted use of surface discourse but also about the currency of the world view that animates the language. One of the consequences for our writing, as it is for any bicultural discussion within New Zealand, is that we sometimes need to use a Māori term to denote a concept, although we do offer a brief English approximation.

As our title talks about creating virtual marae, the term *marae* perhaps also need a little more explanation. Marae are communal (on a family or tribal basis) meeting grounds, and are at the cornerstone of Māori society. In some ways they are similar to North American First Nations' long houses. They are where important social and ceremonial occasions take place, such as farewelling the dead, greeting visitors or discussing issues of current concern. The carved meeting house is a key part of the marae, and so is the open ground in front of the house. In somewhat different ways both areas are constructed as repositories of knowledge and as space for debate. They are accordingly sheathed with traditional protocols that allow a respectful and safe space for participants' engagement. In the discussion that follows we refer to ways in which the protocols and artefacts of the marae have been applied to contemporary media to facilitate Māori participation.

ILLUSTRATIVE CASES, AUTHORS, AND THEORETICAL FRAMEWORKS

The three cases we describe are:

1. e-learning foundational studies at Te Wānanga o Raukawa, a Māori, a tribal university;
2. tribal (Ngāi Tahu) language revitalatisation using digital technologies;
3. digital media used by Māori television and film.

Each demonstrates a way in which digital technology and e-learning have been explored and, to return to the previously cited words of Ohia, 'used with confidence', and perhaps 'predicting what is around the corner'. They are offered as examples of innovative practice rather than illustrations of widespread practice. In each there is a strong connection between technology and Māori development towards sustainability.

As authors we relate in a number of different ways to the study of Māori self-determination and to the specific data discussed in the three cases. Lynne Harata is Māori and relates to Ngāi Tahu, Ngāti Awa and Te Whānau Apanui. The tribal history that backgrounds the second case is her history and as a teacher educator, and tribal developer she has been an active participant in the events and strategies described. She and Janinka recently completed a major study (Greenwood & Te Aika, 2009) of the factors that contribute to Māori success at tertiary[2] level. The data in the first case is drawn from their report. Janinka is a Czech-born childhood immigrant to New Zealand and acknowledges both Māori and European teachers in her adopted country. Her research includes cross-cultural perspectives in theatre and material in the third case is drawn largely from her research. Niki is a very recent immigrant from Northern Ireland who brings into the team her extended practice of research and development of digital technologies in teacher education in the UK and USA.

As a result of our differing backgrounds we have come to this project with a diverse range of theoretical frameworks. In our formative discussions we have drawn on indigenous, and particularly Māori, conceptualisations of knowledge, research and development, as well a number of critical social theories, particularly in the areas broadly labelled as postcolonial, critical literacy, and social ecology. It would be beyond the scope of this chapter to attempt an alignment of these approaches, and possibly a denial of the value of indigenous perspectives which tend to reject discourses. Moreover, indigenous perspectives are wary of discourses that prioritise western concerns, especially if they seem to subsume indigenous concerns (Smith, 2005; Four Arrows, 2008). Therefore in this chapter we give priority to an indigenous Māori approach, that invites discussion of experience, personal and reported, and analysis in terms of self-determination and development. At the same time, because we are all academics within a predominantly western university, we highlight relationships with western socio-critical theories when relevant. In the section that follows we briefly review some of the key relevant literature.

Literature Review

Literature relevant to this chapter includes a considerable body of works discussing digital divides, the history of New Zealand colonisation and Māori recovery, the relationships between language, culture, identity and well-being, the role of education in development, and the key elements of the conceptual frameworks we mentioned above. Here we offer a brief sampling of what seem to us some of the key ideas and discussions.

Discussion in the literature of "digital divides" addresses not only access to computing hardware and the internet but also the degree to which potential users perceive digital technology as useful and usable. Within the global discourse of technology competence and effectiveness, Haythorthwaite's (2007) recent examination of how the digital divide matters for e-learning takes a deficit view in which those in the 'advance party' gain power. She notes that Castells and others have argued that differential timing in access has led to power and information for groups that had early access and this continues. Late access may result in "always running to catch up" and never gaining power. Haythorthnwaite provides evidence of a spectrum of digital divides across gender, age, region (in the USA and Europe) while also clarifying the close connection between poverty and lack of access to digital technology as well as other resources. Our first case invites a reconsideration of this approach.

Davis' (2008) ecological framework identifies the teacher as the 'keystone species' in the evolution of educational ecologies, which in the context of the examples in this chapter might well encompasses family, artists and programme creators. Her ecological perspective clarifies reasons for the reduction of the impact of digital technologies when the teaching and learning ecology does not fit with the ecologies with which it must interact. This includes the school ecology, the wider examination system and the home. On the other hand for indigenous people (and probably others), digital technologies offer communicative and educational opportunities beyond those of school, as our first case suggests.

The developments described in this chapter are perhaps more in line with Dutton's (2004) description of the educational, social and economic challenge of digital technologies in terms of digital technologies as a "double edged sword" with which users (and non users) influence the access and applications of digital technologies through their engagement with them (both increasing and decreasing). Populations who do not use digital technologies have little influence on market-driven evolution. In addition, increasing use of digital technologies results in evolution of relevant resources for users, including relevant languages and cultures. The three cases we discuss show how Māori have grasped this "double edged sword" to build Māori capacity and to spread Māori language, culture and values.

The World Summit on the Information Society in 2004-2005 highlighted the plight of many peoples who were being left out of the rapid developments accelerated by digital technologies. Deer and Hakansson (2005) describe the challenge created by the view of information as a commodity rather that as knowledge, which is "as diverse as individual, social and cultural diversity" (p. 68). At the same time they recognise the role that indigenous cultures play as guardians of biodiversity who have already contributed many important aspects within western culture, including the shaping of democratic ideas. Deer and Hakansson express concerns that "knowledge, information, communication and technology are not context-free, but intimately interwoven with ethical obligations towards the entire web of life." (Deer and Hakansson, 2005, p. 76) In support of this perspective Resta, Christal and Roy (2004) identify new technologies as having the potential to support and sustain native culture and also the potential to accelerate its erosion. They also emphasise the importance of collaborative protocols that prevent usurpation of indigenous knowledge. Similar protocols that ensure participants' co-construction of the narratives are emphasised in New Zealand (Bishop, 2005; Smith, 2005; Macfarlane, 2008; Te Aika & Greenwood, 2009; Greenwood & Te Aika, 2009).

The history of colonisation of New Zealand and its impact on Māori is thoroughly discussed by, among others, Orange (2004), Walker (1990), King (2004), Belich (1996), and from a Ngāi Tahu South Island perspective by Evison (1993). They describe the loss of land, the erosion of rights guaranteed under the treaty of Waitangi, the marginalisation of Māori social, economically and educationally, and the devastating impact of introduced diseases. The oral traditions of the marae as well as the archival records researched and reported by Orange and King testify to an unbroken history of petitions for restoration of land as well as of initiatives by Māori to improve the health and education of their people. To some extent it is useful to apply the lens of postcolonial theory to the processes of marginalisation (Spivak, 1996), other-ing (Said, 1978), and re-inscription (Ashcroft, Griffith, & Tiffin, 1989) as such theorisations allow us to understand colonisation as a global, and continuing process that is predicated on power and the perceived right of the powerful to access resources as needed and to construe others in terms that allow continuing use of power. However, from a Māori perspective, as also from other indigenous perspectives, it is important not to rely on postcolonial literature to address the particular grievances, aspirations and developmental projects of Māori. A global theory carries the threat of a totalising discourse (Smith, 1999) and can threaten to reduce local needs and developments to mere illustrations of global concerns. A growing body of New Zealand literature (Durie, 1998; Pere, 1991; Bishop & Glynn, 1999; Te Aika & Greenwood, 2009) addresses the need to inscribe Māori concepts and development strategies in terms of indigenous values and understandings.

The interconnections and interdependencies between language, culture and identity are repeatedly addressed in New Zealand, as well as global, literature. A seminal work is the Report of the Waitangi Tribunal (1986) on the Māori claim that language is a hereditary right protected by the Treaty. It collates the evidence of tribal elders, examines the unsatisfactory record of monocultural schooling in advancing the interests of Māori, and affirms the integral role language plays in vitalising and transmitting culture, in building connections between people, families, their histories and their ancestral land, and in enabling people to express their values and thoughts in ways that are consistent with their Māori identity. Such affirmations have resonances with the writings of a number of socio-linguists, such as Gee (1992) who details how a society endows its members with a 'Discourse' that predicates conceptualisations, values and social position as well as semantic resources. The unpublished writings and oral testimonies of Rangihau Karetu and Waikerepuru[3] provide a framework for the ways traditional language forms can accommodate contemporary technological innovations, and offer explanation of the relationship between language and ancestral knowledge.

The relation between culture and education is also extensively explored in global and New Zealand literature. Among the New Zealand writings are a number (including Smith, 2000; Reedy, 2000) that address the importance of developing systems of schooling that allow Māori not only to learn their language but also to learn about the world as a whole *in* their language, and so learn the conceptual frameworks that allow them to realise their identity as Māori. In a more general context, Greenwood and Wilson (2006) trace how important it was in the 1970s and 1980s, the early decades of Māori resurgence, to create opportunities for communities as well as schools to learn about Māori ways of seeing the world, and explore ways in which this might be facilitated through the arts. Similarly Macfarlane (2007), focusing particularly on students with challenging behaviours, shows the importance of culture and cultural values in creating a supportive and effective learning space for Māori students. More discussion of the processes of language revitalisation and further related literature will be found in discussion of the second case.

Indigenous research is an umbrella term that covers a number of different indigenous peoples' approaches to research processes and epistemologies. What they have in common is that they assert the existence of different knowledges, they prioritise the value of their own inscriptions, and they actively direct the purpose of research to building their people's capacity and well-being (UN General Assembly, 2007; Durie, 2001). In this chapter we draw on Māori versions of such perspectives, commonly referred to as Kaupapa Māori research. In particular we draw on assertions of the importance of collaborative approaches holistic approaches to knowledge and the overarching goal of self-determination and the development of well-being.

Having laid this brief foundation of literature, we now present the three cases that inform this article, starting with one that is set within formal tertiary education.

Foundations Courses in e-Learning in Te Wānanga o Raukawa

Our first case is a re-examination a case study that was part of a recent study (Greenwood & Te Aika, 2009) of tertiary programmes that are successful for Māori. That study utilised a research methodology firmly grounded in kaupapa Māori approaches, and drew on the voices of tribal elders as well as participants within tertiary institutions. A co-constructive approach to the development of narratives and analysis was used, of which a detailed transcription can be found in the full report of the study. The data and quotations discussed here are drawn from the project report.

Te Wānanga o Raukawa was set up in 1981 by a confederation of three affiliated iwi groups (Te Ati Awa, Ngāti Raukawa and Ngāti Toa Rangatira) to provide a university pathway that would provide nationally recognised degrees in fields that were seen to be important to tribal development, initially Māori language and business studies. It was the first of the several Māori tertiary institutions that now provide degree courses.

Students in the Wānanga are of all ages. Some have recently left school, some are men and women of all ages, including elders, who take the opportunities offered by the Wānanga to improve their skills and gain qualifications[4]. Over half the students study by distance in both New Zealand's large islands, some from their own homes, and some from marae that have genealogical ties with the local tribes. Distance students attend several residential courses. In between these intensive face-to-face classes, students' complete further work through assignments and self-directed study and research with blended e-learning. (Learning with and through digital technologies is commonly refereed to as e-learning. The term blended e-learning is used to signal that the e-learning is designed to blend with the pedagogy; for example, the use of email for communication between the teacher and learner and between learners.)

The decision to include foundation studies in computer competencies and e-learning in every programme was the result of an understanding that Whatarangi Winiata, leader and founder of the Wānanga, gained about the emerging significance of digital technologies. The story is told of some years ago Winiata had needed to urgently contact the Wānanga while abroad in Paris. There was a problem with phone connections from his hotel, and so he was taken to a nearby internet café and logged in. He wrote then to his colleagues of his determination that his people would not miss out on this wave of technological innovation the way they had missed out on so many previous ones. His practical vision included the supply of a computer to every student at the Wānanga with the understanding that the computer would be used not only by the student but also by his or her family and whānau, and so support and develop the whole tribe's digital capabilities. As a result, a computer and internet project was initiated with the goal of adding digital technology to five hundred Māori homes a year.

The Wānanga's approach to e-learning has some similarities to the courses typically found in other institutions and some distinctive features. Perhaps the most significant distinctive feature is the application of a Māori perspective to context, content, learning processes, and interpersonal interactions.

The core philosophy of the Wānanga is to ensure the survival and advancement of the iwi or tribes that it belongs to. Advancement has two complementary aspects: as Durie (2001) describes it, it involves development as Māori and as successful citizens of the world. Consequently, all programmes within the Wānanga are positioned in the dual framework of the acquisition of practical and career-orientated knowledge and skills as well as in the nurturing of Māori ways of seeing the world and interacting. In Māori terms the goals of education are the development of communities as well as individuals and engage spiritual and interactional as well as intellectual domains. To ensure that it is possible to present learning through Māori eyes, all courses are placed in a context that is guided by customary behavioural protocols and values. Such a context is provided on the grounds of the Wānanga itself, on various marae around the country where it has agreed to develop courses, and in the interactions with students on-line and through the help centre.

Content is always related to the applications for family and tribal development. For example, in e-learning, as students are introduced to a new software application or a new process, part of their assignment will ask them to consider how it can be used in their particular environment for the marae or tribal community with whom they work. It is seen as important that knowledge is never seen as disjointed or alien, but continuously integrated with Māori experience and aspirations. The use of Māori terms to identify the components of the computer and key processes is an integral element in ensuring that students can create meaningful bridges between new, apparently western, technologies and their Māori world. A bilingual approach to instruction is thus used in which course materials and web pages have Māori images and language blended with English. A cartoon figure of a motherly elder in a red dress, Whaea Manu, appears on screen during training sessions to greet students and to field their questions. The continuing challenge is to ensure that the new technology, rather than subsuming Māori culture, is imprinted with the symbols and language that make Māori recognisable and distinct as a people.

To ensure that the Māori application is authentic and relevant, the Wānanga has determined to retain control of its teaching materials. For example, Microsoft had offered to collaborate in developing, and subsequently marketing, Māori language materials to accompany its office software, but the Wānanga resolved to operate independently because it did not want to lose control of decisions made and risk a tokenisation of language or cultural concepts for commercial purposes.

Decisions about pedagogy are also made on the basis of Māori protocols and values. Because the foundation studies are the first tertiary experience for most students and because many of them had negative previous experiences of schooling, the development of a safe and nurturing environment and relationships is seen as important. The marae's traditional styles of interaction provide a model. Greenwood and Te Aika's (2009) study cites examples of how students react to the environment. For example, an older student explains: "I love the Wānanga's style as we are learning like our tīpuna (elder) in a traditional way. We study, eat and share together. It's not a one hour class and goodbye. It's learning whānau (family) style." (p. 85).

Learning on the marae is multi-facetted: there are occasions where experts lead, there are others where learning is by doing, and others when ideas and skills are developed by peer interaction; sometimes all happen at once. Students at the Wānanga appreciate the recognition of different learning styles. As one observed: "Some people have their different ways of learning. See, there are some that like to be on their own and study but there's some that like to be in groups and have input from other peoples. I learn from other people." (p. 85).

A significant number of the Wānanga's students are kaumātua (elders), many of whom are motivated to support future generations including

their own grandchildren. Learning materials have been developed to accommodate the needs and preferences of these elders. For example, intensive courses utilise both individual books and a full wall screen that many elders prefer to view. In addition, books are produced in a larger font for those who have reading difficulties, and the use of Māori terms particularly serves this group in making new concepts and strategies more familiar. Initially the Wānanga spread kaumātua throughout the groups, but experience showed that as a whole the kaumātua group had different attitudes to technology than the young people and needed to approach it in their own way. As a teacher explained: "The young people can be very casual in their approach and like to joke around with each other especially when they make mistakes and a kaumātua does not expect that kind of behaviour and the lack of respect that is shown." [Greenwood & Te Aika, 2009, p. 75) An initial concern that some elders would pick up the technology much faster and need to learn in a separate faster group proved ungrounded: The same teacher reported: "The acceptance and respect shown by everyone within a kaumātua group ensures that students who are faster with technology do not clash with others who need to go over everything a number of times." (p. 75).

Similarly the Wānanga recognises the importance of matching the classroom experience to the home experience as much as possible. Many homes would only have internet access through a telephone jack that they would plug into their computer, and without a training experience based on this kind of connection the computer was likely to stay in its box. So intensive classrooms were rigged with a multitude of telephone lines as well as a computer laboratory were setup.

Māori protocols also guide the style of interactions. For instance, the help desk was initially set up on western lines, but the founding CEO Winiata's advice was to ensure they used "whaka papa and reo [relationship ties and language] to support the whole process". The helpdesk co-ordinator explained: "We found that by the time our students rang our help desk they had an issue [that] they have been trying to deal themselves and we are the last port of call. By the time they ring us, the problem is quite full on and they are not thinking right. We've found that if we talk to them in te reo (Māori language) 99.9% of the time it works to calm them down when they are agitated." He further illustrated by citing an incident where the technical problem was resolved relatively quickly, but the restoration of good relationship was an equally important though additional task: "Craig had the longest call – three hours. The issue was actually sorted in the first five minutes. It was an old kuia (elder). That's about being Māori and that relationship thing." (p. 58).

Māori emphasis on relationship ties also informs how the supply of computer hardware is handled. At enrolment, each student is required to have a laptop, printer and internet connection at home. (The college has negotiated packages for their students to enable them to access quality equipment and an Internet connection at cost price, which may be added to the fees and student loan.) Computer leases have three signatures: the student's, the Wānanga's and of a co-signatory from the tribe. The co-signatory, in addition to guaranteeing the loan, commits to give guidance to the student in his/her tribal studies and of kaupapa Māori. Such a guarantee supports the student to honour the trust, complete the course, and share its benefits.

Emphasis on relationships, tribal connections and reciprocity also impact the choices the Wānanga makes when placing courses on other tribal marae. The first priority is always developing the viability of the local affiliation of tribes: the next is to support the autonomous development of groups who share genealogical connections or who have contributed to the affiliations forged by Te Wānanga o Raukawa in the past.. Thus expansion is not determined by desire to grow the institution but to participate in the project of self-determination and capacity building.

The compulsory e-learning foundation course lays the basis for further academic study. The first course consists of nine short modules that include introduction to technology, introduction to computers; set up and care; word processing; introduction to email; internet use and on-line training. The second year course introduces students to desktop publishing; business communications, spread sheeting, power point presentations and on-line training. Throughout, students are asked to apply this knowledge in the assignments and presentations required in other subjects. The courses are structured so that students gain credits in nationally recognised qualifications as well as developing the potential of the new technology for their communities.

NGĀI TAHU KOTAHI MANO KĀIKA: NGĀI TAHU'S STRATEGY FOR LANGUAGE REVITALISATION

Ngāi Tahu is the *iwi* or tribe who asserts chiefly rights to most of the South Island of New Zealand. The history of alienation of its lands in the first decades of colonialism led to its socio-political marginalisation and consequently to severe loss of the indigenous language, te reo Māori (Māori language), and its southern dialects. The details of Ngāi Tahu treaty claims and the establishment of a corporate tribal structure are outside the scope of this discussion. However, in 2000 after a series of consultation meetings with tribal members, Ngāi Tahu embarked on major tribal strategic planning initiatives in key areas of health, education, language revitalisation, housing, employment and development of regional tribal councils. Ngāi Tahu set strategic goals and long term visions to be achieved for all of its tribal members by 2025, as well as intermediary goals to check progress on the journey. Ngāi Tahu were creative and proactive in their thinking around tribal development and have embedded the notion of sustainability and providing for the future needs of tribal members into their tribal pepeha or guiding mantra, "Mo tatou, a, mo ka uri a muri ake nei": for us and the generations to come.

In the area of language revitalisation Ngāi Tahu identified that the intergenerational transfer of Māori language as the main form of communication had not occurred within South Island Ngāi Tahu for about 80 years and in some areas for 130 years. It also identified that there were approximately only three remaining native speakers still living within the tribe that spoke forms of the southern dialect (Te Karaka, 2001).

The challenge was to build areas within the South Island with a high enough density of fluent speakers of Māori for intergenerational transfer to occur and at the same time develop and maintain links with members living in other regions or overseas. Ngāi Tahu as a tribal group consists of approximately forty thousand people with forty five percent living in the tribal region, fifty per cent living in other parts of New Zealand, and five percent living overseas. The tribe also has a high number, nearly sixty percent, of its members under thirty years of age.

Ngāi Tahu began its language strategy by establishing a formal Language Planning Committee in 2000. The committee consisted of tribal members with expertise in a number of areas including, tribal development, language teaching, publishing, and members from all sectors of education from early childhood through to tertiary education and adult life-long learning. The Language Planning Committee also consulted with world renowned language expert, Joshua Fishman from New York, who advised the group to 'focus on the home and intergenerational transfer'. In his opinion this was the only way language revitalisation would be achieved if the parents and children were speaking te reo and Ngāi Tahu dialects in the home (Te Karaka, 2001). Following his advice and after deliberating over other accompanying priorities the Ngāi Tahu (Ngāi Tahu Development, 2001) language strategy KMK was developed, *Kotahi Mano Kāika, Kotahi Mano Wawata*. (Translation:

a thousand homes where the language is alive and transferred from one generation to another to achieve a thousand dreams and aspirations.)

This case focuses on the e-learning components of the KMK strategy, which have been implemented to enhance language access and language engagement for tribal members assisting in Ngāi Tahu identity reconstruction and rebuilding cultural capacity. Resta, Christal and Roy (2004) state that new technologies have the potential to support and sustain native culture and also the potential to accelerate its erosion. Technology in this instance is used as "a tool of empowerment and offers the potential for indigenous people to create their own cultural content and curriculum resources at their own speed, in their time and under their own conditions, using their own knowledge and judgment that defines equity/equality"(Delgado cited in Resta et al., p. 183). It would be fair to say that the e-learning component was initially supplementary to other tribal language revitalisation initiatives. However, as the implementation of the strategy developed, the Language Planning Committee and subsequent advisory groups became more aware of the need to engage the 'tribal masses' not just the converted language followers who faithfully attended language hui (gatherings) but also those who were distanced by geography, time and physical access. The South Island of New Zealand has a large mountain range called the Southern Alps which runs centrally through the whole island and separates the east coast of the island from the west. The application of digital technologies has therefore increased over time to become a major component.

One of the first e-learning initiatives was to set up a website (http://www.kmk.maori.nz) to promote Kotahi Mano Kaika in 2002. On this website Ngāi Tahu families were given access to a range of language resources to assist them in learning te reo Māori and southern dialects. Some of the resources were in the form of sequenced texts called *Te Hu o Moho* with communicative everyday language for use in the home with strong examples of the southern dialect and resources that could be downloaded, thereby overcoming the barrier of distance. The resources include posters for children to learn Ngāi Tahu dialect specific words, and a hip hop and r'n'b album for youth called *Tenei Te Ruru* which was a compilation of youth culture waiata (songs) by various Ngāi Tahu artists. Other waiata, both contemporary and traditional, such as *Te Hā o Tahupotiki*, were produced for a wide range of age groups and were also made available on the site. The songs have downloadable MP3 sound files that can be played on many devices including iPods and PDAs. Access is free not just to Ngāi Tahu tribal members but also to anyone who uses the site. This has proved to be a less expensive way of providing cultural learning tools than costly glossy printed books where numbers had to be restricted to small runs of 2-3000. The biggest advantage of this new e-learning tool for the tribe was enhanced access. It was instant, inexpensive and, with accompanying sound files, provided a self learning tool for all those who used it. The website continues to be developed over time to offer an increasing array of language learning resources for all age groups.

Thus, Ngāi Tahu tribal innovation in e-learning included adapting new forms of technology to benefit transmission of language and culture, to build tribal capacity and to enhance the status of the language amongst its own tribal members and members of the wider community and dialect group. The resurfacing and reconstruction of the southern Ngāi Tahu dialect, which was almost extinct like the native moa bird in New Zealand, has also helped revive southern historic traditions, language, and cultural practices that had been eroded through colonisation and oppressed by the more dominant use of northern dialect as the lingua franca for speakers of Māori language. It combined new ways of knowing with old ways of knowing and, in particular, targeted the children and youth culture of the tribe.

In addition, other e-learning initiatives to increase use of Māori language have been trialled and in particular it was felt that a language critical awareness campaign was needed to persuade the disengaged masses in the tribe about the importance of language revitalisation as a key to revitalising Ngāi Tahu culture and identity (Te Aika, 2004). A pilot campaign utilised some of the strategies udsed in chain mail campaigns to spread the language revitalisation message among young people. Over one weekend 900 hits to the tribal language site resulted from people sending on advocatory messages to their friends. This led to the development of the Generation Reo campaign (http://www.generationreo.com; Tarena, 2009). Key messages in the campaign included:

*Te Reo is a fundamental base of our culture as Ngāi Tahu and as Māori letting people know that the language is in danger and that everyone has a part to play in ensuring it survives into the future. It's about getting as many people as possible-**speaking**te reo Māori. Learning it isn't enough anymore. To survive, te reo needs to be used, to be spoken, to be heard. It sounds simple enough, speak te reo and it will survive. You have a part to play. Do something today. Join the revolution… Generation Reo … it takes one generation to lose a language and three generations to get it back. (Te Runanga o Ngāi Tahu, 2007).*

These chain emails were sent out to registered KMK participants and then they in turn sent the message to someone passionate about te reo Māori. The language campaigns have sound files and are persuasive with Ngāi Tahu cultural propaganda. The branding and colours used on the website are clear and simple. The messages are meant to have an impact on the listener and reader. Ngāi Tahu celebrity voices are used in the campaign and help pull on tribal 'heartstrings' to encourage tribal members to participate. Here is another example of a text used in the campaign:

Every generation needs a cause. Our old people united to fight the Ngāi Tahu claim. Let the cause of our generation be saving the language. Stand up and unite and be a force for change. Saving our language starts with you. Join the revolution Generationreo.com. (Te Runanga o Ngāi Tahu, 2007).

One of the advantages of a website is that the number of website hits can be monitored and the length of interactions monitored as well. For example if the website is only encountered for 30 seconds then the subject matter has not caught the viewer's attention but if it is encountered for 3 minutes or longer then something is likely to have successfully engaged the viewers attention on the website (Tarena, 2009).

Other sections of the *Generation Reo* website are designed to stimulate the visitors to start out on the journey of learning te reo with tips and hints for beginners, as well as frequently asked questions about bilingualism. A promotional DVD was produced in 2008 for parents about raising bilingual children. Families were interviewed and featured in the DVD talking about issues about raising children speaking Māori and speaking the Ngāi Tahu dialect. Only a limited number of DVD's were produced, and so in another initiative to reach the masses the DVD was put on *Bebo* and *You Tube*, as well as the main *Generation Reo* website. This video redistribution of the DVD content has reached wider audiences who continue to engage with the video and write their own reactions and thoughts about it via blogs and commentaries on *Bebo* or *You Tube*. In this way, these e-learning approaches aimed to blend in with the learners' family life and provide cross generational access to digital technologies.

These e-learning applications have helped Ngāi Tahu to further develop their language resources and have provided a teaching and mentoring tool where the viewers and students can also become teachers or critical friends as they publish their own thoughts, commenting on and reviewing the contents of the video and discussing its merits or making recommendations to others. This is a

higher level of interactive engagement and requires higher level thinking skills. The web site viewers are not just a passive recipient but they react and cause others to react in their commentaries. Again this is popular with youth, but not exclusively so.

The community moderates the site effectiveness in their critique and feedback and the virtual whare korero- or (house of talk) comes alive. Not all discussions and reactions are instantaneous; people can enter and depart this virtual marae whenever they like. Everyone has the opportunity to become a speaker, writer and publisher. It is also possible to track the discussion and commentary. Thus the virtual marae is one that is available around the clock.

There have been further suggestions for development of other initiatives to support cultural transmission on the Ngāi Tahu language websites. An example would be a Wikipedia of ancestors to help regenerate knowledge of genealogy connecting descendants with their notable ancestors, and to learn more about the traditions of the ancestors bringing the past into the present (Tarena, 2009).

Māori TV including Digital Media

This cluster of examples involves the use of digital film media for broad education purposes. Earlier we suggested that to some extent computer technology provides a contemporary equivalent to traditional marae-based education processes. In this case, we look at ways in which digital media are being used by Māori for open access educational processes. First we look at how Māori television performs a teaching role in terms of language, content, design and focus on Māori as a people. Then we look more specifically at the work of Don Selwyn, a visionary leader sometimes called the 'godfather' (Greenwood, 1999) of Māori film, theatre and television, and in particular at his film *The Māori Merchant of Venice.*

Māori Television operates as a full time public access 'around the clock' channel and is now in its fifth year of operation. Its significant difference from mainstream television is that it began as the result of a Māori initiative (in fact political insistence), is operated by Māori for Māori (in the first instance) and has an explicit agenda of supporting Māori development and autonomy: an agenda that we can easily construe as educational.

As Māori television has grown, it is watched by an increasing number of non-Māori New Zealanders who enjoy its home-grown and relatively non-commercial focus. In this way its educational role has expanded to one of teaching all New Zealanders something about Māori language, values, perspectives and arts as well as inducting them into a view of New Zealand that is Māori focused.

One of the roles of Māori television that is most overtly educational is the promotion and active teaching of Māori language. The Māori Television website explains: "Māori Television has been established as one of a number of important initiatives to promote and revitalise the Māori language. The aim of our channel is to play a major role in revitalising language and culture that is the birthright of every Māori and the heritage of every New Zealander" (Māori television, 2009a). This project takes a number of forms. There are regular short programmes where vocabulary and grammar structures are introduced and practiced much in the way they might be in a language lab. There are programmes that are completely in Māori language (usually with subtitles) that allow those with Māori language competency to practice aurally, and to experience the cadence and richness of the language as a communicative tool. Language contexts include elders recounting tribal histories, news reporting, debates on topical issues, youth programmes discussing contemporary music, sports and entertainment. There are also a number of bilingual programmes that in a range of ways reflect the way Māori speakers weave both English and Māori languages together in informal conversation. This might be an interviewer speaking Māori with an elder and then turning to the audience and recounting part of the narrative in English. Or it may be a dialogue that reflects the

character of 'Māori English', a particular New Zealand dialect of English which incorporates Māori words that more readily capture concepts that seem difficult to express in English. The television channel acts as a vehicle of cultural reproduction in terms of both Māori language and Māori styles of expression and normalises them, not only for Māori viewers but also for the increasing number of non-Māori. Consequently, more than 1.5 million New Zealanders now tune in to every month (Māori television, 2009b), who, like several of the authors of this article, actively turn to the channel as an escape from imported pulp serials.

Content is as important as language. For example, some items in the Māori news are the same as in the mainstream news, though they may well come with a different interpretative commentary. However, there are other items which would probably never be seen on mainstream news. They are indicative of the different ways that cultures frame their world, and in this way the channel serves to make visible a Māori sense of values. The same occurs with documentaries which may, for example, record tribal histories or dialogues with elders about an aspect of ancestral knowledge, perhaps the history of a place, the use of a medicinal plant, or the right ways to harvest seafood. At the time of its launch the chairman of the channel explained: "From its inception the board of Māori Television has held a strong view that this channel should also be a window into the Māori world, imparting knowledge to all New Zealanders (Walden, 2004). As Gee (1992) reminds us, each society endows its members with a discourse that both empowers them to communicate with each other and constrains them in what they will communicate about, and thus know. The impact of colonisation on Māori comes not only through loss of land and the imposition of laws; it also comes through the discourses that are allowed and encouraged by society, and through the loss of those that are marginalised by it. The effect of Māori television is to give power to discourses that reflect Māori values and knowledge systems. It is not simply a matter of giving them air time, it is also about presenting them as normal rather than exotic, of placing them at the centre rather than at the margins (Spivak, 1996) of society's recognised discourses.

The channel imports a number of foreign programmes. Among these are documentaries reporting the development, and particularly the language revitalisation programmes, of other indigenous peoples. There are also foreign movies, with a special but not exclusive focus on films in other languages than English and those that highlight some aspect of non-Anglo cultural identity. A respected contemporary Māori scholar, Mason Durie, has defined the Māori educational agenda as enabling Māori to be citizens of the global world and of a Māori world (2001). The imported programmes situate the global world within the perspective of a Māori centre. Moreover the privileging of non-Anglo voices within this global context serves to remind viewers, Māori and non-Māori, that other discourses exist outside those that define New Zealand's postcolonial mainstream.

Design is an important component of the way Māori television presents a Māori perspective. Digital technology offers a relatively inexpensive means for creating environmental design and animation. The studio sets feature Māori motifs. The channel's logo is formed by small rubber-limbed warriors who wield their taiaha, battle staves, in the rituals of challenge and welcome before they morph to form the words "ma ratou, ma matou, ma koutou, ma tatou", which the website briefly translates as 'for everyone' and then the name of the channel. The sequence is about Māori branding, but it is also a digital enactment of an important ritual of encounter and it serves to evoke resonances of the spirituality of people coming together for debate and entertainment, a sense of spirituality that is not usually connected with mainstream television.

Encounter is seen as an integral part of the Māori television agenda. There is actual physical encounter. The studios have been designed to recreate the open spaces of a marae "so that, at any time of the day or night, members of the public can view television being made" (Māori television, 2009b). And there are opportunities for any member of the public to 'have their say' by coming in and recording their message at Te Kokonga Kōrero, Speaker's Corner, with the opportunity of it being screened. And, of course, there is virtual encounter on air. Part of that encounter is the profiling of Māori musicians, painters, sculptors, writers and other artists who might otherwise linger in the margins of mainstream media, and who now, to a greater or lesser extent, are part of the overall shape of contemporary New Zealand culture.

The driving force behind Māori television is the result of the commitment of a great many Māori who, over at least thirty years, saw the potential in the media to accomplish Māori developmental goals. The practical platform for the channel is the use of relatively inexpensive and accessible digital technology which makes film-making more dependant on drive than on grants of substantial external funding. Both these factors are exemplified in the work of recently deceased Don Selwyn.

Selwyn was one of the primary 'movers and shakers' in the development of Māori theatre, film and television. He was pulled into theatre by the great Māori opera singer, Inia Te Wiata, who saw the potential of theatre to express stories Māori had to tell. In the 1960s the Māori Theatre Trust was formed. It demonstrated the production values of combining Māori mythology and traditional Māori performance forms with the lighting and sound technology of the modern stage (Greenwood, 2002). However, as the Trust gradually turned from innovative theatre to an ambassadorial and commercial concert party, Don broke away and engaged in grassroots productions of new Māori plays and, with a number of his contemporaries, became involved in New Zealand's new film and television industry. He saw the potential of television "to bring much more global coverage for minority or indigenous perspectives" (Selwyn cited in Greenwood, 2002).

One of Selwyn's projects was the setting up of Tamaki Creative Māori Arts, a training programme for first actors and writers, then production technicians. "One of the things that struck me, being an actor and working in front of the camera was that there were not many Māori behind the camera," he said, "and that's what drove us to start the film school. Now it's not an issue any more because there are so many" (Selwyn cited in Greenwood, 2002). Don further commented that the shared cultural perspective brought by Māori technical crew members was integral to the success of Māori film-making. The availability of a good array of Māori production specialists is an important factor in the current success of Māori television.

Over time the film school developed into He Taonga Productions. Using a platform of low cost digital technology, and a significant understanding of the craft, Selwyn produced a significant number of cutting-edge Māori films. Among these is *The Māori Merchant of Venice*. In the 1940s Pei Te Hurinui, intrigued by both the character of Shylock and the cadences of Shakespeare, had translated the play into classic Māori language. Selwyn was "totally inspired by the ability of [Pei Te Hurinui] who hadn't come through a western education system but had the capacity to handle the classic English language and translate it into classical Māori" (Selwyn cited in Greenwood, 2002). Unable to get a commercial backer, Selwyn created the film with digital camera and a basement studio editing suite.

Within a context of education, the film is a landmark in a number of ways. It records the magnificent cadences of Pei Te Hurinui's classical Māori spoken, simply and strongly, by a cast of native speakers. It frames the racial tension in Shakespeare's play within the context of a cross-cultural New Zealand and provokes us to reconsider our own society. Perhaps most impor-

tantly, it challenges our understandings of what is *normal* by situating the entire action within a Māori framework, where even the immigrants, in this case the Jews, speak Māori and are surrounded by Māori visual motifs. The very act of framing a Shakespearean play within a Māori context (so that non-speakers can only access the text through short subtitles) is an interesting performance of postcolonial deconstruction. By situating the Māori world as the mainstream that assimilates the art work of an exotic foreign culture, it poses a reconsideration of the assimilative presumptions of a colonising culture. Yet at the same time as it questions, it allows a celebration of the imported art within the context of traditional Māori oratory and poetry.

Before he turned to theatre, Selwyn had trained and worked as a teacher. He came to see theatre and film as a more subtle and persuasive way of provoking thought and new understanding. Like Whaturangi Winiata, who envisioned the importance of e-learning as described in our first case study, Don perceived the advent of digital technology as a movement that could assist the development of Māori and enable Māori to speak in ways that are really Māori and to share them with not only New Zealand but also the wider world.

EMERGING THEMES, CONTROVERSIES AND SOLUTIONS

The three cases illustrate a number of recurring themes that we hope have useful messages for indigenous people in other parts of the world as well as our whole global society. Perhaps the overarching theme is that we reject a deficit perspective implied in the term 'digital divide' and instead signal that digital technologies are valuable to the resurgence of indigenous peoples when coupled with the philosophy and culture of the people in a way that builds capacity for the evolution of the application of the digital technologies in inclusive multicultural societies. We now surface four themes that we have identified in all our case studies.

Resurgence of Māori Language and Culture Supported with Digital Technologies

Digital technology has become a significant tool in the renaissance of Māori language, both in its pan-tribal national form and in its tribal dialects. In addition, the language has grown to accommodate the various elements of modern technologies and the activities associated with them. The Ngāi Tahu case shows how a range of informal communities as well as the tribe's strategic planners have utilised formal and informal digital processes to advance language recovery and, increasingly, to communicate within the recovered language. Te Wānanga o Raukawa has developed a way of indigenising Microsoft's instructional guides, using Māori terms for computer parts and e-learning processes, creating learning contexts that apply to family and tribal development, and ensuring that support structures operate in terms of Māori rather than western protocols. Māori TV and film keep the classical language of the elders alive and provide a vehicle for a range of orthodox and less orthodox versions of Māori language to be learned, used and further developed.

Importance of Visionary Leadership

In all three of the cases we have examined, productive pathways were opened up because of the drive of a handful of people who had the vision to see the potential of the media and who strategised to use them for the advancement of their people. Māori society is one that traditionally sets a high value on rangatiratanga, a concept of leadership that carries responsibility as well as authority. "It is by their work that we recognise chiefs," a Māori proverb that tells us: "Ina te mahi, he rangatira!" However, the pathways became functional because

the wider community picked up the initiatives and further extended and adapted them.

Community-Based Constructions of Education

The cases invite us to re/conceptualise education that has expanded beyond schooling and curriculum formats. The Wānanga's practice explicitly situates education as a community/tribal enterprise, open to all and at any ages for a range of purposes to do with their well-being (as opposed, for example, to a construct of the right to education as workforce preparation). The Wānanga's practice also provides evidence of the expectation that more than those enrolled in courses will be beneficiaries of the provision of technology and training in its uses, because it is expected that the formal students will in turn share their computers and their expertise with their families and iwi. Māori television and the film work of Don Selwyn position education as a process of developing new understandings within the public arena.

In our case studies the distinctions between campaign, propaganda, entertainment and education have become a little blurred and they invite us to reconsider the substantive differences between learning and enculturalisation. Although the pictures and narratives placed by young people of Ngāi Tahu on the social web site *Bebo* might be considered as elements of pop culture, they can also be construed as participation in sites of cultural construction in which young people activate the language they are learning and reconnect with relations from whom they have been separated by the modern tribal diaspora.

Sociolinguists, such as Gee (1992) suggest to us that all societies enculturate their members in a wide range of visible and invisible ways, and invite us to recognise that the education that happens in classrooms or structured on-line courses is only a fragment of the ways in which societies society pass on the information and codes needed for their members to survive and thrive. Traditional Māori education, utilising intergenerational transfer and largely centered in the processes of the marae, might be conceptualised in similar terms to those quoted above from Don Selwyn when he was talking about the potential for film and television. In contrast, schooling and curriculum might be seen as western constructs and relatively modern approaches to knowledge and its promulgation.

Development of a Virtual Marae

As we have explained in the introduction, traditionally the marae was the centre of community activities, and the repository of knowledge in a range of oral, visual and physical forms. The development of digital technologies, particularly the web, text messaging and digital recording and editing, is presenting the opportunities of a virtual marae, specifically a virtual space which can be used by its people to recreate the opportunities of the marae across time and space. The Wānanga in the first case study partnered with marae to support e-learning on multiple locations and then also developed a blended approach that seems to us to extend their e-learning into a shared virtual marae. The second case study moved from the use of print and DVD and onto the web where interaction emerged and so it evolved a web-based virtual marae. Māori TV's virtual marae is a blend between TV and the web, with very strong links to many marae, including those within the Te Papa museum in Wellington.

As an educative media and virtual space these digital technologies allow users to control the timing and extent to their engagement. As a tool of decolonisation it allows participants to place Māori perspectives at the centre of their communication and self-configuration, including expressions of their own identity as Māori expressed in Māori language, Māori cultural artefacts and practices.

FUTURE RESEARCH DIRECTIONS

As we examine possible future research directions, we are very aware that the initiatives we have described are themselves the start of a substantial platform for Māori research and development in digital technology and e-learning. We would like to track some of the emergent developments further. For example, the language revitalisation projects described above are already being followed up with tribal plans to use digital technology to develop websites that will support cultural reconstruction, particularly allowing participants to share the tracing of genealogies, histories and other important elements of Māori knowledge. We would like to track the development of what might become a digital cultural encyclopaedia.

In addition, we would like to contribute to the on-going exploration within New Zealand of how mainstream schools and tertiary institutions can use people–centred and culturally appropriate processes such as those discussed in the case study of Te Wānanga o Raukawa in their classroom teaching. Work by McFarlane (2007) and Greenwood and Wilson (2006), discussed previously are examples of previous contributions to this exploration. In particular, we are interested in exploring how e-learning interfaces can be developed to integrate what we already know about cultural difference and the various cultural needs of learners. We also wish to support culturally appropriate distance education in schools and in tertiary education. For that we plan to work with Māori students in our teacher education and professional development programs, including online learning.

CONCLUSION

Around the world digital technologies have been used both for the benefit and detriment of indigenous people. The focus of this paper has been to examine some of the ways Māori innovators have explored, adapted and acculturated a range of digital technologies in broadly educational ways. The key features we have examined are the ways groups have used such technologies to serve the needs of their people, and wider community, and the ways they have retained Māori cultural values within the adaptation of the technologies.

We would be disappointed if our account was read in any way that universalised such practices, or that suggested that technology holds a simple answer to historic and continuing colonisation, political and commercial. What these examples do illustrate is how innovators have used the media that in other hands often threatens to further erode indigenous culture in order to in-inscribe popular culture and commercial technologies with indigenous values.

We have likened Māori use of digital technologies to the creation of 'virtual marae': meeting houses in cyberspace where Māori can meet, greet, acknowledge and support each other, where they can learn according to their own perceived needs rather than according to the choices of an education system, and where they can debate and inform contemporary issues. Like the material marae the virtual marae can be a place where Māori values and protocols have pre-eminence. It can also create a surrounding of Māori images, oratory and performing arts.

To a relatively minor extent now, such virtual marae become part of physical marae, as they do in the rural marae that have become outposts of The Wānanga o Raukwa's programme or in the physical premises of Māori television. Perhaps they might do so more over time.

Younger generations are assimilating participative web tools as part of their everyday interaction with mass and global culture. Through the initiatives such as those described above they are now also bringing Māori dimensions to this arena. It will be interesting over time to see how much these interactions might impact on the values of the popular media or how much Māori values might be re-colonised by popular values. What

happens will be influenced, as the innovations we have described have been, by the strength of the younger generations' commitment to Māori self-development and self-determination.

REFERENCES

Andreotti, V. (2008). *Development vs poverty: Notions of cultural supremacy in development education policy.* In D. Bourn (Ed), Development Education: Debates and Dialogues. (45-63). London: Institute of Education, University of London.

Arrows, F. (Jacobs, D.) (2008). The Authentic Dissertation: Alternative ways of knowing, research, and representation. New York: Routledge.

Ashcroft, B., Griffith, G., & Tiffin, H. (1989). *The empire writes back: Theory and practice in post-colonial literature.* London, New York: Routledge. doi:10.4324/9780203426081

Belich, J. (1996). *Making peoples: A history of the New Zealanders: From Polynesian settlement to the end of the nineteenth century.* London: Allen Lane.

Bishop, R. (2005). Freeing ourselves from neocolonial domination in research: A kaupapa Māori approach to knowledge creation. In Denzin, N. K., & Lincoln, Y. S. (Eds.), *The Sage handbook of qualitative research* (3rd ed.). Los Angeles, London: Sage.

Bishop, R. (2005). Freeing ourselves from neocolonial domination in research: A kaupapa Māori approach to creating knowledge. In Denzin, N., & Lincoln, Y. (Eds.), *The Sage handbook of qualitative research* (3rd ed., pp. 109–138). Thousand Oaks, CA: Sage Publications.

Bishop, R., & Glynn, T. (1999). *Culture counts: Changing power relations in education.* Palmerston North: Dunmore Press.

Cant, G. (2005). Ngāi Tahu and its eighteen papatipu rūnanga in a contested post-colonial New Zealand. In Cant, G., Goodall, A., & Inns, J. (Eds.), *Discourses and silences: Indigenous peoples, risks and resistance* (pp. 199–208). Christchurch: University of Canterbury.

Davis, N. E. (2008). How may teacher learning be promoted for educational renewal with IT? In Voogt, J., & Knezek, G. (Eds.), *International handbook of information technology in primary and secondary education*. Amsterdam: Springer. doi:10.1007/978-0-387-73315-9_31

Davis, N. E., & Fletcher, J. (2009 in preparation). *E-learning for adults in New Zealand with literacy, numeracy and/or ESOL needs. Report for the New Zealand Ministry of Education*. Christchurch: University of Canterbury College of Education.

Deer, K., & Hahanasson, K. A. (2005). Towards an indigenous vision for the information society. In T.J. van Weert (ed). *Education and the knowledge society: Information technology supporting human development*. pp 67-80. Boston/Dortrecht/London: Kluwer Academic Publishers.

Durie, M. (2001). *A framework for considering Māori educational advancement. Opening address*. Hui Taumata Mātauranga.

Durie, M. H. (1998). *Te mana te kawanatanga: The politics of Māori self-determination*. Auckland: Oxford University Press.

Dutton, W. (2004). *Social Transformation in the Information Society*. Paris: UNESCO [online] http://portal.unesco.org/ci/en/ev.php-URL_ID=12848&URL_DO=DO_TOPIC&URL_SECTION=201.html

Evison, H. (1993). *Te Waipounamu: The greenstone island*. Wellington: Aoraki Press.

Gee, J. P. (1992). *The social mind: Language, ideology and social practice*. New York: Bergin and Garvey.

Giroux, H. (1988). *Teachers as intellectuals: Towards a critical pedagogy of learning*. Granby, MA: Bergin & Garvey.

Greenwood, J. (1999). *Journeys into a third space: a study of how theatre enables us to interpret the emergent space between cultures*. Doctoral thesis, Griffith University, Brisbane, www.gu.edu.au/ins/lils/adt/

Greenwood, J. (2002). *History of bicultural theatre: Mapping the terrain*. Christchurch: Christchurch College of Education.

Greenwood, J, & Te Aika, Lynne Harata (2009). *Final report of the Hei Tauira project: Teaching and learning for success for Māori in tertiary settings*. Wellington: Ako Aotearoa.

Greenwood, J., & Wilson, A. M. (2006). *Te Mauri Pakeaka: A journey in to the third space*. Auckland: Auckland University Press.

Haythornthwaite, C. (2007). Digital divide and e-learning. In Andrews, R., & Haythronthwaite, C. (Eds.), *The Sage handbook of e-learning research* (pp. 97–118). Los Angeles, London: Sage.

Kawharu, I. H. (Ed.). (1989). *Waitangi: Māori and Pakeha perspectives of the Treaty*. Auckland, New Zealand: Oxford University Press.

King, M. (2004). *The Penguin History of New Zealand*. Auckland, New Zealand: Penguin.

Macfarlane, A. (2007). *Discipline, democracy and diversity: Working with students with behaviuour difficulties*. Wellington: NZCER Press.

McLaren, P. (1997). *Revolutionary multiculturalism: Pedagogies of dissent for the new millenium*. New York: Routledge.

Ohia, M. (2004). *Introduction: Critical success factors for effective use of e-learning with Māori learners*. Retrieved from http://elearning.itpnz.ac.nz/files/Hui_Report_finall_Effective_use_of_Māori_eLearning.pdf

Orange, C. (2004). *An illustrated history of the Treaty of Waitangi*. Wellington: Bridget Williams Books.

Pere, R. (1991). *Te wheke*. Wellington: Ao Ako Global Learning Ltd.

Reed, A. W. (1935). *The Māori and his first printed books*. Dunedin: Reed.

Reedy, T. (2000). Te reo Māori: The past 20 years and looking forward. *Oceanic linguistic, 39*(1), 157–169.

Resta, P., Christal, M., & Roy, L. (2004). Digital technology to empower indigenous culture and education. In Brown, A., & Davis, N. E. (Eds.), *Digital technology, communities and education. World Yearbook of Education 2004* (pp. 179–195). London: Routledge Falmer. doi:10.4324/9780203416174_chapter_11

Said, E. (1978). *Orientalism*. New York: Vintage Books.

Simon, J. (1986). *Ideology in schooling of Māori children: Delta Research Monograph, No 7*. Palmerston North: Massey University Press.

Smith, G. (2000). Protecting and respecting indigenous knowledge. In Battiste, M. (Ed.), *Reclaiming indigenous voice and vision* (pp. 209–225). Vancouver, BC: UBC Press.

Smith, L. (1999). *Decolonizing methodologies: Research and indigenous peoples*. Dunedin: University of Otago Press.

Smith, L. (2005). On tricky ground: Researching the native in the age of uncertainty. In Denzin, N., & Lincoln, Y. (Eds.), *The Sage handbook of qualitative research* (3rd ed., pp. 85–108). Thousand Oaks, CA: Sage.

Smith, L. T. (2005). On tricky ground: Researching the native in an age of uncertainty. In Denzin, N. K., & Lincoln, Y. S. (Eds.), *The Sage Handbook of Qualitative Research* (3rd ed.). Los Angeles, London: Sage.

Spivak, G. (1996). Explanation and culture: Marginalia. In Landry, D., & MacLean, G. (Eds.), *The Spivak reader* (pp. 29–51). New York: Routledge.

Tahu, K. (2006) *Kotahi mano kāika, kotahi, mano wawata: One thousand homes, one thousand aspirations*. Retrieved from http://www.kmk.maori.nz/ Accessed 16 February 2009

Te Aika, L. H., & Greenwood, J. (2009). Ko tātou te rangahau, ko te rangahau, ko tātou: A Māori approach to participatory action research. In D. Kapoor & S. Jordan (Eds). *Education, participatory action research, and social change: International perspectives. Palgrave Macmillan*.

Television, M. (2009a). *Mission statement*. Retrieved from http://www.maoritelevision.com/

Television, M. (2009b). *Website: frequently asked questions*. Retrieved from http://corporate.maoritelevision.com/faq.htm

Tribunal, W. (1986). *Report of the Waitangi Tribunal on the Te Reo Māori claim*. Wellington: Government Printer.

UN General Assembly. (2007*). Declaration on the rights of indigenous peoples*. http://www.iwgia.org/sw248.asp

Waldon, W. (2004) Official launch of Māori television http://www.maoritelevision.com/newsletter/issue5/index.htm

Walker, R. (1990). *Ka whawhai tonu matou: Struggle without end*. Auckland: Penguin.

ADDITIONAL READING

Dutton, W. (2004). *Social Transformation in the Information Society*. Paris: UNESCO [online] http://portal.unesco.org/ci/en/ev.php-URL_ID=12848&URL_DO=DO_TOPIC&URL_SEC-TION=201.html

Greenwood, J., & Wilson, A. M. (2006). *Te Mauri Pakeaka: A journey in to the Third Space*. Auckland, New Zealand: Auckland University Press.

Kotahi Mano Kaika. (n.d.). Retrieved from http://www.kmk.māori.nz/ (KMK is the acronym for "Kotahi Mano Kāika") Retrieved from http://www.māoritelevision.com

One Sqaured. (n.d.). Retrieved from http://www.ngāitahu.iwi.nz

Reo, G. (2007). Retrieved from http://www.generationreo.com

Te Aika, L. H. & Greenwood, Janinka (2009). Ko tātou te rangahau, ko te rangahau, ko tātou: A Māori approach to participatory action research. In D. Kapoor & S. Jordan (Eds). *Education, participatory action research, and social change: international perspectives*. New York, Palgrave Macmillan.

ENDNOTES

[1] Māori language has been included in this chapter (with short explanations). Readers are also encouraged to use a dictionary such as www.maoridictionary.co.nz produced by Te Tumu o Te Whare Wānanga o Ōtākou

[2] In New Zealand tertiary education refers to the post-compulsory sector. Tertiary institutions, therefore, include universities, polytechnics, wānanga, and a a number of other institutions run by private providers.

[3] John Rangihau, Timoti Karetu and Huirangi Waikerepuru are renowned Maori scholars and leaders in the revitalisation of language and culture, A significant part of each of their scholarly contributions has been through traditional processes of oral presentation and debate in communal gatherings.

[4] We note from American readers that a term such as certifications rather than qualifications, might be more readily understood within their local context. We retain the term qualifications because within a New Zealand context certification would have a more narrow meaning, pertaining only to a lower level qualification, whereas *qualifications* embraces degrees as well as diplomas and certificates. The book is targeted at an international readership: however, we are aware that to be meaningful the global often needs grounding in the local.

Chapter 5

From Igloos to iPods:
Inuit Qaujimajatuqangit and the Internet in Canada

Cynthia J. Alexander
Acadia University, Canada

ABSTRACT

Inuit in the Eastern Arctic of Canada reclaimed their homeland on 1 April 1999 when the newest territory in Canada, Nunavut, was created. Inuit are using new media technologies to preserve and promote their language, traditional knowledge, and ways of being. In this chapter, the reader is offered an exploration of the challenges northerners face in the digital era, including affordable, reliable access to the Internet. However, the author shows how the resilience that characterizes Inuit culture extends to their innovative adoption of new media technologies. The author offers insight into one web development project, a partnered initiative with Inuit, which enables Inuit youth to learn from their Elders, and for users around the world to learn from Inuit via an interactive online adventure. The case study of The Nanisiniq Inuit Qaujimajatuqangit, or IQ Adventure, provides an interesting example of how harnessing the power of new media can support Indigenous peoples' decolonization efforts.

From the skies above, you zoom down from space towards Earth, listening to the sounds of wildlife and Inuit music. You navigate through a 360 degree shot, thereby finding yourself on a hill above the capital of Nunavut, Iqaluit. An Inuksuk stands before you and you hear a voice...the former Commissioner of Nunavut and Inuit cultural teacher, Peter Irniq, explains he is here to help guide you through a fantastic journey:

But see...the Earth is in Trouble!

And we haven't much time...

In less than 200 years we have lost so much.

We Inuit have been displaced from our traditional lands,

And we almost lost our language.

DOI: 10.4018/978-1-61520-793-0.ch005

Copyright © 2011, IGI Global. Copying or distributing in print or electronic forms without written permission of IGI Global is prohibited.

Our culture was forced aside... but we have survived.

But there's little time left.

Now we know... the ice is melting.

The waters will rise...in the ocean

And our animals, of the sea and of the land, are disappearing.

Learn well... come on!

Now we know that our Earth is in trouble!

Will humans survive? Hurry! Make a start!

AJUNNGI! We can do something!

You zoom in on the ship that has appeared through the window of the Inuksuk. You accept the challenge.

INTRODUCTION

Immersed in actual footage of a real voyage through the Eastern Arctic, the virtual explorer sets out on quest via an interactive film on the new website, *The Nanisiniq Inuit Qaujimajatuqangit* (hereafter referred to as the IQ Adventure, from which an image of the homepage appears in Figure 1) that the author co-designed and –developed with a number of partners, including the Government of Nunavut. Throughout the journey, the user tackles issues and makes choices at key decision-points in the five policy-based scenarios which represent a range of contemporary issues in the Arctic—from climate change to consensus-based decision making to sustainable community development. Each scenario provides an opportunity for the user to apply information and knowledge shared by Inuit experts who play themselves in the film (including senior policy makers, hunters, and art-

ists). Importantly, the user is challenged to apply Inuit's ancient knowledge system to *contemporary* challenges; each time a challenge is successfully completed, the user receives a stone to build a virtual Inuksuk, which by the end of the voyage represents the insight that the user has gleaned via the film about Inuit knowledge systems, decision approaches, and values that can be called upon to help navigate through life.

The website illustrates the power of new media technologies to counter the one-way information flows to the North that reinforced colonial systems, from judicial to health to education, which have negatively impacted Inuit well-being. By necessity, over millenia Inuit have developed an evolving system of values, knowledge, skills and approaches that enabled them to survive on the tundra; this foundation of knowledge and way of being, and the predisposition to innovate and adapt to survive underpins the ways that Inuit have begun to harness new media technologies to reclaim their intellectual and cultural legacies.

The need to do so is urgent given that Inuit of Nunavut, the majority of whom are under the age of 25 years, face the kind of deep-rooted challenges that accompany persistent colonial-

Figure 1. Main Page of the Nanisiniq Inuit Qaujimajatuqangit or IQ Adventure website © [2006], [C. J. Alexander. Used with permission)

ism. In the Arctic, where small changes have big effects, the disruption of Inuit's way of being is so profound that lives are lost, reflected in the fact that suicide rates among Inuit youth are among the highest in the world. In the chapter, the author examines the role of new media resources in the preservation and promotion of Inuit knowledge among Inuit youth[1], with the objective of contributing to the development of pride in self, culture and community.

The case study of the development of the IQ Adventure website illustrates the critical role that information and communication technologies (ICTs) play in decolonization efforts by indigenous peoples. Inuit are key actors in designing the post-colonial digital landscape by sharing with the world, in unprecedented ways, their values, knowledge and experiences. In doing so, Inuit's digital designs are contributing to new forms of identity and community among the world's indigenous peoples. The examples from the interactive film on the *Nanisiniq Inuit Qaujimajatuqangit* (IQ) Adventure website that have been chosen to share in this chapter focus on Inuit's foundational principle of interdependence among humans, and vis-à-vis nature, which has been imperative for Inuit to survive and thrive. The user's coach during the interactive film, Peter Irniq, the former Commissioner of Nunavut, has made it one of his priorities as a cultural teacher, Elder, politician, and human rights activist, to inform and inspire Inuit and non-Inuit alike about an ancient system of knowledge and values, *Inuit Qaujimajatuqangit* (literally translated as that which has long been known by Inuit), commonly refered to as IQ, and its relevance in the 21st century as an evolving system.

ICED OUT: COLONIALISM AS CONTEXT

Inuit lived for thousands of years as a self-governing people. We had our own learning systems, *justice system, and lived our lives based on a philosophy of how to live a good life, at peace with the land.*

Our lives changed abruptly only a short time ago. Missionaries came to our land and changed our names, and our lives. We were kidnapped from our parents and put in residential schools where many of us suffered unspeakable abuse. We were taken off the land and into communities. Commercial forces depleted the wildlife on which we depended to live. But we persevered, and we were patient in our efforts to reclaim our land.

Dr. Peter Irniq, Nanisiniq Inuit Qaujimajatuqangit Adventure Website

Ancient values and practices were communicated to Inuit at a very early age through stories, songs, direct modeling of behaviour and legends; survival in the Eastern Arctic was dependent on remembering them. This system and the age-old methods for communicating its values have been interrupted by outside influences and colonial institutions. The 'discourse of colonialism' is "the name for that system of signifying practices whose work it is to *produce and naturalise* the hierarchical power structures of the imperial enterprise, and to mobilise those power structures in the management of both colonial and neo-colonial cross-cultural relationships" (Slemon, 1986, p. 6). The 1992 Nunavut Land Claims Agreement, enacted in 1993, reclaimed 136,000 square miles of land for Inuit; doing so was the first step towards post-colonial relations in Canada, followed by the creation of the Territorial Government of Nunavut on 1 April 1999. After the tenth anniversary year of the creation of the newest territory in Canada, Inuit of Nunavut are seeking to infuse their language and culture in the value system that underpins public policy processes in every field, from education to health to justice. Beyond the need for Inuit to reclaim their language and culture, there is much for the rest of the world to learn from ancient

Inuit values and processes about living respectfully with each other, and vis-à-vis nature. Inuit's long-standing, time-honoured way of being has intellectual, spiritual and applied significance for Inuit, resonance for Indigenous peoples everywhere, and relevance for all humanity.

Inuit live in four geographic communities in northern Canada, including Nunavut[2], which covers approximately 20 per cent of the landmass of Canada. There are 28 communities located across the tundra of Nunavut, none of them connected by road. The population of Nunavut is approximately 29,000, of whom 85 per cent are Inuit. Inuit lived on the land for millennia, in snow houses or igloos in the winter, and in tents in the summer; they survived in the harshest environment in the world. The residential school experience has impacted the stability of Indigenous communities, and contributed to the higher levels of poverty[3] experienced by Indigenous communities than exists elsewhere in Canada. Most Inuit women and girls suffer violence and abuse; specifically, "Nunavut's crime rate in 2004 was eight times the Canadian rate… Between 2001 and 2004, the number of women and children who left their homes in Nunavut because of violence increased by 54 per cent. Over the same period nationally, the increase was 4.6 per cent" (Pauktuutit, 2006, np). Suicide rates for Inuit are 11 times higher than the Canadian average. Jack Hicks has researched the problem and "believes that many of Canada's aboriginal societies are 'in something of a state of shock'" and asks: "What will future generations think of how the federal and territorial governments addressed a terrible need of services and resources in the 1990s, this decade and the next? I think their judgements will be 'indifference' and 'neglect'. …You could have done much more and you didn't" (as cited in *Human Rights Tribune*, 2008, np.). In June 2007, Canada voted against a resolution at the Human Rights Council in Geneva, which requires states (ie. Canada) to undergo periodic United Nations' reviews of its human rights record; Canada was the only country that voted against the resolution. In addition to the impoverished material conditions that define the lives of the majority of Inuit and other indigenous peoples in Canada[4], they have suffered the loss of language and cultural discontent that leaves youth vulnerable in the face of the economic, political and other challenges.

In 1977, Thomas Berger stated: "What happens in the North will…tell us what kind of a country Canada is; it will tell us what kind of a people we are" (Coates, Lackenbauer, Morrison, & Poelzer, 2008, p. 112). On 11 June 2008, the Prime Minister of Canada apologized to First Nations, Inuit, and Métis[5] peoples for their treatment in the residential schools system, to which children were removed and isolated from the influence of their homes, families, traditions and cultures with the goal of assimilating them into the dominant culture, based on the assumption that aboriginal cultures and spiritual beliefs were inferior and unequal. The Prime Minister explained:

The government of Canada built an educational system in which very young children were often forcibly removed from their homes, often taken far from their communities. Many were inadequately fed, clothed and housed. All were deprived of the care and nurturing of their parents, grandparents and communities....

First Nations, Inuit and Métis languages and cultural practices were prohibited in these schools. Tragically, some of these children died while attending residential schools and others never returned home. The government now recognizes that the consequences of the Indian residential schools policy were profoundly negative and that this policy has had a lasting and damaging impact on aboriginal culture, heritage and language. (Canada, 11 June 2008, np.)

Bluntly stated, as it has infamously been said, the objective was "to kill the Indian in the child." As we approach the end of the first decade of the

21st century, policy makers, academics and citizens alike pay scant attention to how some communities in Canada, the United States, and other developed nation-states live in so-called Third World conditions. Three federal systems—Canada, the United States and Australia—are among those nations that have marginalized the indigenous peoples of the land for hundreds of years; further, at the same time that these states have called for democracy in other parts of the world, their own citizens lack basic human rights, including fresh water, basic housing, education, and health services. The United Nations Declaration on the Rights of Indigenous Peoples that was adopted by the U.N. General Assembly on September 13, 2007 by an overwhelming majority vote of 144 to 4 was opposed only by Canada, the United States, Australia and New Zealand. Canada and the United States, however, continue to claim that the Declaration should not apply to them. This is the first time that Canada has sought to be exempted from a human rights standard adopted by the U.N. General Assembly. In Canada as in other federal systems, including the United States and Australia, indigenous peoples have experienced colonialism as an historic wrong, and also, as a persistent system of unequal power relations in contemporary life. In August 2009, the Prime Minister of Canada visited Nunavut. Canadians were shocked by a photograph of two boys sleeping on the pavement in Iqaluit, the capital of Nunavut. Nunavut Tunngavik's director of social and cultural development, Natan Obed, observed that there is a lack of capacity to deliver public services, and that culturally-specific programs are needed for Inuit, who are dealing with historical trauma (as cited in Paperny, 17 August 2009, p. 2). Beyond the socio-economic challenges that Inuit face, science reveals irrefutably that the Arctic is melting much more quickly than was predicted even a decade ago. Adaptable, resilient, and innovative, Inuit persist. Despite the range of policy challenges, the preservation and promotion of their culture remains one of the Government of Nunavut's top priorities. In this context, the relationship between decolonization and new media technologies can be broached. Issues of decolonization underpin efforts to create culturally-infused new media systems. Indigenous peoples' visions of change are grounded in a recognition of, and respect for, indigenous language, culture, knowledge and belief systems. Although the field of indigenous studies is rapidly expanding, there has been a dearth of studies that examine the relationship between digital equity and Indigenous political systems in North America. The question of digital equity with respect to Indigenous peoples in Canada or elsewhere requires an appreciation of the challenges decolonization initiatives face. Inuit are creating *post-colonial* media systems.

DECOLONIZATION AND DIGITAL DESIGNS

I will coach you during our journey, so that you find the strength and vision you will need. You will learn wisdom from the Elders. Our journey won't be easy.

You will have to learn to listen very carefully to hear their message, and all that is said. And sometimes, you will have to wait to understand what you have heard.

Dr. Peter Irniq, Nanisiniq Inuit Qaujimajatuqangit Adventure Website

New media technologies provide an opportunity for Inuit and other indigenous peoples to renegotiate identities. The need to do so continues in the face of resistance to cultural reclamation processes that are underway around the world. As Odin (1997) contends, hypertext and post-colonial strategies "represent two manifestations of the topology of postmodern information culture where grand narratives are being replaced by local narratives and local knowledges" (p. 627). New media technologies provide a mode of information and communication sharing that supports the sharing of ancient legends, knowledge, and belief systems in new ways for new users, as illustrated by the IQ Adventure website. Inuit, like other indigenous peoples around the world, are demonstrating Boulou E. D'beri's (2005) suggestion that "emerging practices of memory through new media technology, illustrate the 'shift' of modernity, particularly vis-à-vis certain monopoly of knowledge and cultural expressions"; that is, new media technologies provide resources for Inuit and other Indigenous peoples to "re-activate their silenced-cultural practices" (p. 2). The infusion of Inuit-values, languages, knowledge systems, and decision-making processes throughout the IQ Adventure website is indicative of the way in which Inuit can transverse geography and generations, and confront *persistent* colonial stereotypes. New media constructions like the IQ Adventure website need to be read as more than cultural statements; that is, they should also be seen as *political* statements that contribute to the deconstruction "of the ethnographic discourse, that of the lazy Inuit, inept of doing or producing anything good in the present as well as in the past; or that of an Inuit manufactured as the uncivilized Indian, who is aggressive and acts savage-like" (E.D'beri, 2005, p.7). A particularly controversial new book, *Disrobing the Aboriginal Industry* (Widdowson & Howard, 2008) has led to an exceptionally heated national discourse about Indigenous traditional knowledge, self-government, and assimilation. For exmaple, in his favourable review of the book in the conservative paper, the *National Post*, Jonathan Kay writes, quoting Widdowson and Howard:

"In archeological terms, various aboriginal groups were either in a Paleolithic or Neolithic stage," he writes. "Isolation from economic processes has meant that a number of Neolithic cultural features, including undisciplined work habits, tribal forms of political identification,

animistic beliefs and difficulties in developing abstract reasoning, persist despite hundreds of years of contact."

In other words, the encroachment of Western values in aboriginal communities isn't the problem—it's the solution (Kay, 3 February 2009, p.2).

Moved to respond, Mary Simon, the President of Inuit Tapiriit Kanatami, the national voice of Inuit in Canada, stated:

We've already experienced cultural assimilation, and look what it has produced—unparalleled educational failures, the Residential School tragedies, poverty, suicides and the resulting social destruction. These are the very issues that Inuit struggle to overcome every day.

Jonathan Kay's pejorative reference to presumably aboriginal "cultural traditions" as "hidebound" and "religious fairy tales" is a mean-spirited, narrow-minded and perhaps even racist attack on practices that ensured Inuit survival in the Arctic for thousands of years Societies adapt and evolve, and for the past century Inuit have been doing that more than most. (Simon, 13 February 2009, p.2).

Indigenous peoples continue to confront the strong currents of colonialism in the media, in politics, in education systems, and in society. Just as indigenous leaders like Mary Simon are working against the persistence of racial stereotypes and policy neglect, so too are indigenous scholars concerned enough to respond to assertions such as those that Widdowson and Howard have made. For example, Kiera Ladner, Canada Research Chair in Indigenous politics and governance at the University of Manitoba, explains that Widdowson perpetuates a "fantasy of the master race," where the "civilized" ruled over "savages"—a view that's "decades, possibly centuries," out of date (as cited in Shimo, 24 February 2009). As Marie Battiste (2005), a noted Aboriginal scholar, states: "Cognitive imperialism and racism have been sources of discrimination and disempowerment for Aboriginal peoples in schools" and it should also be noted, in mainstream media systems (p. 17). The promotion of indigenous knowledge systems via new media technologies provides both intellectual and political muscle in the ongoing battle against colonialism. Battiste (2005) advises: "Decolonizing education must engage with Indigenous knowledges, languages, and heritages to create an 'honourable' education system" (p. 15). Digital equity is about political, economic, and *cultural* inclusion; that is, it is about using information and communication technologies (ICTs) as the means through which skills are improved, quality of life is enhanced, educational objectives are realized, economic well-being is secured, *and additionally*, historically marginalized cultural knowledge systems and ways of being are promoted, considered, understood and hopefully, respected, within national and global communication systems.

In Canada[6] as elsewhere, indigenous peoples' cultural distinctiveness continues to be threatened, ever-the-more-so given the increasing flow of southern values via television and more recently, via the Internet. In this context, it is challenging to resist Eurocentric values and norms, approaches, privileges and frameworks. For example, patriarchy is one facet of the dominant society that has unwoven indigenous culture and unravelled indigenous families, communities, and governance systems. For Inuit, as for other indigenous peoples across Turtle Island[7], men and women were equal, and shared balanced—interdependent—roles within the family, governance and spiritual practices. The displacement of women parallels the overthrow of indigenous communication, learning, healing and other systems that placed interdependence as a foundational principle to their way of being. Aboriginal author Paula Gunn Allen states:

Since the coming of the Anglo-Europeans beginning in the fifteenth century, the fragile web of identity that long held tribal people secure has gradually been weakened and torn. But the oral tradition has prevented the complete destruction of the web, the ultimate disruption of tribal ways. The oral tradition is vital: it heals itself and the tribal web by adapting to the flow of the present while never relinquishing its connection to the past (p. 45).

New media technologies *can* be designed to serve post-colonial objectives. Indeed, multimedia systems are particularly suited to indigenous oral traditions by privileging visual and oral modes of communication, and in doing so, displacing text as the dominant discursive mode. The loss of an indigenous peoples' knowledge system is an irreplaceable loss for humanity. As Governor General Michaelle Jean stated at the first-ever National Summit on Inuit Education in April 2008: "I think that the more adults and Elders are empowered, the more youth are going to take that power and make it theirs." A year later, the National Inuit Education Accord, a new multi-stakeholder policy initiative signed by Inuit and federal government representatives on 2 April 2009, represents an historic undertaking that holds great promise to realize post-colonial Inuit education. Technology has played a key role in graduating the first 21 Inuit students in the country's only part-time M.Ed. program focused on education and leadership; that is, the M.Ed. program is the first graduate degree program offered in Nunavut, enabling Inuit students to study part-time through face-to-face course in two communities in Nunavut, Iqaluit and Rankin Inlet, and critically important, through online learning. Inuit leadership within the school system is necessary to reclaim knowledge, voices and vision in the development of a post-colonial education system. By the time such a system is up and running, the generation of Elders who grew up on the land will have died, and with them, the legacy of Inuit knowledge. Hence, preserving their voices is imperative.

Since the 1950s, Inuit made the involuntary transition from a world of igloos to one of iPods. They continue to experience the traumatic effects that have resulted from a period of compressed colonialization—moving from the ice-age to the space age in two generations—including loss of land, language, culture, and Indigenous knowledge systems and ways of being. What role do deeply entrenched societal and institutional values and socio-cultural and economic cleavages play in digital developments, if any? Techno-enthusiasm should be nuanced:

However sophisticated a system may be, technological change offers no easy 'solution' to alleviate complex social and political problems or ameliorate socio-political and economic cleavages. Indigenous peoples, women, and other underrepresented citizens in our political system have not yet found quick-fix 'technological solutions' to the challenges of economic dependence, weak political influence, and social inequity. We should not expect any technological silver bullet to deal with such complex and recalcitrant issues" (Alexander & Pal, 1998, p. 6).

WEBS OF KNOWLEDGE: INUIT QAUJIMAJATUQANGIT (IQ) AND THE INTERNET

We call this ancient body of knowledge "Inuit Qaujimajatuqangit."

Come on, try to say it: Inuit Qaujimajatuqangit."

Listen again. "Inuit Qaujimajatuqangit."

You'll be able to learn some of this ancient knowledge, and your eyes will be opened... if you listen carefully to the Elders.

This is going to be a great journey for you, I just know it. Ajunnginiaqtummarialuujutit! You're up for the challenge!

Dr. Peter Irniq, Nanisiniq Inuit Qaujimajatuqangit Adventure Website

Inuit of Nunavut developed a knowledge system over thousands of years of living in the Eastern Arctic. IQ was communicated to young Inuit through stories, songs, direct modeling of behaviour and legends that spoke of the success associated with remembering them. The traditional learning processes that Inuit used to communicate their values and knowledge to youth were interrupted, and almost extinguished, by outside influences and the imposition of institutions that were foreign to their belief systems. John Ralston Saul observes: "Everyone in the Arctic recognizes that education is the key to what happens next....But what kind of education? The Canadian model is based on largely urban assumptions. Applied in the Arctic, it is largely irrelevant. ...But the purpose of education is not to empty the Arctic of its citizens. Nor is it to undermine their self-confidence as Northerners. The Southern system in place couldn't help but do both" (Ralston Saul, 2008, p. 289). The process of legitimating and valuing Indigenous knowledge is vital to the evolution of a democratic society. Inuit are seeking ways to infuse these beliefs and values, language and learning processes, and their knowledge system, into their lives, communities and organizations.

Inuit Qaujimajatuqangit is traditional Inuit knowledge. Words of advice have often come from Inuit Elders who learned these values from their Elders before them. Inuit as a people have a long-standing code of behaviour based on enduring values and time-tested practices. The core values upon which the time-honoured Inuit belief system lies are ones we all recognize:

- **Connection Values:** sharing, generosity, family, respect, love, listening, equality, significance and trust.
- **Work Values:** volunteer, observe, practice, mastery, teamwork, cooperation, unity, consensus and conservation.
- **Coping Values:** patience, endurance, improvisation, strength, adaptability, resilience, resourcefulness, moving forward, take the long view, survival, interconnectedness and honesty. (Nunavut, Department of Human Resources, 2005, np)

Inuit are seeking ways to build these beliefs into what they do today so that once again these beliefs infuse the value system across Nunavut. In 1999 the first government cabinet of the newest territory in Canada, Nunavut, formulated a five year plan, *Pinasuaqtavut* or "that which we have set out to do", also known as the Bathurst Mandate. In the Bathurst Mandate, the Government of Nunavut (GN) made a formal commitment to use Inuit traditional knowledge—known as *Inuit Qaujimajatuqangit (IQ)*—as its foundation and set out a vision for community development. *IQ* is an ancient, and importantly, evolving body of knowledge, ways, beliefs, and public philosophy that has, thus far, been poorly represented in the English language and strongly resisted in archaic administrative systems. Knowledge of this value system is also potentially of great benefit to non-Inuit who seek a respectful way to live on this planet.

With the purpose of developing and sustaining a community-driven research initiative, in June 2005, the author co-organized an international symposium, Nunavut @ Five, to provide an opportunity for senior Inuit and northern policy makers, teachers, and human rights activists to reflect on the first five years of the creation of Canada's newest territory, Nunavut, and articulate their policy priorities. During the symposium, Inuit participants emphasized the imperative for Inuit youth to reclaim their legacy, by creating more opportunities for youth to learn from Inuit Elders. From this ongoing dialogue and deep collaboration, a community-driven research initiative emerged with a priority placed on the need to articulate *Inuit Qaujimajaatuqangit* (IQ) for Inuit youth in a way that demonstrates its value as a 'living' decision system that evolves over time.

New media systems can be designed to reconstruct knowledge sharing between generations of Inuit. The IQ Adventure website exemplifies the opportunity that ICTs afford Inuit youth and others to go to Elders for wisdom and guidance. Given that the generation of Inuit who grew up on the land and who are the keepers of knowledge are dying quickly, such a website provides access to Elders, in their own voices, for current and future generations. In the absence of being able to go to Elders in person, ICTs provide opportunities for Inuit, and for all of humanity, to draw upon their knowledge.

Values, assumptions, preferences and choices are deeply embedded in technical systems. Within the design of the IQ Adventure website, we sought to encompass the complexity of Inuit's relationship with the land that has evolved. The digital design was itself an experience in decolonization, by sharing Inuit Elders' recollections of their cultural legacies, and creating, through the co-development of scenarios in the interactive film, alternative visions of how to face *contemporary* challenges by drawing on the wisdom of an *ancient* culture. Native American scholar, Taiaiake Alfred (2005), explains that the European colonizers' "simplistic approach of assimilation resulted not in the acceptance of their ways but in cultural pollution", the result of which Alfred calls 'cultural blanks' (p. 11). Alfred's latest book, *Wasáse (2005)*, is titled after a way to live as *Onkwehonwe*, a term that is symbolic of the social and culture force dedicated to "altering the balance of political and economic power to recreate some social and physical space for freedom to emerge. …and do what we must to force the Settlers to acknowledge our existence and the integrity of our common connection to the land" (p. 19). Digital designs provide a vehicle for indigenous peoples' struggles to decolonize, with the goal of inspiring and empowering youth and future generations.

Distinctively, perhaps, Inuit are generous in sharing their Elders' insights, not only with their own youth, but with non-Inuit everywhere. What is becoming increasingly evident is that efforts of Inuit and other Indigenous peoples to share their narratives--in their own voices—via digital resources contributes to decolonization of knowledge and communication systems. Further, partnered, collaborative, community-led initiatives are contributing to the evolution of post-colonial methodology.[8] Such work privileges Indigenous voices and values, political thought and ways of being, methodologies and approaches. Among Indigenous peoples, beyond resistance to the persistent forces of colonialism, there is not a singular approach or vision in digital design. For example, it is crucial to note that diverse Indigenous peoples have their own sensibilities and practices about whether and how to communicate and importantly, with whom to share their intellectual property. Over millenia, survival for Inuit depended on interdependence, and that principle underpins the 'open access' of their digital designs. Inuit choose to share the wisdom of their teaching stories with the world, a choice that is explained, at least in part, by Inuit leader and Nobel Prize nominee Sheila Watt-Cloutier: "When I think about education, the first thing that comes to mind is the need we Inuit have to teach the world about our culture and our arctic environment. Why does this first come to mind? Because of the great responsibility we bear to educate the world of the crucial function of our culture and our environment for this planet's well-being" (2004, np).

The goal of the collaborative IQ Adventure website (facilitated in southern Canada by faculty and staff at Acadia University and the Inuit-led media firm, drumsong communications inc., partnered with the Government of Nunavut, and funded by the Government of Canada) was to develop an online learning resource that enabled the learner-user to understand, respect, and potentially, practice, Inuit knowledge, values and practices as 'living IQ.' The challenge was to bring the subject alive, to make it a world in which one can immerse oneself, experiencing it as young Inuit traditionally did—experientially—while traveling on the land, learning by doing with the insight gained from stories, songs, and legends. The goal has been to re-empower Inuit Elders who hold the knowledge of IQ, by creating a virtual meeting place for users 'to go' to the Elders for wisdom and guidance. Dialogue with Inuit inspired the development team in the South to help rebuild the ancient cycle of knowledge transfer. From a design perspective, the challenge was to *contextualize* IQ, and this was accomplished by developing a script and scenarios for a virtual learning environment that would enable learner-users to learn—through an immersive quest-based experience—how an ancient body of knowledge can inform decision-making in the 21st century.

WEB DEVELOPMENT STRATEGY: CREATING INUIT-INFUSED NEW MEDIA

Digitization of 100 Hours of Existing Film Footage of 50 Inuit Elders from across Nunavut

The project's co-developer, film-maker and artist John Houston[9], spent the first eight years of his life in Cape Dorset, Nunavut, growing up listening to legends and stories from Inuit Elders. As an adult he has travelled across Nunavut recording their memories about life on the land, and listening to the legends Inuit grew up hearing and were willing to share. What Houston heard from Inuit provided the foundation for his Arctic film trilogy, beginning with Songs in Stone, then Nuliajuk: Mother of the Sea Beasts, and Diet of Souls. He traveled across

Nunavut to listen to Elders recollections about an ancient tale about *Kiviuq*, the legendary Inuit epic hero who is called, in various dialects, Kiviu, *Kiviuq*, Qiviuq, Qooqa or Qayaq. *Kiviuq* was a shaman, and missionaries among the Inuit forbade his name to be spoken, or set down in the syllabic writing they introduced. Kira Van Deusen writes that the legend of Kiviuq "contains history and practical life skills, but also works as a powerful tool for spiritual development. Kiviuq's story is very long, with many complex and multi-layered morals. It compares better with Homer than with Aesop" (2009, p.12). In 2004-05 Houston and his drumsong film production team concluded a year-long journey across Nunavut to film dialogues with 50 Inuit Elders talking about Kiviuq. The *IQ Adventure* website project team gained exclusive access to this material, and digitized 100 hours of the film footage for the IQ website. Through the legend of *Kiviuq*, Elders share insights into the core of *Inuit Qaujimajatuqangit* (IQ)—spirituality— which is based on a close and deep relationship with the land that underpins the Inuit philosophy of interdependence between humans and nature.

The project development team digitized the Houston *Kiviuq* archival film collection and created, in Inuktitut and English, a searchable database of the 50 Elders' interviews (a screen shot appears in Figure 3); further, some of the film interviews with Elders have been embedded within the *Nanisiniq Inuit Qaujimajatuqanginnik* interactive film.

From the Elders, users hear the relevance of this ancient tongue, and learn about a legend that has the power and significance of Homer's Odyssey in Greek culture. For example, in the searchable Elders' Database, Mariano Aupilardjuk shares his recollection of the importance of the legend of Kiviuq. Aupilardjuk states: "I feel that the stories told to me of Kiviu by my mother are real for me and that they have helped the Inuit to be able to survive. I think all the Inuit feel that the Kiviu stories are real and that they have a use in Inuit life.[10]

Elder Rachel Ujarasuq (a still image from her video interview appears in Figure 3) explains: "It's those lessons that teach you about kindness and love and affection for other people that have been around since long before contact with Europeans. And those are the tools that were used to pass on those values, especially when it comes to a story about an orphan boy or an orphan [as Kiviuq was]. It's always been taught that you have to be kind to other people -people less fortunate than you are. You eat with them. You laugh with them. And that's how you pass on these types of values."[11]

The life lessons of Inuit were passed through generations to stories and legends such as Kiviuq.

Figure 2. Screen Shot from The Elders' Kiviuq Database © C. J. Alexander, 2006

In his recollections, elder Sampson Quinangnaq (pictured in Figure 4) stated: "I can say for sure that the Inuit never had anybody who was able to write anything down like the white man. So therefore, all they had was their words to pass on to their own children, over time. They keep on telling their stories over and over so people could remember them."[12] Sampson was born in 1924 in the Back River country of the territory that is now called Nunavut; like many other Inuit in the 1950s and '60s, Sampson was forced by authorities to relocate, in his case, to Baker Lake in 1959, a move he did not like because there was much less wildlife to depend on for sustenance. Sampson appears in Houston's film *Kiviuq* which was screened at a number of film festivals internationally throughout 2007 and 2008. On 20 May 2006, Sampson Amarok, the man called 'wolf,' passed away, affirming the urgency of the importance of recording Elders' knowledge. Via the Internet, film-goers and 'net explorers can listen to Sampson and the other Elders represented in the searchable database, in Inuktitut or English.

For Inuit youth, having access to Elders, to the legend of Kiviuq, and to their ancestral language via the Internet is part of current efforts to reclaim and assert their shared identity. Nunavut's Department of Education offers a number of reports and draft papers, available online, which state that pride in Inuit culture and language are fundamental for adults and children. New media learning environments do not replace 'on the land'

education initiatives such *Reclaiming Our Sinew*, a 24-week full-time holistic program to develop practical skills, self-esteem and cultural pride. Nunavut's Department of Education is exploring how to integrate Elders into the school system. Given that time is short to learn from Elders, the IQ Adventure website's searchable database of Elders' recounts of the legend of Kiviuq alone is indicative of the role of ICTs in the development of Inuit-specific curriculum resources. Recording the memories of those who grew up in snow houses is an activity that is part of the multiple paths to healing that Inuit are pursuing, all of which center around the importance of learning about traditional knowledge, culture and practices. For healing to happen, as indigenous women concluded at a national roundtable, "Elders and seniors must be honoured and recognized, women must be returned to their rightful place within aboriginal culture as leaders and life givers, and families must again become the central focus of any work aimed at healing women or addressing family violence" (Canadian Heritage, Family Violence Initiative, 2002, np).

Digitization of 200 Works of Inuit Art from the Houston North Gallery Collection

Inuit have been artists for millenia. In the contemporary world, art is a mode of expression for

Figure 3. Screen Shot of Rachel Ujarasaq from The Elders' Kiviuq Database © 2006, C. J. Alexander. Used with permission.)

Figure 4. Screen Shot of Samson Quinanqnaq from The Elders' Kiviuq Database © 2006, C. J. Alexander. Used with permission.)

Inuit to express their emotional attachment to the land, and via a symbolic plane, to express their individual and collective identity. For example, Inuit artist Irene Avaalaaqiaq's extraordinary wall hangings and drawings have allowed her to recount her grandmother's oral stories, and share the legends and lifestyles "of those Inuit who once lived in the manner of their ancestors. …Oral traditions of storytelling and the passing on of moral tales find a new mode of telling" in her "highly dramatic and powerfully expressive wall hangings and drawings" (Nasby, 2002, p. 84). For the last fifty years, art has also provided an important opportunity for Inuit to earn a living in a way that leaves a light ecological footprint. A July 2009 fact sheet on Inuit artists notes that according to Census Canada, in Nunavut Inuit artists represent 1. 86% of the territory's labour force which is more than double the national average of 0.77 percent (Canada Council of the Arts, July 2009, p. 1). Inuit art has received national and international recognition, and remains an important mode for expressing Inuit identity.

The IQ Adventure website development team digitized 100 Inuit carvings and 100 Inuit graphic prints from the Houston North Gallery. The IQ Adventure's co-developer, John Houston, is an artist himself. His father, James Houston, travelled to the Arctic in 1948 to make drawings of Inuit. When he was first given a gift of carvings from his Inuit neighbours, he recognized their talent and subsequently introduced Inuit art to the world, taught them print-making, and supported the creation of the famous West Baffin Eskimo Co-op in Cape Dorset, Nunavut, where many Inuit have created their masterpieces.[13] John Houston now serves as the director of the Houston North Gallery, and as a partner on the website development team he provided the opportunity to create a searchable database of the gallery's collection for the IQ Adventure website. Such an undertaking in itself alone is an important enterprise, given first, that the Houston North Gallery's Inuit graphics collection is considered the largest outside of museums, and second, that the Gallery's Inuit sculpture spans 58 years, from recent masterworks through to the beginnings of contemporary Inuit art.

Inuit art has played an important role in keeping legends, traditions, and traditional ways of being alive. Historical relationships are recorded in Inuit art, as exemplified in the (1976) print by Napachie Pootoogook, entitled *The First Policeman I Saw*. Nancy Doubleday (2005) explains: "Inuit art has achieved recognition on the world stage as a result of its unique aesthetic, its origin and, to some extent, its political utility. …The articulation of Inuit identity in art and political life… flows from the same, shared spring of relations between Inuit culture and 'land' and is captured within the Innuqatigiittiarniq' [the Inuit word for the concept of the healthy interconnection of mind, body, spirit and the environment] "(p. 169). One of the artists who IQ adventurers meet and learn from in the interactive film, sometimes refered to as the virtual learning environment (VLE), is Siassie Kenneally, known internationally in the art world as one of the three cousins (along with Suvinai Ashoona and Annie Pootoogook) of Cape Dorset. Siassie Kenneally is an artist who grew up on the land; she lived in a camp established by her father that was located in the midst of an abundance of wildlife, including beluga whales, polar bears and seals. From her father and older brother, she learned to protect herself and others in this environment, to hunt at an early age, and how to traverse the ice flows in her boat at the age of 12; she has built upon this knowledge base, inventing new hunting techniques.

Figure 5 depicts a work in progress by Siassie Keannelly. The web development team created hot spots by embedding links at certain locations in the piece of art for the user to discover. By clicking on the links one has the opportunity to hear Siassi sharing a story of community, history and IQ related to the experience she had growing up on the land. One of the stories Siassie shares is

Figure 5. A still image of Cape Dorset artist Siassie Keannelly's work in progress to which she refers during her interview in the interactive film on the IQ Adventure website © 2006, C. J. Alexander. Used with permission

the Cape Dorset community development model of survival through adaptability.

Siassi is distinctive as an Inuk woman, in that she is a young person who has learned how to survive and thrive on the land, knowledge that has been lost to far too many young Inuit men and woman. Her knowledge of the land is reflected in her art, and her goal is to share traditional Inuit ideology through her art. Siassie travelled to Acadia University to share and record her understanding of the changing cultural and environmental context in which she lives. Upon discovering the hotspots embedded in Siassie's extraordinarily detailed artwork representing the region where she was raised, the user hears the artist recount the skills she learned from her father, the mis-adventures she shared with her siblings, the encounters she experienced with wildlife, and the values she still counts upon to survive a difficult life. For Siassie and other Inuit artists, art has provided them with a way to survive, as explained by Coach Irniq in his briefing to the user before concluding the scenario set in Cape Dorset, Nunavut:

Inuit have the highest number of artists, per capita, in Canada. As we have adapted, art has served us well. We have preserved our culture, including our legends, through our art.

We Inuit have traveled around the world through our art, and our art has been purchased worldwide. The Government of Canada gives our art as gifts to dignitaries.

We adapted, as best we can, to the market economy. Art has given us the money we need to try to provide for our families, and because we believe in sharing, for our communities. (Dr. Peter Irniq, Nanisiniq Inuit Qaujimajatuqangit Adventure Website)

Inuit art explains a way of life. It offers a visual history life on the land. Inuit myths and legends were kept alive through art when missionaries and school masters forebade the practice of Inuit spirituality, including drumming and the sharing of legends such as Kiviuq. Like the Elders' Interview database, the art database stands alone as a resource and is embedded in the interactive film for users to draw upon to learn more about IQ. As Coach Peter says, "By the time you finish this journey, you will have learned who Kiviuq is" (2006, *Nanisiniq Inuit Qaujimajatuqangit Adventure Website*). The art database is embedded within the interactive film for users to access during their quest, and is supplemented by video-interviews with artists and other Inuit in what we call the Community Voices (CV) database. For example, we filmed Annie Pootoogook, a contemporary artist with a growing international reputation, during our sea voyage and share her insights in the CV database. Importantly, Annie chose to talk about the role that art has played in her own healing, and some of her work depicts domestic abuse; her work departs from the cliché forms of Inuit art by chronicling life in the midst of a rapidly shifting culture, caught between life on the land and the flow of ideas, information, and material goods from the south. Among the other artists who users 'meet' is Kenojuak Ashevak, perhaps the world's most famous Inuit artist and an Elder who is in her 80s, who was filmed by the project development team at the West Baffin Eskimo Co-op in Cape Dorset that she helped co-found.[14] Kenojuak was filmed in action, drawing as she has for over 50 years in the Co-op. As she sketched, we filmed her talking about her memories of making art, from the moment that James Houston handed her the first piece of paper she had ever seen, to the importance of the Co-op in sustaining Inuit culture and communities, to her role as a mentor to generations of Inuit artists. Our interview with her includes her views on the role that art has played in Inuit life—as a source of income, healing, and identity, among other things—in the last fifty years or so. It our hope that the user will find inspiration on how to face

his or her own life challenges by paying attention to and learning from Kiviuq's adventures, from exploring the art database, and hearing from artists during the IQ quest.

Being Up There: The Virtual Learning Environment

The interactive film provides artistic representations, philosophical insight, story-telling, and policy analysis to nurture a distinctively integrated, holistic understanding of *IQ* as *living policy*. It guides the user through an immersive experiential learning adventure—a virtual learning experience (VLE)—about the relationship between the Inuit and the land, understood through a multi-media journey that provides the learner-user with the opportunity to interact with Inuit Elders. The quest presented within the VLE is for learners to understand, appreciate, and respect, and perhaps incorporate, *Inuit Qaujimajatuqangit* into their lives. The principle goal was to share IQ via the interactive online learning resource with Inuit youth, given that half of the population of Inuit is under the age of 15 years, and that Inuit comprise 85 per cent of Nunavut's population. In the 20th century, one of the most powerful modes of storytelling in our culture was film; in the 21st century, the learning potential of VLEs lies in the power of the media to create immersive, interactive experiences.

Providing a contextualized story for digitized content such as the databases helps to realize the potential to engage and inform youth in new, compelling and perhaps, addictive ways. In the privacy of his or her own cyber-space, someone who may be too shy to initiate a conversation with an Elder can set out on a journey with 50 Elders ready to be called upon. In a jolts-per-minute culture, the learner can lose himself, and time, in a story in which he actively plays a role, makes decisions based on new-found understandings, and realizes, in a safe, private space, the consequences of his actions. Gaining knowledge about Inuit ways of being, becoming articulate about Inuit stories, growing secure in his sense of place in the landscape, the user may go back into his classroom, home, and community, with tales to tell, successes to recount, and curiosity to delve deeper…and in so doing, Inuit youth will contribute to the preservation, and promotion, of the legacy of the Elders. Having initiated the conversation, he will bring back to the webspace stories to share with other users heard from his Elders, sketches he's made of seals and owls, experiences he has had in trying to follow, imitate, and recount IQ in his own daily life.

The VLE provides the user with the opportunity to make a number of crucial decisions related to the relevance and appropriateness of IQ-based actions, or inactions, and will reveal the implications of different decisions throughout the journey. The quest presented to the learner-user within the VLE is to make decisions at crucial points, calling upon Elders in the database to assist and 'draw' on art to inspire and inform, in ways that will show an growing understanding of, and respect for, Inuit knowledge and way of life. The VLE was designed to share insights "and experiences of *particular* worlds and *particular* relationships" (Bogost 2007, 241). The following passage, exerpted from the original script John Houston, Peter Irniq, and the author wrote, is suggestive of the experience the user embarks upon:

Having magically been transported on board the MSS Explorer, which is anchored in the deep water harbour in Iqaluit, you find yourself in a room. You are exploring the 360 environment when the phone rings. You answer it. It's a call from Peter Irniq, the former Commissioner of Nunavut. He welcomes you aboard, glad that you've taken up the challenge. He's agreed to be your mentor on this journey, guiding you along your way, encouraging you to succeed. He asks if you'd like this. When you say yes, then he appears as a hologram right there in front of you, standing with you in Homebase.

Coach Peter explains the variety of resources you'll have available, and tells you about the magical portals that will take you to different times, to meet different Elders, and to explore different places. He explains that you will encounter some magic. Art may come alive. Spirits may throw curve balls in your way. But he assures you that he'll be there throughout the journey.

(Dr. Peter Irniq, Nanisiniq Inuit Qaujimajatuqangit Adventure Website)

For non-Inuit youth, the opportunity exists to get closer to the tundra and to traditions, to legends and language through art, through Elders' stories, through the filmed sequences of the Eastern Arctic into which the databases are embedded. For a primary school student, she will bring tales of creatures and quests that *Kiviuq* pursued, and in so doing, will make Nunavut a part of her reality. For teens seeking adventure, the quests provide endless hours of opportunities to learn how to progress through the VLE, and in so doing, learn about an ancient culture whose stories, way of being, may find a place in a young Canadian in Toronto or a Japanese youth in Kyoto. The resource provides insight into Inuit spirituality, and following Kiviuq via this route into Inuit intellectual systems. The immersive experience is designed to provide some appreciation for Inuit's ancient foundational principle of interdependence within communities, across inter-generations, and vis-à-vis nature. Such connections might resonate with non-Inuit youth across Canada, and around the world. Historically, there has been a fascination with the Inuit and their land, and we hope that this resource will interest and engage life-long learners in a compelling new way.

Unlike traditional websites where much of the information is text based and supported by images and other media, the VLE is designed around the rich media content. Scenes are interactive, users select various and differing paths and adventures. As stated earlier in this chapter, users are guided by Coach Peter Irniq, mentored by members of the community, and informed and inspired by Elders through a quest. The project development team sought to create an immersive environment in which users would learn by observation, by listening, and by participating as if they were

Table 1. Still Images from the IQ Adventure website of 360 Degree Panoramic Photographics taken during the 2006 Voyage on the MSS Explorer. © 2006. C. J. Alexander. Used with permission

A still image from the interactive film at the outset of the journey. The user has just been briefed by his mentor for the journey, Dr. Peter Irniq, who provides briefings and de-briefings before and after each of the user's quests. The user can user his mouse to explore a 360 degree panoramic view of Nunavut's capital, Iqaluit, and is invited to click on the ship, visible in the distance, to board the ship MS Explorer.

The image to the left is a still from the 360 degree panoramic photograph of the Government of Nunavut's Legislative Assembly. Hotlinks are embedded on Inuit symbols in the Legislature, with explanations provided by P. Irniq.

The bridge of the MS Explorer, the vessel for the virtual voyage from Iqaluit to Cape Dorset, Nunavut. During the Sept. '06 voyage in honour of James Houston, John Houston, Alethea Arnaquq Baril, Dave Sheehan C. Alexander, and other members of the development team filmed the scenarios for the interactive film, which features community members in Iqaluit, Kimmirut, and Cape Dorset.	Digges Island, location of the justice scenario, featuring Inuit culturalist, lawyer, designer of traditional Inuit clothing, and musical performer, Aaju Peter. The user faces a challenge, and must make a decision informed by *IQ* before s/he can move forward in the quest.	Pritzler Harbour, location of the consensus-based decision-making simulation, which the user must successfully complete before moving forward in the journey. In this survival scenario, the user must determine how to survive on an isolated, unihabited island.

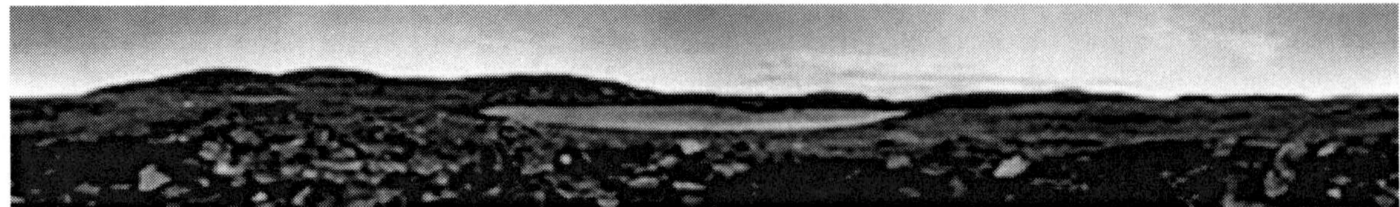

Jimmy Manning was filmed at the ancient sacred site, Malik Island (near Cape Dorset where, among other roles, Jimmy has worked as the print-shop manager at the West Baffin Eskimo Co-operative, and where Kenojuak Ashevak, Annie Pootoogook and many other Inuit artists have created their masterpieces). On Malik Island, at the ancient tent circle, Jimmy concludes the voyage in the interactive film with reflections on sustainable community understood in the context of Inuit's 4000 years serving as stewards of the land.

visiting the actual sites and communities, with the ship anchored in the distance at each locale. Scenes were carefully crafted to allow the user to learn by doing, making mistakes, receiving corrective feedback and positive reinforcement of his or her acquired knowledge. The opening scene of the VLE is reflective of this approach.

In addition to the use of actual footage we shot of a real voyage we took by sea through the Eastern Arctic, the developers also used 360 degree virtual reality photography that enable the user to pan up, down, and around specific sites for each of the scenarios in the interactive film (please see examples offered in Table 1).

In each of the five scenarios the user has to make some choices at specific times that will reveal whether he has learned about how to apply IQ; each time he successfully completes a challenge, the user receives a stone to build his or her own virtual Inuksuk, as explained by Coach Irniq, below:

If you are successful in an activity, you will get a stone. Once you have successfully completed all the challenges, you will have enough stones to build your own inuksuk.

Myself, I have built many inuksuit in many foreign lands. The inuksuk you build yourself will be your compass, telling you where you're heading... and where you've been. It also provides you further insight into the world of Inuit Qaujimajatuqangit. (Dr. Peter Irniq, Nanisiniq Inuit Qaujimajatuqangit Adventure Website)

The quest is a search to understand Inuit values, language, and culture; ultimately, it is a quest to understand Inuit identity through the story of Kiviuq and IQ. When s/he successfully completes the adventure, the Inuksuk will stand as a sign of the knowledge and values that s/he gained that can be called upon to navigate through life.

In the interactive movie the user participates in five core scenarios that are focused on five themes: environmental stewardship (ie. examining the relationship between Inuit culture and hunting, IQ about climate change/global warming); justice; learning from five Elders in Kimmirut who were interviewed on film (in Inuktitut) as they butchered a seal, skinned it, sewed, and cooked; consensus-based decision making; and understanding what sustainable community means through a visit to Cape Dorset. Figure 6 shows a map of the journey. In each of the scenarios the user interacts with notable Inuit, including: Cape Dorset artist and hunter, Jimmy Manning who goes out on a zodiac through the Lower Savage Islands with the user and offers eye witness accounts of the effects of climate change on the fragile ecosystems and on the Indigenous peoples, Inuit, who have served for millenia as the original enviromental stewards of the land; with Aaju Peter, an Inuk lawyer, who the user has to face in the justice scenario (see Figure 7); and Kenojuak, an Inuit artist of international acclaim, who was one of the first Inuit to advance the idea of an art co-operative in Cape Dorset, an excellent example of Inuit living lightly on the land in the modern world.

Developing the IQuest: Example from the Environmental Stewardship Scenario

Inuit may be the original ecologists. Given the challenge that global warming presents, they

Figure 6. Map of the voyage of the MSS Explorer that the user embarks upon in the interactive film on the IQ Adventure website. © 2006, C. J. Alexander. Used with permission

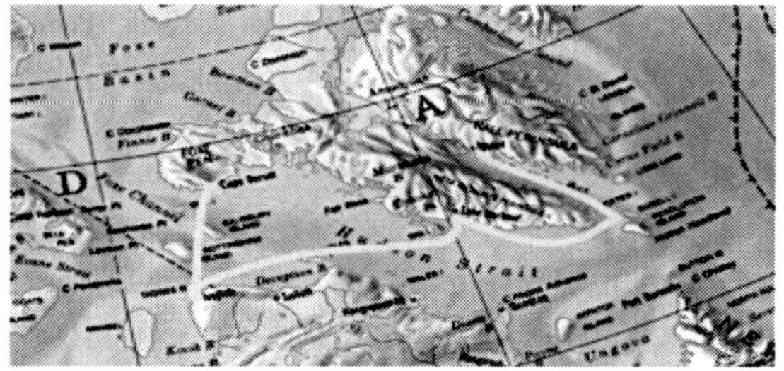

Figure 7. Draft of the flow chart created for Justice scenario in interactive film. © 2006, C. J. Alexander. Used with permission

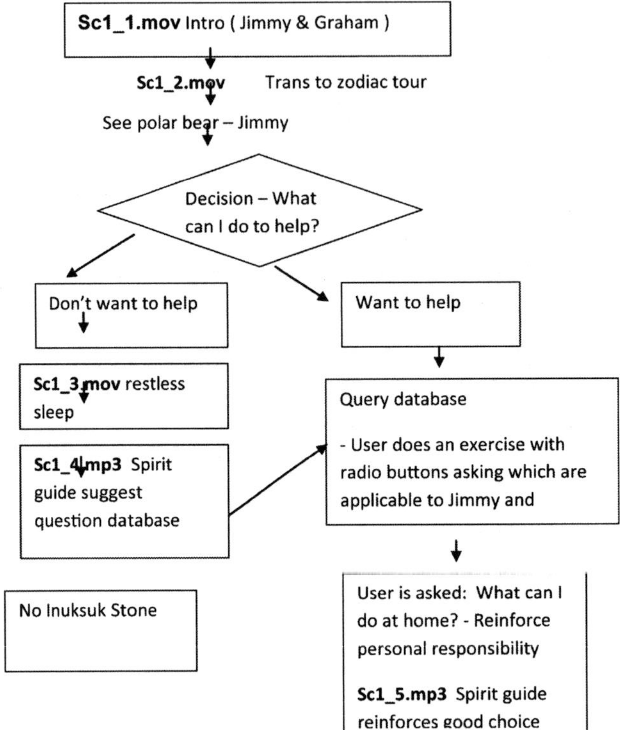

may offer some of the insights and alternatives we need. As the voyages embarks and the ship sails through the Lower Savage Islands, to gain insight into their intrinsic sustainable worldview and way of life, the learner has a dialogue with a Western scientist, Professor Graham Daborn, and an Inuk hunter, Jimmy Manning. This dialogue reveals that the scientist and the Inuk hunter share many concerns about environmental changes in our midst. They also agree that they have a need and interest in working together, with the scientist acknowledging that there is much to learn from Inuit about the manifestations of climate change, and other environmental challenges, and environmental stewardship.

Out in the zodiac with Jimmy Manning, an Inuk hunter, and Graham Daborn, a marine ecologist, the user begins to learn more about the impacts of climate change on the fragile Arctic ecosystem. The learner discovers that scientists predict that Canada's polar bears will be extinct within 25 years as global warming shrinks the ice cover on which they depend for feeding and giving birth. On the other hand, he learns about the significance of hunting to the Inuit way of life. He learns that Inuit are seeing 'more' polar bears in parts of Nunavut, and that they are encroaching upon and therefore, endangering life in communities. He starts to recognize that we all live in a very vulnerable eco-system; on the other hand, as Inuit remind us, nature adapts…the question is whether humans will survive. Back on the ship, the learner struggles to make sense of information and understanding what he has gleaned. In his sleep, a spirit challenges him to make a difference. He tosses and turns. Upon awakening, he has to figure out: What can he do, personally, to effect change, given the environmental disaster that is unfolding in our midst? Table 1 exemplifies the simple decision matrix that the development team created for this particular scenario. Each scenario has a challenge and/or a choice that the user must complete successfully on his or her quest. Only upon the successful completion does the user receive a stone for his virtual inuksuk. He can go deeper into the resources to glean more insight from the databases, listen more carefully to the Elders and community voices, and explore other resources in homebase.

Each of the scenarios is distinctive in form and substance, explained briefly below:

- The *environmental stewardship* scenario takes place out in a zodiac at sea, providing the user with a chance to explore the surroundings with Jimmy Manning and consider the impacts of climate change. The learner gains more insight into humans' interdependence with the land, and of their role as environmental stewards.

- The *consensus-based decision making* scenario leaves the user stranded on an island with other adventurers from the ship. The challenge is how everyone will survive without supplies and without radio contact with the ship. It is a scenario about how people interact, and the quest is to learn the Inuit way of solving problems. In this scenario, the learner learns something about the role of mutual respect in a society that depends on interdependence for survival.

- The *justice* scenario focuses on how Inuit resolved disputes, a system very different from the emphasis on retribution that has dominated colonial legal systems. Something is stolen, and the user has to figure out how to face the consequences. Inuk hunter, seal-skin traditional clothing designer, musical performer, activist, and lawyer, Auju Peter plays a key role in this scenario. In this scenario, the learner explores the role of healing, humour, and integrity in an interdependent culture.

- In the *education* scenario, the learner has the opportunity to observe a community in action, and like a young Inuk, understands how by watching others he or she can learn a great deal. Learning by doing has always been an effective pedagogy among Inuit. The learner also learns, by observation, the equally important roles that men and women play in an interdependent community.

- In the *sustainable community* scenario, the learner has the opportunity to take a trip through Cape Dorset with two internationally renowned artists, one a carver and the other a print-maker. The learner has the opportunity to listen and understand how individuals in a community have cooperated to survive forced relocation and immersion into a new economic system; further, he learns that interdependence still defines Inuit community life.

In each of the scenarios, the learner also discovers something new about the fundamental principle of interdependence. Before and after each scenario, in "homebase" aboard the ship, the user is briefed by Coach Peter Irniq. One of the "360 navigable environments" is the homebase on board the ship which the user will have access to throughout his journey. Using his mouse, the learner can zoom around and zoom in to access a variety of resources that will help him or her in the journey. In addition, Coach Peter phones and emails the user throughout the journey, sending information and ideas that will help him successfully complete the challenges. Some documents, including those from the government that are relevant to the challenges, are accessible at home base. Importantly, homebase will be the point of entry into three "portals" that will take him into three different temporal worlds:

1. a portal entry into the adventures of Kiviuq, the great Inuit hero and shaman, told through the video-recorded stories of Elders;

2. a portal entry that bring selected pieces of Inuit art to life;
3. a portal entry into the scenarios.

The user therefore goes to 'homebase' to prepare for and to access each of his adventures. The game-like, fun learning environment:

- Exemplifies IQ principles and values with learning scenarios.
- Deepens understanding of Inuit values and knowledge, and challenges the user's ability to apply that understanding in simulated, real-life situations.
- Explores the role of art and music in Inuit culture.
- Facilitates a connection with Inuit Elders, their communities and the land.
- Encourages a closer relationship with the land.
- Introduces the Inuit concept of fluid leadership.
- Offers an introduction to the Inuit model of consensus-based governance.

The final experience for the user in the interactive film takes place on Malik Island.

On the final trek of the virtual voyage, the explorer 'meets' with Jimmy Manning once again. Jimmy shares his reflections on the Thule people who lived on Malik Island a thousand years ago, and he takes the user through an ancient Thule home, sharing insight into how they lived on the land. S/he listens to Jimmy reflecting about the changes in our midst, and what we can learn from those who lived in harmony on the land. The learner is challenged to consider what sustainable community means in his own environment, and in the world.

Nakamura astutely asserts that the "Internet has created and defined digital visual capital, a commodity that we mark as desirable by conferring on it the status of a language unto itself; we speak now of digital literacy as well as visual and the ordinary sort of literacy"(2008, 15). The visions and voices of Indigenous peoples are being carried across cyberspace. As Inuit youth learn how to build igloos and inuksuit (the plural of inuksuk), they themselves are turning to their Elders, video-recorders in hand, to learn more about Inuit ways of being, knowledge system, language and culture.

FUTURE RESEARCH DIRECTIONS

Approximately 50 years ago, Inuit were living as they had for millenia, and then were suddenly uprooted from their camps, language, culture, knowledge systems, and ways of being. By the end of the twentieth century, they reclaimed their land, created a public form of government in Nunavut, the newest territory in Canada, and set out to infuse IQ throughout the public sector. Like the canary in the coal mine, in one of the world's most fragile eco-systems, Inuit are facing the rapid effects of global warming and climate change. While *Nunavummuit*, the Inuktitut word for those who call Nunavut home, have celebrated their tenth anniversary year as a territory, the federal government has failed to live up to its commitments to the Nunavut Land Claim Agreement of 1993, leaving many Inuit homeless, hungry, unemployed, and ill. Yet Inuit *are* resilient. In the midst of a housing crisis, public officials have placed language and culture as top priorities on the policy agenda. New media technologies are playing a key role in securing Inuit knowledge systems for current and future generations of Inuit, and for humanity. However, online resources such as *The Nanisiniq Inuit Qaujimajatuqangit* or IQ Adventure website may be out of reach for northern youth, given the high cost and instability of Internet access across rural and remote communities of Canada. The digital divide has not disappeared and therefore, merits ongoing analysis. Among the questions to be broached are the following: What is the role of government in providing affordable, reliable

Internet service across remote regions? What are the implications of the lack of such service for northerners in terms of access to education resources in Inuktitut? The relationship between the Internet and citizenship merits much more attention. For example, the following questions merit consideration: What are the implications for democracy if some citizens do not have the kind and quality of Internet service that citizens in urban centers enjoy? What is the role of the state in funding the development of online resources that help ensure that Indigenous languages and cultures that might otherwise become extinct, are revived and promoted? A concern about citizenship among marginalized, oppressed communities of interest leads to questions such as the following: How, if at all, is the Internet connecting Indigenous peoples in specific countries such as Canada or the United States? How, if at all, is international indigeneity supported by the Internet? From an international comparative perspective, are new media technologies supporting decolonization strategies of Indigenous peoples and if so, how so? Then too, as new learning resources become available that can help fill the gap in understanding about Indigenous peoples', in their own voices, it is not necessarily a given that school teachers and school administrators will take-up the opportunity to draw upon such resources. For example, will teachers whose time may already be stretched too thin find new online resources related to subjects they themselves may not have studied, too challenging or time-consuming to incorporate into their lesson plans? What is the nature of adoption, or rejection, of such new learning resources in schools that serve primarily Indigenous students, and in schools that serve primarily non-Indigenous students? Will such resources provide Inuit youth with an anchor to their traditional knowledge systems that buttresses their individual and collective pride? Further, it will be interesting to see whether, and if so how so, users' interest in such new learning resources provides new pathways of understanding and respect for Indigenous peoples.

CONCLUSION

This chapter's focus on Inuit-infused digital designs addresses the specificity of Indigenous peoples' struggles in the modern Canadian state, but Inuit's digital compass is reflective of and relevant to other marginalized communities' post-colonial and anti-racist initiatives. The case study of Inuit's stake in cyberspace is important within discussions about ICTs in the global context, if for no other reason than the fact that, in the face of climate change and global warming and the retreat of the polar ice cap, the Arctic has become a new stage for international economic and political battles over energy resource development, new transportation routes, and sovereignty claims. Globalization presents new challenges to Indigenous peoples' search for autonomy in Canada and elsewhere in the Circumpolar North. In the face of such rapidly changing ecological, political and economic environments, Indigenous peoples in the Arctic of Canada may find themselves, once again, iced out in political and policy decision making, despite the advances they have made in self-government.

In this chapter, I offered some insight into the creation of the interactive film, one particular resource on the IQ Adventure website that I co-developed. The website, a community-based initiative, was inspired and driven by the need for cultural preservation and promotion as articulated by Inuit. As a multi-media resource, the website was developed as a response to the 'state of emergency' that exists in terms of recording and sharing Inuit Elders' knowledge before they, and the knowledge they hold, is gone. New media technologies are critical resources in efforts to deconstruct colonial discourses and practices, and to develop post-colonial knowledge systems that draw on ancient wisdom and ways of being. The case study of the development of the IQ Adventure website illustrates the critical role that information and communication technologies (ICTs) play in decolonization efforts by Indigenous peoples.

Inuit are key actors in designing the post-colonial digital landscape by sharing with the world, in unprecedented ways, their values, knowledge and experiences. In doing so, Inuit's digital designs are contributing to new forms of identity and community among the world's Indigenous peoples.

The intergenerational grief that continues to grip individuals and their communities resulting from colonialism, including the residential school experience, leaves Indigenous peoples particularly vulnerable in a period of upheaval in their midst, including climate change and global warming. Yet Inuit have survived in harsh conditions for thousands of years, and are an adaptive and resilient people. In the 21st century, in a resource-rich developed nation such as Canada, it may be surprising that Inuit experience persistent poverty, a direct result of colonization. The challenge is to spark the interest of youth who, as a result of colonization, do not have the anchor of their cultural legacy. New media environments can be designed as *culturally–infused* media resources, created by and for Indigenous peoples as a means to help address the ongoing intergenerational effects of colonialization. Community-based initiatives to develop culturally-appropriate new media resources support post-colonial movements. Inuit Elders are uniquely positioned to provide profound insights into bigger societal questions, such as the relationships between humans and vis-à-vis the environment.

Arctic peoples' cultures and perspectives have been marginalized. From a political and policy perspective, they have not been treated as equal citizens. Their basic needs have not been met. Inuit of Nunavut have reclaimed the land for which they have served for thousands of years, as the primary custodians. As they continue to reclaim their cultural identity, knowledge and belief systems, Inuit are reaching out to the world in the hope of sharing the lessons they have learned about living with other humans interdependently, while serving as environmental stewards of the land. Interestingly, Inuit are unambivalently positioned to stake their claim in cyberspace at the same time that nation-states are jockeying to advance their strategic positions as the ancient ice in the Northwest Passage quickly melts. It will be important to continue to explore the role that connectivity and new media technologies play in articulating and advancing Inuit's socio-political, cultural, and economic needs and interests in such a rapidly changing environment.

REFERENCES

Alexander, C. (2001). Wiring the nation! Including first nations? *Journal of Canadian Studies. Revue d'Etudes Canadiennes, 35*(4), 227–239.

Alexander, C., & Pal, L. (Eds.). (1998). *Digital democracy: Policy and politics in the wired world.* Toronto: Oxford University Press.

Battiste, M. (25 October 2005). *Reconciling aboriginal diversity in learning systems.* Keynote presentation. URL: www.usask.ca/education/people/battistem/keynote-OSSTF-2008.pdf

Boulou E. D'beri. (2005). *The new practices of memory.* URL: mokk.bme.hu/centre/conferences/reactivism/FP/fpBB

Canada. Department of Indian Affairs and Northern Development. (11 June 2008). *Statement of apology to former students of indian residential schools by prime minister stephen sarper.* [http://www.ainc-inac.gc.ca/ai/rqpi/apo/index-eng.asp]

Coates, K. S., Lackenbauer, P. W., Morrison, W. R., & Poelzer, G. (2008). *Arctic front: Defending canada in the far north.* Toronto: Thomas Allen Publishers.

Doubleday, N. (2005). *Sustaining arctic visions, values and ecosystems: Writing inuit identity, reading inuit art in cape dorset, nunavut. Presenting and representing environments.* The Netherlands: Springer.

Galabuzi, G. (2006). Canada's *economic epartheid: The social exclusion of racialized groups in the new century.* Toronto: Canadian Scholars' Press.

Government of Nunavut. Department of Human Resources. (2005). *Inuit qaujimajatuqangit.* [http: http://www.gov.nu.ca/hr/site/beliefsystem.htm].

Human Rights Tribune. (8 January 2008). *Native suicide surge, a post-colonial trauma.* URL: http://www.humanrights-geneva.info/Native-suicide-surge-a-post,2630

Kay, J. (3 February 2009). "The left's aboriginal blind spot." *National Post.* URL: http://www.nationalpost.com/opinion/columnists/story.html?id=f80486d4-3866-4932-a3a5-4ec927795139&p=3

Pauktuutit, Inuit Women of Canada. (1 October 2006). Keepers of the Light: Inuit Women's Action Plan, Ottawa. http://www.pauktuutit.ca/pdf/publications/pauktuutit/KeepersOfTheLight_e.pdf

Saul, J. R. (2008). *A fair country: Telling truths about canada.* Toronto: Viking Canada.

Shimo, A. (25 February 2009). *Tough critique or hate speech?* Macleans.ca URL: http://www2.macleans.ca/2009/02/25/tough-critique-or-hate-speech/print/

Simon, M. (13 February 2009). Assimilation is no solution. *National Post.* URL: http://www.nationalpost.com/life/footprint/story.html?id=1285763

Slemon, S. (1987). Monuments of empire: Allegory/counter-discourse/post-colonial writing. *Kunapipi* 9.3, p. i-i6.

Smith, C., & Ward, G. (Eds.). (2000). *Indigenous cultures in an interconnected world.* Toronto: University of British Columbia Press.

Van Deusen, K. (2009). *Kiviuq: An inuit hero and his siberian cousins.* Montreal: McGill-Queen's University Press.

Widdowson, F., & Howard, A. (2008). *Disrobing the aboriginal industry.* Toronto: McGill-Queen's University Press.

ADDITIONAL READING

American Indian Digital Divide Task Force. Retrieved from www.indiantech.org

Branswell, H. (26 February 2004). Lifespan up to 10 years shorter for first nations, inuit peoples. *The Globe and Mail.*

Conference Board of Canada. (2001). *Aboriginal digital opportunities: Addressing aboriginal learning needs through the use of learning technologies.* URL: http://www.conferenceboard.ca/aboriginal

Harvard Kennedy School. (2010). *Harvard Project on American Indian Economic Development.* Retrieved from http://www.ksg.harvard.edu/hpaied/

Hurley, D. (September 2003). *Pole star: Human rights in the information society.* Retrieved from http://www.ichrdd.ca/english/commdoc/publications/globalization/wsis/PoleStar-Eng.html

Kuh-ke-nuh Network of Smart First Nations (K-Net). (nd). *Using broadband communication technologies and services in keewatinook okimakanak first nations.* Retrieved from http://knet.on.ca

Stackhouse, J. (10 December 2001). The wired warrior's digital dream. *The Globe and Mail.* URL: http://www.globeandmail.com/servlet/ArticleNews/printarticle/gam/20011210/FCSTAC12Y

Valaskakis, G. (1996). Citizens of the electronic village: smartening up or dumbing down? *Couchiching Online. URL:* http://www.couch.ca/history/1996/valaskakis.html

ENDNOTES

[1] The author is grateful to the Social Sciences and Humanities Research Council of Canada for funding her research on digital equity in Canada that informs this chapter. The generosity of time, insight, patience, and willingness to collaborate of Inuit of Nunavut continues to inform and inspire the author. The website was developed with funding from the Canadian Culture Online program in the federal Department of Canadian Heritage. The diverse partnerships that underpinned the website development were critically important to the creation of such a distinctively Inuit space in the digital universe.

[2] Nunavut means our land in Inuktitut.

[3] The health of people is tied to their living conditions: "Poor living conditions including overcrowding, lack of clean water and safe waste disposal all contribute to higher levels of disease… Life expectancy in Nunavut is ten years below the national average" (Government of Nunavut 2004, p.3).

[4] As Teillet reminds us: "If we want aboriginal people to buy into the Canadian dream, we must understand what we have offered them to date" (Griffiths 2007, p. 189).

[5] The term 'Métis' refers to a person of mixed Aboriginal and European ancestry, who identifies as Métis.

[6] Grace-Edward Galabuzi's research reveals that "[w]hile Canada embraces globalization and romanticizes the idea of multiculturalism and cultural diversity, persistent expressions of xenophobia and structures of racial marginalization suggest a continuing political and cultural attachment to the idea of a White-settler society" (Galabuzi: 2006, xii).

[7] Turtle Island is a traditional name for North America among Indigenous peoples.

[8] Linda Tuhiwai Smith writes: "The word itself, 'research', is probably one of the dirtiest words in the indigenous world's vocabulary. When mentioned in many indigenous contexts, it stirs up silence, it conjures up bad memories, it raises a smile that is knowing and distrustful. It is so powerful that indigenous people even write poetry about research. The ways in which scientific research is implicated in the worst excesses of colonialism remains a powerful remembered history for many of the world's colonized peoples" (1999, 1). For this reason, the IQ Adventure website has been developed on a foundation of community-driven research over the past decade. For example, Acadia University received a grant from the Social Sciences and Humanities Research Council of Canada (SSHRC) (C. J. Alexander served as Principal Investigator) to invite Inuit and other northern policy makers (elected and appointed), educators, elders, community activists for a three day retreat and symposium, called the International Nunavut at Five Symposium, to discuss their research priorities. It was at that symposium that *Nunavummuit* (literally translated as those who call Nunavut home) identified the preservation and promotion of IQ, for current and future generations of Inuit youth, as a top priority. Subsequently, another SSHRC grant enabled Acadia University and other university researchers to travel to Iqaluit, the capital of Nunavut, to meet with elders, youth, principals and teachers, the Premier and other politicians, and civil servants. Acadia University has hosted Elders and Artists in Residence from Nunavut, who have met with students, faculty, and worked closely with the IQ Adventure website development team. In 2006-2007 the federal Department of Canadian Heritage provided a major grant to develop the IQ Adventure website with a community-driven partnered

research team (including Acadia University, the Department of Education and the Department of Culture, Language, Elders and Youth (CLEY) in the Government of Nunavut, John Houston's Inuit-led film-company, drumsong communications inc, and Adventure Canada). We have learned much about community-driven initiatives and through the creation of the www.InuitQ.ca website, we have identified additional culturally-appropriate processes that will guide future research and development initiatives.

9 John Houston has produced the following films (copyrighted through two entities, drumsong communications inc., and Houston Productions Inc): Songs in Stone: An Arctic Journey Home (1999); Nuliajuk: Mother of the Sea Beasts (2002); Diet of Souls (2004); Kiviuq (2007); James Houston: The Most Interesting Group of People you'll ever Meet (2008); The White Archer (2010). Two publications of interest by Houston are: "Songs in Stone: Animals in Inuit Scupture", *Orion Nature Quarterly* (Autumn 1985); "Art and Soul", in *Equinox Magazine* (no. 81, 1995).

10 The quotation is an excerpt from Mariano Aupilardjuk's interview in The Elders' Kiviuq Database on the IQ Adventure website, and is identified in the project team's database archival records as <01 Tape KIV-D-21>.

11 The quoyation is an excerpt from Rachel Ujarasuq's interview in The Elders' Kiviuq Database on the IQ Adventure website, and is identified in the project team's database archival records as <05 Tape KIV-D-36a>.

12 The quotation is an excerpt from Sampson Quinangnaq's interview in The Elders' Kiviuq Database on the IQ Adventure website, and is identified in the project team's database archival records as <12 Tape KIV-D-16>.

13 In the 1960s, after Inuit were forced out of their camps and into settlements, the federal Government of Canada established arts and crafts centers as a way to engage Inuit in the cash economy.

14 Another Inuk who users get to know on the quest in the interactive film is Jimmy Manning, who plays a key role in the environmental stewardship scenario and in the concluding scenario. Jimmy is a hunter, artist, and the studio manager of the West Baffin Eskimo Co-op. In a January 2009 interview, he worries "about the co-op's future—now that Houston's original stars, for one, Kenojuak Ashevak, famed for her wildly imaginative, colourful prints of humans transforming into animals—are aging" [Source: http://mediacentre.canada.travel/content/travel_story_ideas/cape_dorset_coop]. Creating art is about spirit, but it also is a critical source of income for many Inuit.

Chapter 6
The Digital Abyss in Zimbabwe

Jill Jameson
The University of Greenwich, UK

ABSTRACT

Just as refugees fleeing to escape Zimbabwe have struggled to cross the crocodile-hungry waters of the Limpopo, so are Zimbabweans battling to find ways to traverse the abyss of a digital divide affecting their country. In 2008-09, Zimbabwe was rated third worst in the world for its national information communications technology (ICT) capability by the World Economic Forum, being ranked at 132/134 nations on the global ICT 'networked readiness index'. Digital divide issues, including severe deficits in access to new technologies facing this small Sub-Saharan country, are therefore acute. In terms of global power relations involving ICT capability, Zimbabwe has little influence in any world ranking of nations. A history of oppression, economic collapse, mismanagement, poverty, disease, corruption, discrimination, public sector breakdown and population loss has rendered the country almost powerless in ICT terms. Applying a critical social theory methodology and drawing on Freirean conceptions of critical pedagogy to promote emancipation through equal access to e-learning, this chapter is written in two parts. In the first place, it analyzes grim national statistics relating to education and to the digital divide in Zimbabwe, situating these in the wider context of Africa; in the second part, the chapter applies this information in a practical fictional setting to imagine life through the eyes of an average Zimbabwean male farm worker called Themba, recounting through narrative an example of the impact on one person's life that could result from, firstly, a complete lack of educational and ICT resources for adults in a rural farming situation and, secondly, new opportunities as a migrant to become engaged with adult and higher education, including ICT training and facilities. Access to education, to book publications, to ICT facilities, in dialogue with others during a long process of conscientization, are seen to open up democratising and liberating opportunities for Themba in South Africa. The powerful transformation that takes place Themba's life and propels him towards many achievements as an e-learning teacher is inspired by Freire's critical pedagogy: it provides a message of hope in an otherwise exceptionally bleak educational and technological situation, given the current difficult socio-economic and political situation that has resulted in a digital abyss in Zimbabwe.

DOI: 10.4018/978-1-61520-793-0.ch006

Copyright © 2011, IGI Global. Copying or distributing in print or electronic forms without written permission of IGI Global is prohibited.

INTRODUCTION

The great snaking curves of the 'grey-green greasy' (Kipling, 1993) Limpopo river coil themselves around the southern border of Zimbabwe, marking a key geographical and political transition into South Africa. To the north, the Zambezi river flows furiously along the border with Zambia, tumbling rapidly down from Mosi-oa-Tunya ('the smoke that thunders') at Victoria Falls. For a landlocked country like Zimbabwe, these rivers have both a literal and representational importance, separating out its national identity from other countries and marking timeworn boundaries that have witnessed both turbulent and calm waters, in friendship and hostility, struggle and survival, freedom and change with and against other southern African nations. It is to the Limpopo that many refugees and economic migrants from Zimbabwe have fled in past decades, taking scanty belongings, often escaping at night, in a struggle to reach South Africa before being caught and possibly imprisoned by the Zimbabwean African National Union-Patriotic Front (ZANU-PF) elements of government, and before cholera, HIV, poverty or old age have rendered them helpless.

More than three million people have left in similar ways: the Zimbabwean diaspora has grown exponentially, with around a quarter of the population now scattered across the world. In 2007 it was reported that more than 17,000 Zimbabweans a week were flooding through the Limpopo border into South Africa (Meldrum, 2007). Often, these escapees have had little money, few resources, and no one to meet them in South Africa. Many times, they may have had little hope of reaching any satisfactory destination after a dangerous journey. The end of that long journey may also often have been characterized by brutal treatment from hostile South Africans angry about increasing competition from these generally well-educated, skilled (Nherera, 2000) and desperate Zimbabwean immigrants, hungry for scarce resources and jobs. Zimbabwean refugees and migrants often have faced a menial, deskilled life in their new country, beset by many hardships (Bloch, 2006). But still they have come, crossing the vast chasm of the Limpopo river valley, on their many hopeful journeys.

Just as the rivers north and south separate Zimbabwe from other nations, so technological and developmental divides mark out its difference in recent decades characterized worldwide by progress in globalization, increased communications and market expansion fuelled and supported by new technologies. Zimbabwe has been more or less left behind with regard to technology during this time of massive global technological development (Nherera, 2000). Although Africa as a whole is rated as 'the least wired region in the world' (ITU, 2007d), the situation affecting Zimbabwe is acute even within Africa. While a stark 'digital divide' in comparison with other continents affects the whole of Africa in both relative and absolute terms, even within this overall challenging situation for access to information and communications technology (ICT) across Africa, the situation in Zimbabwe is now one of desperate poverty combined with an extreme lack of access to advanced technologies, particularly in the rural areas, in which 70% of the country's population lives (Matsika, 2007). In effect, Zimbabwe faces a digital divide of such magnitude that it forms an ever-deepening abyss separating this beleaguered country from more technologically advanced nations.

The struggle to cross this chasm of inequality in technological developmental terms is bound up with the desire for emancipation of the Zimbabwean people from the many political, economic and social difficulties with which they are faced. A long history of colonial and post-colonial political domination, oppression, mismanagement, economic collapse, corrupt politicians, electoral scandals, human rights abuses, impoverishment and disease in Zimbabwe has had traumatic effects on the population, rendering virtually powerless the educational and information and communications technology capabilities of the country, its economy and its industries (WOZA, 2009). The chasm between Zimbabwe's current level of ICT capability and those nations that have achieved higher levels of ICT development has occurred as a result of a difficult national history arising from a number of complex factors, as this chapter will discuss, using an implicit critical social theory philosophical approach to examine the digital divide issues affecting Zimbabwe.

The key aspect of critical social theory informing the chapter is an imaginative fictional portrayal of a critical emancipatory pedagogic approach based on the work of Paulo Freire (1921–1997), as seen through the eyes of a farm worker in rural Zimbabwe who has had little access to education in his adult life and no access at all to ICT. Freire's *Pedagogy of the Oppressed* (1970), developed originally from adult education literacy work with rural peasants and villagers in Brazil, and deeply connected with the cultural and contextual understandings of rural adult workers suffering both poverty and oppression, is appropriate to this Zimbabwean farming context. The fictional narrative in part two is informed by a consideration of the Freirean processes of *'conscientization'* (the development of critical consciousness) that can occur with ongoing opportunities enabled by educational dialogue, ICT and other learning facilities in adult education and subsequent progress in more advanced levels of education.

Freire (1970) advocated liberation from oppressive conditions through the development of *praxis*, a combination of critical reflection and action. Freire's liberatory critical pedagogy validates the humanizing potential of an 'authentic struggle' to be fully human and free from oppressive conditions. This emancipation involves a dialogic, problem-posing approach to education, based on mutual respect, which can only be undertaken by the oppressed to free both themselves and their oppressors from the dehumanizing conditions of the 'banking concept' of education. The 'banking concept' dehumanizes both oppressors and the oppressed, as it is a reductive approach based on the

idea that education is 'an act of depositing, in which the students are the depositories and the teacher is the depositor' (Freire, 1970, p. 53). Freire argues that the problem with the 'banking' approach is that it lifelessly dehumanizes both teachers and taught, because the 'teacher talks about reality as if it were motionless, static, compartmentalized and predictable.... to "fill" the students with the contents of his narration' (p. 52). The creative potential involved in the transformational approach of *critical pedagogy*, by contrast, emphasizes that both teachers and students can be, in a mutually respectful, life-enhancing way, involved in a problem-posing process of developing knowledge and understanding in education 'through invention and reinvention, through the restless, impatient, continuing, hopeful inquiry human beings pursue in the world, with the world, and with each other' (p. 53). The dichotomous relationships between teacher and student and between oppressed and oppressor are therefore transcended through critical pedagogy in a mutual process of respectful inquiry, in which both teachers and students are humanized and validated. Through this process, those who were formerly oppressed are freed through conscientization and critical praxis to become 'beings for themselves' (p. 55). This approach is reflected in Themba's story.

BACKGROUND

The post-colonial era in Zimbabwe began in 1980, when the country became an independent nation, following 90 years of oppression and inequality under colonial rule (Addison and Laakso, 2003; Meisenhelder, 1994). At the time of independence, Zimbabwe had a strongly productive economy, being a net exporter of food and numerous crops, and the country, formerly the British colony of Southern Rhodesia, was then popularly labeled both the 'jewel' and 'breadbasket' of Africa (Power, 2003). However, the robust economy and excellent standards of technological and agricultural achievement that the new independent nation inherited in 1980 were built on profoundly unjust inequalities (WOZA, 2010). Many social problems were embedded within Zimbabwe's initially successful colonial inheritance (Meisenhelder, 1994). The country's wealth was based on a historic marginalization and disenfranchisement of the indigenous black African population in favor of white, mainly European, Zimbabwe-born descendants of the minority group of colonialists. In 1980, around 70,000 white people owned the majority of the fertile land in the country, while the large majority of the population, nearly 12 million black people, remained relatively disenfranchised and dispossessed (Chan and Primorac, 2004). The country's economic success was for the most part based on a robustly productive commercial agricultural sector that had around 5,600 white farmers who between them owned approximately 15.5 million hectares of well-managed commercial farmland (Addison and Laakso, 2003). There was, therefore, general support for a negotiated land reform program to be agreed amongst all new political stakeholders in the country following the transition of power to the new ZANU-PF led government in 1980. Ideally, this would legally compensate existing landowners and gradually transfer appropriate ownership to black Zimbabweans, simultaneously up-skilling the new farmers to enable agricultural success to continue (Chan and Primorac, 2004; Meisenhelder, 1994).

Implementation of this general country-wide desire to redress the historical and divisive disproportionate allocation of land was initially delayed during the first two decades of independence (Meisenhelder, 1994). When the land reform program was then implemented more vigorously in 2000, it was regrettably mishandled, being bound up in defensive political actions by the then ZANU-PF government in reaction to the formation in 1999 of the new Movement for Democratic Change (MDC) opposition party and defeat by the ZANU-PF government in the 2000 Constitutional Refer-

endum (Chan and Primorac, 2004; Isaacs, 2007). A program of violent farm seizures that ousted many white farmers and dispossessed black farm workers arguably led inexorably to a catastrophic decline in the economy and a rapid brain drain or 'reverse transfer of technology' (Logan, 1999) of professionals. The country was plunged into a deepening crisis of escalating hyperinflation and social unrest (Isaacs, 2007; Chikwanha, Sithole and Bratton, 2004) during the years 2000-09, which led ultimately to economic collapse and the abandonment of the Zimbabwean dollar as an independent currency:

Over the past few years, the Zimbabwean economy has been beset with crises, characterised by an unsustainable fiscal deficit, an overvalued exchange rate, and rampant inflation (which stood at 1,000% in 2006). The government's controversial land reform programme has reportedly been the cause of significant damage to the commercial farming sector rendering the country a net importer of food after having traditionally been the source of jobs, exports, and foreign exchange. Financial support from the International Monetary Fund was also suspended due to arrears in repayments on loans (Isaacs, 2007).

The Organisation for Economic Co-operation and Development (OECD) reported in 2004 in the OECD African Economic Outlook 2003/4 for 22 countries (OECD, 2004) that Zimbabwe's isolation had increased in that period as a result of the government's unwillingness to address the economic and political crises affecting the country, notably as a result of the decrease in agricultural production, the lack of foreign currency, the credit crunch affecting the region and the insecurity of land tenure. As a result, the OECD reported (2004) that around 2.5 million Zimbabweans had left the country, since a crisis situation had developed in Zimbabwe. The vast number of Zimbabwean migrants in the diaspora has since continued to rise (Bloch, 2006; Meldrum, 2007).

Post-Independence
Education in Zimbabwe

During this period, the 'remarkable gains' made in the first decade of Zimbabwe's post-1980 independence regarding progress in educational achievement in Zimbabwe faltered, unemployment rose dramatically and access to education and employment, in particular for girls, declined, owing to a range of difficult socio-economic conditions (Al-Samarrai and Bennell, 2007; Peresuh and Ndawi,1998; Meisenhelder, 1994; WOZA, 2009). Despite the initial highly positive post-independence educational initiatives to reverse the unequal and unjust pre-independence years of 'quality education for "whites only" with minor exceptions to build up a black ruling elite class' (WOZA, 2010:1), a serious decline in educational standards in Zimbabwe subsequently occurred, reaching crippling levels of incapacity, with 9,000 teachers failing to report to work in April, 2008 as a result of the hyperinflationary economic conditions and difficult political situation (Africa Research Bulletin, 2008). In February, 2009, the group Women and Men of Zimbabwe Arise (WOZA) reported that the state of education in Zimbabwe was that of *'a dream shattered'*: a catastrophic decline had occurred to Zimbabwe's 'once vibrant education system'. WOZA's research report noted that 'along with our children's lives', education 'has been destroyed by political interference':

Who could have believed that we could sink so low as to reach the situation we are in today? Only private tertiary institutions function, government universities, colleges and schools are closed. The rot began many years ago; by 2001 the number of school-aged children not in school was already higher than in 1991, literacy rates had started to fall by 2002 and qualified teachers had begun leaving Zimbabwe by 2003. But the worst devastation has come in the past four years. Schools are closed because there are no teachers; teach-

ers are not there because they have been chased away from their classrooms by the meagre wage offered by government. Large numbers of them have opted to swim across the crocodile-infested Limpopo River to seek a living wage in South Africa (WOZA, 2009, p. 1).

So poor has the Zimbabwean education system become that in January, 2010, more than 800 members of the WOZA group engaged in street protests in Bulawayo, protesting peacefully and handing over to government a new WOZA report on the education system in Zimbabwe– *Looking Back to look Forward* (WOZA, 2010). WOZA particularly recommended the introduction of more vocational and technical education, given the need for employment-related opportunities and skills, based on the never-released government-commissioned 1999 *Nziramasanga Report*, which called for an increased 'vocationalisation of the curriculum' (WOZA, 2010). Amongst the essential features of vocational education is the provision of effective ICT in education, the lack of overall facilities for which is the focus of the next section.

ZIMBABWE: A SPECIAL CASE OF THE DIGITAL DIVIDE WITHIN AFRICA

The above complex situation makes Zimbabwe a special case regarding the digital divide within Africa, in which there exist, more generally, wide disparities in ICT capability. In 2006, the ITU World Telecommunication/ICT Indicators Database for Africa (ITU, 2007a) recorded that although Africa had around 14% of the world's population, with 221 million total telephone subscribers, on average only 5.6% of the continent's population were online fixed and mobile subscribers. The 22 million internet users in Africa in 2006 provided an average internet penetration rate of just 5% of the continent's population, in strong contrast with Europe's internet penetration, then seven times higher than in Africa.

In addition, the ITU also reported that Africa had the world's lowest penetration of fixed telephone lines, with an average of around three main lines per 100 people across the continent, while more than 20 countries in Africa had a national average of less than one main telephone line to serve 100 people. Of 221 million telephone subscribers, around 198 million were mobile cellular subscribers, since Africa has the highest ratio of second generation mobile to total telephone subscribers of any region in the world and also the highest mobile cellular growth rate. Africa therefore has its own digital divide, since in 2005-6, around 75 per cent of the continent's fixed lines were to be found in only six of Africa's 55 countries. A major barrier to the uptake of fixed broadband in Africa has thus been the lack of fixed telephone lines. Hence the second generation mobile broadband market has been dominant in Africa, although increasing availability of third generation (3G) high speed wide area wireless network services is expected to change this situation during the next few years, if barriers such as the cost of 3G services and mobile devices can be overcome (ITU, 2007a).

Further ITU statistics (ITU, 2007c, ITU, 2007d) report that a vast information gap separates rich from poor globally, particularly with respect to Africa. For example, fewer than four of 100 people living in Africa used the internet in 2005, in comparison with around 50 per cent of the general population, or one out of every two people, living in the G8 (Group of Eight) nations of developed countries (Canada, France, Germany, Italy, Japan, Russia, the United Kingdom, and the United States, plus European Union representatives). Hence, although the G8 nations collectively comprise fewer than 15 per cent of the world's population, around 45% of the world's internet users reside within these countries. The ITU also reported that, predicting from 2004-2005 growth rates, although around 25 per cent of people living in the developing world would be likely to be online by 2010, an estimated 30 per cent of all villages worldwide would still have no access to telephone services. This divergence in access to information, knowledge and communications via the use of internet resources means that African populations tend to be not only poor in terms of material resources, but also in the knowledge resources, educational and cultural information from which the world's richest countries have derived significant benefits and global competitive advantage. Thus, the rich get richer and the poor get poorer. This is, to an extent, unsurprising, given the historical disadvantages suffered by the African continent during the colonial period and the exploitation of the continent's resources by a variety of colonial powers in former eras.

There is, however, some evidence that the global ICT situation affecting Africa is gradually changing, notably due to developments in the mobile phone sector. By around the end of 2007, more than one out of four African people had a mobile phone (ITU, 2007a), while, by the end of 2008, the ITU recorded that there were 246 million mobile subscriptions in Africa. These figures relate to the 43 African countries served by the ITU Regional Office in Addis Ababa. Within these African countries, there are a vast range of differences as regards overall ICT Development Index profiles (IDI), according to the ITU. For example, the IDI profile achieved by the Seychelles (1), Mauritius (2), South Africa (3), Gabon (4), Cape Verde (5) and Botswana (6) highlights these as the highest ranking six African countries in terms of ICT development, while the D.R. Congo (35), Eritrea (34), Chad (33), Guinea-Bissau (32), Congo (31) and Zimbabwe (30) were the lowest ranked countries (though note that not all countries were measured for the IDI; see ITU, 2009, p. 51).

Overall, the African continent has in recent years experienced the highest growth rate in the world for mobile telephone subscriptions (ITU, 2009). The distribution of mobile phone usage has also spread out across a range of African countries during the past decade, with Nigeria, Kenya, Ghana, Tanzania, Côte d'Ivoire and

Ethiopia now catching up with South Africa as regards the number of mobile cellular subscriptions in the region (ITU, 2009, p. 4). Hence the digital divide in terms of mobile telephone access has been substantially reducing over the past few years, with the penetration of mobile phones into Africa sharply escalating from around five per cent in 2003 to more than 30 per cent in 2009.

Over a similar time period, the use of the internet in the ITU-supported African countries rose dramatically from three million internet users in 2000 to 32 million internet users by the end of 2008 (ITU, 2009). Continuing relatively low growth in fixed telephone lines and fixed broadband internet access is still, however, also a feature of the continent. The vast distances involved in implementing telecommunications structures across the African landmass, combined with poverty, poor governance of ICT and telecommunications strategies, a 'brain drain' of talented staff, relative lack of industrial and commercial competition and numerous financial and technical difficulties with ICT infrastructure have meant that the more temporary, insecure expedience of mobile access rather than fixed line or fixed broadband access has dominated ICT markets in Africa. Hence the ITU report noted in 2009 that:

Notwithstanding the emergence of Africa as one of the most dynamic regions in terms of ICT growth, the region's absolute figures as well as penetration rates remain low. Two decades ago, achieving a teledensity of one per one hundred inhabitants represented a major milestone, but today's benchmarks of achievement are much higher. While the rest of the world has forged ahead with strong ICT investments and the adoption of new technologies, and although Africa has made impressive gains, it remains far behind the ICT penetration levels of the world, and even those of developing countries.... The gap is more pronounced in the case of broadband Internet services, and fixed telephony.... (ITU, 2009, pp. 1-2)

The digital divide therefore remains a serious problem across Africa as a whole regarding broadband, internet uptake and the knowledge, information and communicative power that relate to such services. It could be argued that there are many other more pressing immediate problems that should be addressed rather than ICT development for people in the African continent, such as access to food, sanitation, clean water, shelter, electricity, improved living conditions and medical care. It could also be argued that access to ICT through education is, relatively speaking, a luxury that should only be considered for development long *after* many other socio-economic, housing, health, medical and legal initiatives. Pragmatism suggests that in the worst circumstances this is correct. If people are dying of starvation, being persecuted for their political views or have no access to clean water, food, shelter or electricity, technology access *is* a relative indulgence. Yet once such basic living conditions for survival are satisfied, further arguments that ICT access is a 'luxury' deny the stark realities of the current global era, in which essential information is now located as much on the internet as it is in books and papers.

In effect, rapid global expansion of the information society and knowledge economy has radically changed the world during the past twenty years: access to ICT is increasingly becoming a necessity for economic and social survival. Lack of access to ICT now means lack of access to information, to knowledge, to education and skills, to resources, qualifications, careers advice, social and employment networks and jobs. It would be grossly unfair to deny the potential of African peoples living in poverty to achieve the levels of economic and social growth, self-empowerment and self-determination that are possible through the democratizing and humanizing process that effective interactive access to ICT information can facilitate over time. Improved access to education, general use of ICT telecommunications for daily life and for the numerous employment and training opportunities opened up by ICT developments are vital to the quality of life of people across the African continent in a longer-term sense (ITU, 2007c, 2007d; ITU, 2009).

The fact that Africa now has the highest growth in mobile phone subscriptions in the world, even in rural areas where there are no formal geographical addresses, is an indication of the desperate need and desire of the 'information poor' to gain access to communications and information. As an example of this, the ITU report quotes the President of Rwanda regarding the growing importance of mobile communications in Africa: "In ten short years, what was once an object of luxury and privilege, the mobile phone, has become a basic necessity in Africa." (ITU, 2009:24). In a further example of the ways in which technology can act as a tool for development, a correspondent from Zimbabwe noted the following response in an online BBC debate about the digital divide (BBC, 2002):

The debate on priorities for getting clean water, food, etc. or e-mail, Internet access first of all is obviously misdirected. Why not look at IT as a possible tool, (amongst many others), to get clean water, better health care, better education, better governance to the people of Africa. Narrowing the digital divide can one be one of those tools for achieving those much-needed outcomes. People in the northern hemisphere think they need to dump food, money, etc. to help the poor. Yet here in the south we want roads, vehicles, to get that urgently needed aid to the people but even more we want education, jobs, political and economic stability for growth and independence from western aid. Yes, as badly as we need roads we also need the information highways to empower people of Africa to build a better future. (Klara Tisocki, Harare, Zimbabwe: BBC, 2002)

Yet as the figures above indicate, despite strong growth, internet usage is still particularly low in Africa, with only around five per cent of the continent's population able to access online resources, in comparison with more than forty per cent of the population in Europe, the Americas and the continent of Oceania (ITU, 2007a). Furthermore, the expense of internet access is often relatively higher in poorer countries, and access to e-learning educational resources is more difficult, factors which in themselves contribute to a wide gap between rich and poor countries. Sadly, generations of Africans have, therefore, been partially or completely deprived of ICT and mobile telecommunications access at the same time as there has been an exponential digital explosion of information and communications technology access and the growth of new kinds of web 2.0-enabled 'knowledge work', such as web searching, social networking, blogging, interactive video, multi-user games, the use of enterprise technology, self-publishing and rapid news, information and social communication dissemination services such as RSS and Twitter across the developed world. Exclusion from engaging in new kinds of communication exchange and knowledge working means that ordinary people in Africa, notably in countries facing multiple socio-economic, health, education, human rights, legal and political challenges, such as Zimbabwe, have found themselves in a new digital dark age regarding serious deficits in access to the power of communications and information technology, while people in more developed countries are thriving from the social and economic benefits released by such resources (Gebremichaela & Jackson, 2006).

The fact that access to ICT is important for poorer countries in Africa is not just an indulgent post-colonial techno-deterministic fantasy dreamed up by those living outside Africa. Information poverty, social exclusion and lack of access to information, education and skills to achieve employment in the global knowledge economy are key aspects of the 'digital divide' affecting many of the world's poorest countries in Africa. The complex factors involved in such exclusion from the opportunities provided by ICT are particularly acute in Sub-Saharan Africa, as Gebremichaela and Jackson report:

.... more than three-quarters of the Sub-Saharan population have been marginalized by advancements in ICT adoption and usage, divided along the lines of age, gender, rural and urban areas, level of education and unemployment, ignorance, illiteracy, poverty, and other forms of marginalization. (Gebremichaela and Jackson, 2006, p. 271).

Two recent reports from the group Women and Men of Zimbabwe Arise (WOZA) (WOZA, 2009, WOZA, 2010) on the state of education in Zimbabwe highlight the importance that Zimbabweans attach to relevant vocational and technical education opportunities such as the provision of ICT in education. Based on their research with 1,337 WOZA Zimbabwean members, WOZA reported that the state of education in Zimbabwe was now so desperately poor that it was 'a dream shattered':

We as a people have always valued the education of our children more than anything else. For that we sell our precious beasts, work two jobs, travel across borders, go into exile, buy less food and once joined the liberation struggle. We hang the hard-earned certificates on the unpainted walls of our small homes, and ululate at the graduations. An educated son or daughter is our pride and joy, the fulfilment of years of struggle..... We fought for the right to free education, and thought we had gained that right. Now it has been taken away; now there is no education in Zimbabwe; now our children are starving for food and starving for the learning, which will enable them to become productive adults. (WOZA, 2009, p. 2).

In 2010, summing up the appalling state of education in the country, WOZA recommendations for Zimbabwean education proposed that there should be a complete overhaul of the educational system to introduce more vocational subjects:

- A revamping of the [Zimbabwean education] curriculum to ensure its relevance to the children who learn, producing a school leaver who has skills with which to enter the formal economy or provide adequate self-employment.
- This means introducing more vocational subjects – both commercial and technical – what Nziramasanga called 'vocationalisation of the curriculum', and providing opportunities for children to be attached in work places during their senior years. (WOZA, 2010)

It is clear from the views expressed by WOZA members in this research that the Zimbabwean women interviewees strongly desired and sought vocational education including technology in education opportunities. As expressed in 'Dreaming of a New Zimbabwe People's Charter', the community participants in WOZA are determined to engage proactively to create a better education system for their own future, stating: "We must keep in mind..... that we deserve better and we must not be afraid to believe that we have the right to a brighter future and we have the right to contribute to building it."(WOZA, 2009, p. 5).

METHODOLOGY

This chapter is comprised of two parts. In the first part, a factual report on Zimbabwe's bleak position regarding ICT capabilities, 'networked readiness' and World Values global data is presented in the context of wider African and global research information. This sets the scene for the second part, which provides a fictional narrative account of the educational development of an adult male Zimbabwean rural farm worker called Themba. Themba's escape from oppression and poverty into a new and more empowered life in South Africa is provided as a fictional example of a way out of the 'abyss' in which Zimbabwe finds itself: it is proposed that Zimbabwean teachers in the diaspora such as Themba might wish to return to their roots in Zimbabwe to assist in rebuilding the educational and ICT capabilities of the country in positive ways, deploying the optimistic approach of critical emancipatory pedagogy.

In the second half, underlying the fictional narrative, the chapter adopts an implicit critical social theory methodological approach to digital divide issues affecting Zimbabwe. Critical theory is generally applied in the social sciences and humanities to critique social phenomena for the purpose of achieving emancipation from oppression (Horkheimer, 1982; Brookfield, 2005). The use of the theory in the social sciences aims to scrutinize and reflect critically on existing social relations in order to change society by encouraging the liberation of social groups from the multiple limitations imposed by oppressive social and political structures, ultimately to enable more balanced and equal social participation. In education, the theory draws on the work of Paulo Freire (1970), whose critical pedagogy aimed to develop self-awareness, reflection and knowledge amongst both teachers and students to liberate them from conformity to oppressive controlling forces exemplified by Freire's 'banking' concept of education. The educational emphasis of critical pedagogy privileges critical reflection and social learning over the rational-technical 'training' focus of instrumental accounts, while the critical social theory approach of ICT technological analysis foregrounds issues relating to social and political forces of power and domination in the use and application of ICT rather than technologically deterministic perspectives (Fuchs, 2008a, 2008b; Smith, 1997, 2002).

PART ONE: GLOBAL ICT AND WORLD VALUES FINDINGS

Information and technology access has been directly linked with economic growth and development, as noted in the *World Information Society Report 2007* (ITU, 2007a), which observes that: 'Access to ICTs increasingly determines access to wealth and income, and will, in turn, determine the leaders in tomorrow's knowledge economy.' (p. 9) Williams (2008, p. 1) observed that 'economic data at the national level shows that investment in ICT results in a higher rate of long-term economic growth (Roller and Waverman, 2001)'.

The 2008-09 *Global Information Technology Report* (GITR) World Economic Forum (WEF) 'Networked Readiness Index' (NRI) rankings published by in partnership with INSEAD Business School (WEF, 2009) placed Zimbabwe at the very low position of two points from the bottom of the global league table, scoring 2.49 points, in comparison, for example, with Denmark (ranked first with 5.85 points), Sweden (ranked second with 5.84 points) and the United States (ranked third with 5.68 points), as illustrated in Figures 1-2. It can be seen from these figures that Zimbabwe's 'network readiness' in ICT is desperately poor in comparison with all of the top twenty wealthier countries in Figure 1 and even with such neighbouring Sub-Saharan African countries as Lesotho, Tanzania and Mozambique, which have overtaken its capability amongst the 'bottom 20' league of nations in Figure 2.

Despite its adoption of a dedicated national ICT policy in 2005 (Isaacs, 2007), Zimbabwe is therefore ranked at 132nd of 134 nations globally for information and communications technology (ICT) on the Networked Readiness Index (NRI), according to the GITR report. The NRI ranks the ability of countries to make use of ICT opportunities, assessed according to three dimensions: firstly, the environment for ICT; secondly, the readiness of key interest groups such as government and businesses in the relevant country to

Figure 1. Top 20 countries in ICT as listed by the GITR Report

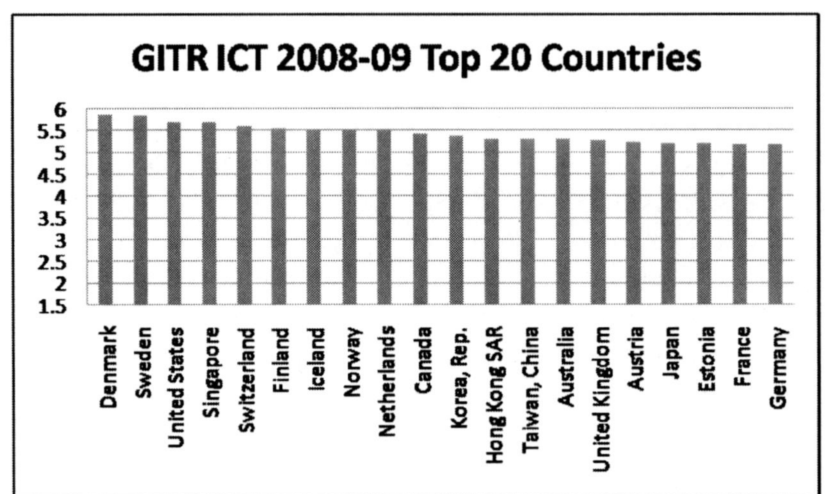

use ICT; and thirdly the actual ICT use amongst those interest groups. In the case of Zimbabwe, exceptionally low figures for ICT usage include the fact that there are only 2.5 fixed telephone lines per 100 inhabitants, only 6.5 mobile lines per 100 people and fewer than 9.5 internet users per 100 inhabitants.

Zimbabwe's placement in the 'usage' component of the GITR report is even worse, at 134, indicating that it is the least developed country in the world as regards business usage of ICT (134), government usage of ICT (134), availability of new telephone lines (134) and freedom of the press (134) while its capacity for innovation in ICT (132), and extent of business internet usage (132) rankings are almost as low. In reaction to the publication of the GITR, *The Zimbabwe Times* commented in March, 2009, that the newly appointed Information Communication Technology Minister, Nelson Chamisa, faced serious challenges in trying to improve Zimbabwe's ICT capability, teledensity and overall rating according

Figure 2. Bottom 20 countries in ICT as listed by the GITR Report

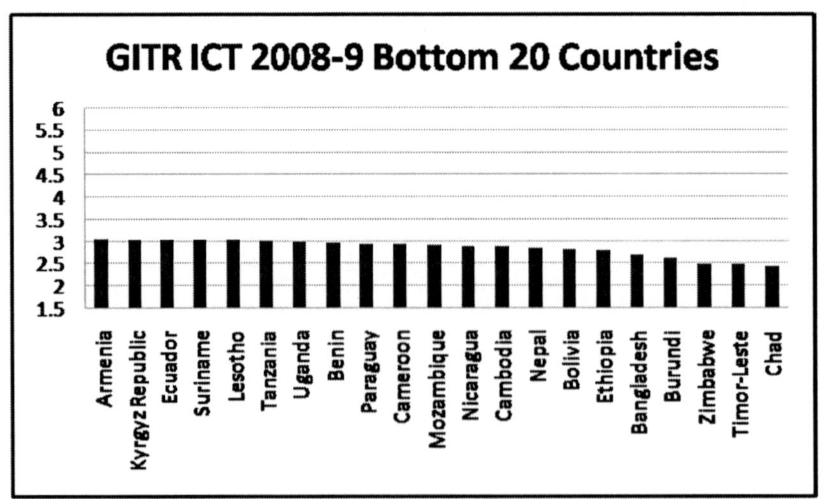

to global ICT criteria (Zimbabwe Times, 2009). Furthermore, although it had been agreed by the coalition Government of National Unity (GNU) that Chamisa, a Movement for Democratic Change (MDC) minister, was appointed to this role, a dispute affecting the remit of his ministry in 2009 led to a serious clash regarding which governmental ministry controls the remit for ICT. The ZANU-PF side of the coalition government has now tried to take away the ICT portfolio from Chamisa and place this within the Ministry of Transport, which is controlled by ZANU-PF. This dispute is likely to continue to affect ICT development for some time, further impeding any actions to improve ICT development and global competitiveness in Zimbabwe (Zimbabwe Times, 2009).

As a result of this sterile climate in national development and business competitiveness in ICT, Zimbabwe was positioned 133rd out of 134 countries on the *Global Competitiveness Index* 2008–09 within the GITR report (WEF, 2009). The GITR WEF report also identifies that Zimbabwe has the lowest international ranking (134/134 nations) for the category 'ICT use and government efficiency', meaning that overall in governmental usage terms Zimbabwe is now the least developed nation in the world for ICT. This deficit situation highlights the huge gap identified by the GITR between Zimbabwe's digital development and the position of 131 other countries higher up the ICT pecking order, with only the countries of Timor-Leste (East Timor) and Chad being positioned below Zimbabwe in global ICT 'networked readiness' (ibid).

As to the wider significance of Zimbabwe's placement two places from the bottom of the GITR Networked Readiness Index, the GITR report's *Preface* notes that ICT competitiveness is increasingly linked to economic development and that the Networked Readiness Index identifies factors that enable countries to benefit in full from technological advances (WEF, 2009). According to the GITR, ICT is 'a critical enabler of growth,

development and modernization' in terms of world-wide national competition:

Recent economic history has shown that, as developed countries approach the technological frontier, ICT is crucial for them to continue innovating in their processes and products and to maintain their competitive advantage. Equally importantly, ICT has proven instrumental for enabling developing and middle-income economies to leapfrog to higher stages of development and fostering economic and social transformation… All over the world, ICT has empowered individuals with unprecedented access to information and knowledge, with important consequences in terms of providing education and access to markets, of doing business, and of social interactions, among others. Moreover, by increasing productivity and therefore economic growth in developing countries, ICT can play a formidable role in reducing poverty and improving living conditions and opportunities for the poor. (WEF, 2009)

Yet the ICT situation affecting Zimbabwe was not inevitable: Dzidonu, Rodrigues and Okot-Uma observed as early as 1989 that Africa had many opportunities to make the most of the potential for communications that new technology promised (Dzidonu, Rodrigues and Okot-Uma, 1989). Furthermore, Zimbabwe did not always have a low-ranking position in ICT, as the Media Institute of South Africa (MISA) has reported:

Zimbabwe once had the second [fastest] growing Information and Communication Technology (ICT) sector in sub-Saharan Africa after South Africa but has since suffered due to collapse of the economy. The sector has also suffered due to the neglect of policies like the National Telecom Policy, National Postal Services Policy and the Universal Services policy. (MISA Zimbabwe, 2009)

MISA Zimbabwe has also noted with 'great concern the appalling state of fixed and mobile telephone networks in Zimbabwe', reporting during 2008 that:

The sole fixed telephone network, run by the state owned TelOne, is in an appalling state of affairs with erratic coverage in the urban areas and is virtually non-existent in the rural areas. This has inadvertently led to a major increase in the use of mobile telephones by the majority of Zimbabweans from all socio-economic and geographic backgrounds. The three mobile telephone networks, (Econet Wireless, Telecel, and the state owned Net One) have, however, failed to cope with the market demand for their services in Zimbabwe's hyperinflationary environment. (MISAZimbabwe, 2008)

and, further, that:

The use of Information and Communication Technologies (ICTs) has become unaffordable, as the majority of Zimbabweans still have no access to foreign currency as they are paid in Zimbabwean dollars. For this reason, access to tools such as the internet, fixed and cellular telephones networks will be a privilege for a few, going against the emphasis of the World Summit on Information Societies (WSIS), Tunisia 2005.... The deprivation of the people of Zimbabwe's right to communicate is in MISA Zimbabwe's opinion, an impediment on a basic right granted to communicate and express themselves as guaranteed in Article 9 of the African Charter on Human and People's Rights. (MISAZimbabwe, 2008b)

In effect, the ITU reported that there were only two economies: Turkmenistan and Zimbabwe, which experienced falls in their Digital Opportunity Index (DOI) scores, between 2004-2006 (ITU, 2007b). In Zimbabwe, this situation has occurred as a result of the relative dispossession of the general population in economic and industrial terms, as it has become more and more difficult to live on a daily basis in the country. The gradual depletion in resources and the resultant removal of the right to communicate has therefore led to a wider dispossession in information and social terms that means Zimbabweans are now increasingly isolated from the rest of the world.

Given that Zimbabwe's access to ICT and related income development, access to information, knowledge, employment and social networking opportunities has been so stunted and that the country has been suffering from multiple deprivation in health, living standards, security and human rights, it is hardly unsurprising that Zimbabwe has also been recently rated as 'the unhappiest country in the world', according to the reported levels of subjective well-being (SWB) in the USA-funded World Values Survey of a representative sample of Zimbabwean citizens (Inglehart, Foa, Peterson and Welzel, 2008). The findings from the survey indicate that freedom of choice, tolerance and equality are linked with greater levels of happiness amongst nations. The World Values Survey ranking reveals that, as a result of a lack of personal choice within an intolerant and unequal society, Zimbabwean people are self-reporting severe levels of dissatisfaction, as illustrated in Figure 4, which displays the results for the bottom 20 countries as rated for Subjective Well-Being (SWB), in contrast with those countries listed in Figure 3 at the 'happiest' end of the SWB scale. These results confirm earlier work carried out in the Afrobarometer Survey in 2001 and 2005 (Afrobarometer, 2002, 2006). The Afrobarometer independently measures the social, political and economic environment in eighteen African countries, through face-to-face interviews carried out in four stages with a representative sample of the adult population. Results reported are:

Asked whether they are "satisfied with the way democracy works" in their country, Afrobarometer respondents are lukewarm. The proportion expressing satisfaction with democracy's performance (58 percent) lags behind the proportion that say they support democracy (69 percent...).

Figure 3. Top ten countries for subjective well-being, world values survey

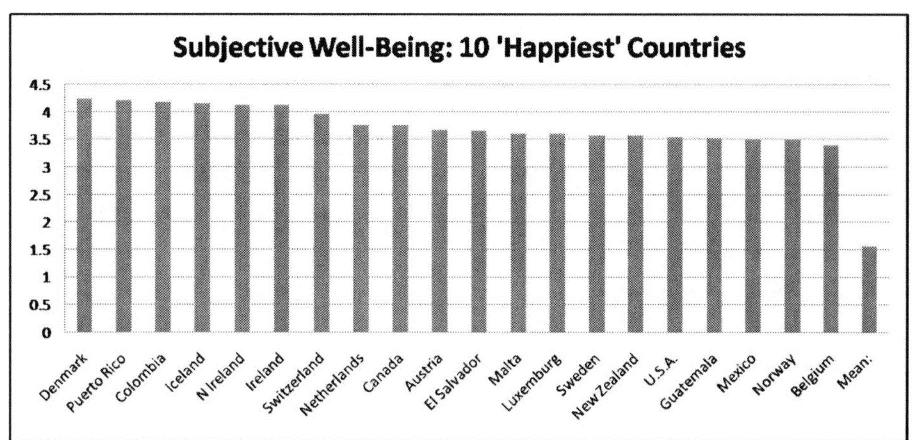

And only 21 percent are "very satisfied." The lowest levels are recorded in Zimbabwe, where only 18 percent are at all satisfied (Afrobarometer, 2002).

Clearly, the concept of 'democracy' as government by and for the people, is not working well in Zimbabwe, with around 29% of people (the highest response category) surveyed by Afrobarometer in 2005 reporting that "Zimbabwe is not a democracy" (Masunungure, Ndapwadza and Sibanda, 2006). In addition, the economy of the country has teetered on the edge of collapse for nearly a decade, more than half the population is dependent for survival on UN food aid, more than 90% of Zimbabweans are unemployed and a major cholera epidemic has recently (2008-09) killed around 4,000 people (Zimbabwean Times, 2009). Within this bleak overall situation, digital divide issues echo the greater abyss affecting the happiness, security, wealth and life expectation of people in Zimbabwe.

The subjective well-being (SWB) Index, illustrated against the per capita gross domestic product (GDP) in different types of societies across the world, was reported in the World Values Survey conducted by Inglehart, Foa, Peterson and Welzel, 2008. The index of well-being is based on participants' reported levels of life satisfaction and happiness, using the mean results from all available World Value surveys conducted during 1995–2007.. In this Index, Zimbabwe is

Figure 4. Bottom ten countries for subjective well-being, world values survey

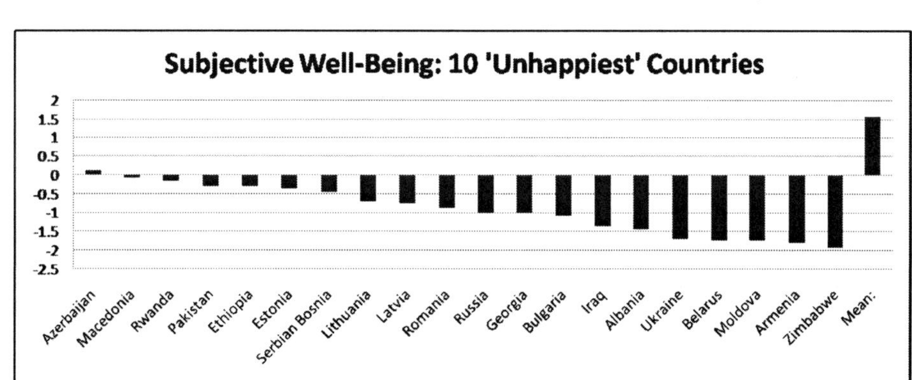

positioned at the lowest point, indicating that the reported subjective well-being of Zimbabweans is the lowest in the world, while the country's per capita GDP is also amongst the lowest in the world. Thus in terms of both self-reported levels of well-being and income per capita, Zimbabwe is performing extremely badly, in addition to its severely marginalized position in relation to ICT. Since both equitable access to and capability in new technologies are linked to global competitiveness and economic growth, as well as to the creation of new jobs and prosperity nationally, and since prosperity, at the minimum level of ensuring livelihood, is linked to well-being (Inglehart, Foa, Peterson and Welzel, 2008), it makes sense that some of these different aspects of Zimbabwe's performance may to some extent be correlated. However, this is as yet not a proven relationship but an observed one (see Inglehart et al, *op cit*). (See Figure 5).

What is interesting also is that, at the other end of the spectrum, a number of countries that perform extremely well in ICT networked readiness also report high levels of subjective well-being amongst the population: for example, Denmark, Iceland, Switzerland, Ireland, Sweden, the Netherlands, the USA, Canada, Luxembourg and Norway are all represented at the upper end of the ICT and SWB scales(Inglehart et al., 2008). It is therefore an interesting proposition to consider whether high levels of subjective well-being and feelings of sustainable self-empowerment may be to an extent coincidentally (or not) related to reasonably effective access to and use of ICT resources and that countries that are relatively effective in access to digital resources are also more favorably situated regarding a well-being/lack of well-being divide. Inglehart et al. (2008) investigated the happiness of numerous countries at different stages of development, involving nearly 90% of

Figure 5. Global Subjective Well-Being (SWB) Index, per capita gross domestic product (GDP) across different types of societies, as reported in the World Values Survey 1995-2007

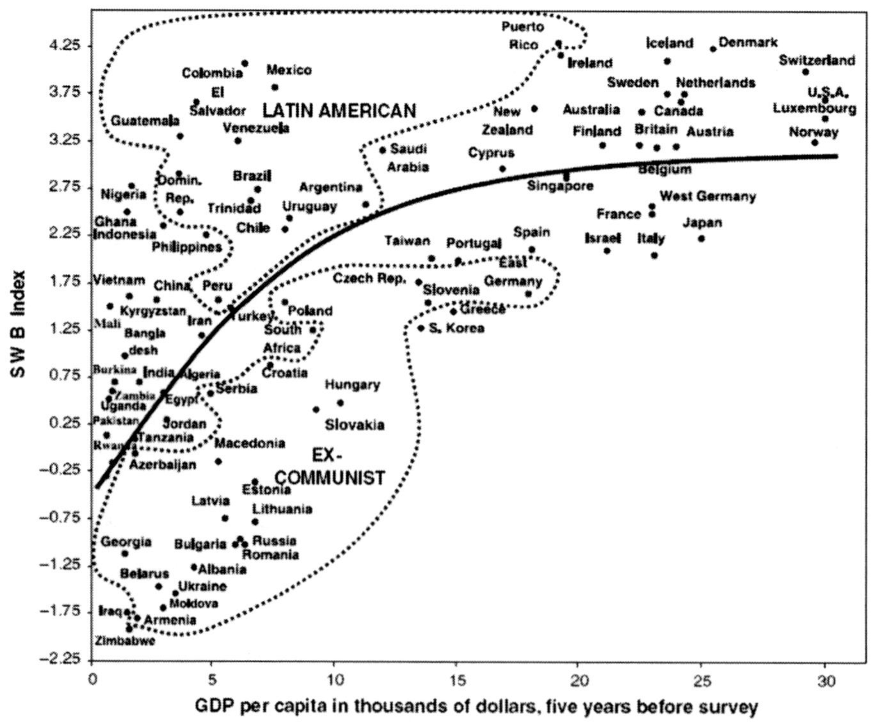

the world's population. These research reports are comprised of data from both the World Values Survey and European Values Study in several waves of data collection from 1981 to 2007. Hence this is unlikely to be a technologically-skewed result, as it would be if only countries with access to technologies were involved, using a limited data collection. To complicate the picture, however, groups such as the New Economics Foundation (NEF) would argue more from an overall ecological perspective, saying that self-reported happiness is not necessarily correlated with great wealth of material resources or super-efficiency but may be connected more with an effective, responsible *balance* of sustainable well-being and ecological efficiency. For example, countries such as Costa Rica, which has a moderately good, improving score in WEF ICT NRI rankings, is ranked as the happiest place in the world (1/143 countries) in the NEF Happy Planet Index (HPI), which measures a combination of health, life satisfaction and ecological responsibility, while Zimbabwe yet again scores the lowest rank (143/143) (NEF,2009). The NEF analysis tends both to enrich and complicate the findings of Inglehart et al. (2008), but, for Zimbabweans, the outcome is pretty much the same.

PART TWO: NARRATIVE

The Story of Themba

Given the above situation, it is possible to imagine some of the impact that the lack of access to ICT might have on members of the general population in Zimbabwe. We can imagine a hypothetical Zimbabwean male, perhaps called Themba, 'trusted one'. Let us imagine that Themba was born in what was then Southern Rhodesia around 1959 in a rural area. When he was young, Themba was educated in a rural mission school. Though he learnt how to read and write and is well educated in these skills, he left school early around the age of fifteen in 1974, as did many of his fellow pupils, without qualifications. Themba is now getting on in years at the age of 50 in 2009. He has an ailing mother, which is one reason why he hasn't left Zimbabwe, despite all the difficulties that have beset the country. Themba has worked on a commercial potato farm near the rural area of Zimbabwe in which he was born for around thirty-five years, since before the ZANU-PF government came to power in 1980. He remembers the old days of the colonial past very well: he was unhappy then at the lack of power of black people in the white-dominated government of Southern Rhodesia and he experienced first-hand the joy everyone seemed to share when Robert Mugabe and the ZANU-PF black majority government came to power in 1980. At last, he thought, African people in the new Zimbabwe would get their fair share of power.

Themba was, therefore, glad that the power of the whites was curbed by the black government in those early years, but otherwise he has forgotten the initial euphoria of 1980s independence and he does not think much about politics or power, just about survival. Themba, at the moment, regards himself as politically neutral: he has not yet recognised that there may be no such thing as 'political neutrality' or 'objectivity'. He just wants to have a good life, to live well with his family, earn a decent wage and be healthy and happy. However, he knows that, to ensure his own continuing existence during past decades, it has been better to declare that one is ZANU-PF when anyone from the government has been around, which has been seldom: only before election times, really. Themba gets on all right on the farm: he does not think much about the power structures in society that mean the white farmer who owns the land also pays his wages. The white farmer does not get in Themba's way, as long as it is obvious that Themba is working hard. So Themba has lived for many years, more or less peacefully, watching his children grow up, his wife growing older and fatter, and his mother's health gradually decline.

Themba does not have a mobile phone, though many farm workers do, as he is the kind of man who thinks more of feeding his family than of getting new things for himself. He has used the fixed line telephone of the farmer once or twice when there was an emergency or an important event, but he does not have any access to computers or television. He knows what computers are, as he has seen the farmer calculate the wages on this and print out documents, but he has never used one and he doesn't really know much about them.

Nevertheless, Themba is relatively privileged amongst Zimbabwean farm workers as he is one of the ten per cent of the population who still has a steady job, even in 2009, since, as luck would have it, the farm he works on is one of the very few big successful commercial farms left that has not yet been seized for compulsory redistribution to 'war vets' or other favoured recipients by the government. Themba is an excellent worker, diligent, good-hearted and very skilled at his farm work, but he is worried that he is now getting old and is feeling somewhat vulnerable on account of this. As one of 150 black workers on the farm working for a young black foreman, Themba is part of a large community. He also has five people in his family, including his wife, his aging mother, his wife's sister and her two young children, a girl and boy, all of whom he supports with his wages. Themba also has four grown-up children of his own, all of whom left a long time ago. His children saw no future for themselves in rural farm working in Zimbabwe and so, disillusioned with the government of Zimbabwe and the prospects for their future, they left, one by one, for South Africa, Mozambique and Botswana. He has not seen them for five years or so.

Since Themba lives in a rural area that is relatively protected by the commercial operations of the farm, he and his family have access to food, clean water and safe sewage systems. Themba has, therefore, not been affected by the cholera that has killed more than 4,000 people elsewhere in Zimbabwe. He has heard of this and is deeply worried by it. He is also fearful about the farm invasions that have happened elsewhere in Zimbabwe, as he knows that a remaining purge of all the 200 or so remaining white-owned commercial farms is now taking place. Themba has heard about the ways in which around 450,000 or more experienced, skilled black farm workers plus their families, altogether comprising around a million people or more, have lost their livelihoods and homes, along with the dispossessed white farmers, when those farm invasions took place. Themba knows that those who replaced the farmers and farm workers have not subsequently maintained the land or the crops, which have fallen into wastage while the country has starved. He knows that if there is a violent take-over of the farm on which he works, he will most likely lose his job and his farm workers' dwelling. He does not know where he and his family will go. He is so worried that he tells his family to be prepared in case a sudden violent take-over happens. He tells them to make up a small container with a few of their most precious possessions and to have this ready at all times in case the worst happens.

So, one night, when he is lying in his bed trying to sleep, Themba is deeply alarmed when he hears gunshots and shouts from the nearby farm buildings. He sees bright flashes of white torchlight scattering giddily through the darkness of the trees. He rushes outside in the starlit night and sees from far off that the white farmer and his family, including his wife and two daughters, have been dragged from the house. The white family is standing shivering with fear outside the farmhouse as the farm invaders circle them aggressively, shouting, pointing and laughing raucously. The invaders seem to be drunk with power. They yell a succession of rude words out loud, ordering the family about and spitting at their feet, saying that they have come to take the farmer's land and property. He sees the leader of the war vets strike the farmer hard. The farmer's nose starts bleeding. Crimson spurts down his terrified ashen face, but he does not react. Themba knows what will be next: the farm workers will be dragged into the dispute. They might even be injured or killed. Though he has no animosity towards the white farmer, he also is not particularly close to him and he doesn't want to risk the possibility of imprisonment, torture or death if he stays to protect the farmer, as others have done at other farms: others who were then injured or lost their lives.

Themba calls softly to his family to grab their things quickly to leave. They must take little to go as far as they can as soon as possible. He gets his wife, her sister and the two children ready, calling out quietly in the darkness for them to hurry up. They are fine to go, but Themba is worried about his mother, who cannot walk properly as she is old. His heart is breaking, as he realises he may have to leave her behind. Themba adores his mother. He has been with her almost every day of his 50 years of life. She is 71 and he does not know how he could bear to leave without her. But his mother pre-empts him. 'Themba,' she says, softly, 'you must go. Take care of the women and children and yourself, be strong and survive. I will be all right here. I cannot walk, I am too old. Go without me and go now. I love you, my son: God go with you!'

Themba feels the tears running fast and hot, silently down his cheeks. He cannot bear it, to leave her here, so vulnerable and alone. But he knows she is right, so he consoles himself with the idea that perhaps things will be all right for her – there are other women who will stay on the farm, the new owners might look after them, it might be OK. He realises that because neither he nor she has a mobile phone, and since both are unlikely to get one, they may never be in contact again. He once wrote a postcard to his mother years ago when he was young and went on a trip to Mozambique - she still has that displayed by her bedside - but she cannot read very well now that she has cataracts and the Zimbabwean postal system to farming families in rural areas is very slow and unreliable. Themba also does not know much about email, nor does he use computers, so

he knows he cannot communicate with her in the way he has heard the farmers do between each other with the computer network.

Themba is very sad to leave his mother, as this may be the last time they ever communicate. There is so little time. He takes one last lingering look at her face as she stands there, so frail and old in the starlight, her long, shabby night robes ruffling against the light winds. Then he has no choice: he has to run, as the war vets are rounding people up and are beginning to come towards the workers' area of the farm. He kisses and hugs his mother goodbye and gives one last, tender stroke to her wrinkled face. Then hastily he runs, with the women, the girl and the boy, quickly, silently, away from the farm and into a clutch of trees nearby, where they hide briefly before making a dash for it towards the hills behind the farm. They are lucky. It is still very dark, no-one sees them and they escape..........

Some weeks later Themba and his family arrive in Johannesburg, exhausted, hungry, dirty and desperate to find a place to stay, having walked for days, taken lifts, caught a couple of overloaded dusty buses and then trudged again, on and on, grabbing whatever meals and transport they could, towards their destination. They are told about the Catholic mission that allows Zimbabwean refugees and migrants to stay. They are successful in finding a small, safe place for the family to sleep in the mission. Gradually, they settle in and are given shelter, some blankets, a little food. The government of South Africa lets them stay, as by now cholera has taken over the exit and entrance points to Zimbabwe at Beit Bridge and no-one is being sent back any more. The violence against Zimbabweans has also thankfully died down since the government intervened to stop this, and the mission has been safeguarded for refugees. So, unlike so many others before them, Themba and his family are safe, free to go about their business and to create a place for themselves in South Africa as Zimbabwean refugees.

Once his family is settled, Themba goes to try to find a job in an employment agency. But it is hard. He has no skills except as a farm labourer on the potato farm, at which job he is excellent, but there is no such work in the city. He also has no references and he is only ten to fifteen years from retirement age. So he asks the woman behind the desk what he can do. She offers him a free course in computing skills, saying that the local community college is organising this for refugees from Zimbabwe like himself with government funding and he might like to try this. She says that if you have computing skills there are lots of jobs. Themba jumps at the chance to gain some skills in using computers and he starts attending the course the very next day. At first, he finds it all very alien and intimidating. He wonders if he will get used to using a mouse and keyboard, as it is so strange. He does not know what any of the gadgets are called and the screen is confusing. However, gradually, he picks up knowledge and skill in the use of computers, with the help of his teachers and fellow students.

He gains his first elementary qualification in computer usage for business administration and proudly goes back to the mission to his family to show them this. Then quite quickly he begins to find his way with computer use for business on the follow-on course and he realises that you just need to work at it to understand how to use computers more. He gains an intermediate qualification, then a more advanced level. Quite soon, Themba finds himself at the college 'graduation' ceremony for vocational certificates for business administration. He gains a job in this field with his new skills, and at long last he starts to earn money again. Themba and his family have been living at the mission for so long that they are quite sad to leave, but now that he is earning a decent wage, they decide they want to live in a small house of their own rather than the mission. So Themba and his family move into a little house in a run-down area near the city. It is quite pleasant no longer to have to stay in the mission. Themba shows his

wife the opportunities available in the local college so that she too can gain some skills. His wife does this and within no time at all, she gains some qualifications in secretarial work and a new job as a secretary. She and Themba joke to each other that, although they sadly miss Zimbabwe and the old ways of living in the rural area, in other ways, they have never had it so good: a whole new life is opening out for them.

Meanwhile, the children and his wife's sister are also gaining skills in using computers. They need to study out of hours, preferably at home, and gradually Themba realises that he and his wife can now afford to buy a small computer with internet access for home usage. He does this, and feels a tremendous sense of pride that at his late age he has managed to set up his family with resources that they need. Themba enrols in night school and starts to do a degree in computing, which he passes with flying colours. He gains a new job to teach computing, enrols in night school again to do a post-graduate teaching qualification, and soon enough his career has taken off, even at his rather late age, and he is soon earning a good wage. As the years go by, however, he starts to miss Zimbabwe more and more and to plan a trip back home.

One day, several years later, while surfing the internet at home in the evening, reading up about Zimbabwean news, which he frequently does, missing his homeland, Themba comes across an archived newspaper article that recounts the story of the take-over of the farm in rural Zimbabwe on which he used to work. The journalist who wrote the story had obviously been there at the farm on the night that Themba and his family ran away, as the events are all explained just as vividly as Themba remembers them. He is, once again, shocked to relive the painful memory of the experience of leaving so suddenly. But what is different this time is that Themba has moved on. His perceptions and thoughts have changed. He has read more and more on the internet about Zimbabwe and the events that happened there.

He has browsed through blogs, read many free newspaper articles and joined in group meetings, diaspora events and internet discussion boards on Zimbabwe.

In doing so, he has read up also about the country's opposition parties, about the bravery many leaders demonstrated in fighting for civil liberties in Zimbabwe, which ended up changing things for the better. He has begun to learn of the struggle of other peoples who have been oppressed at different times in different countries. He begins to draw parallels between Zimbabwe and other countries, to see that the way in which ZANU-PF has developed propaganda has some similarities with Stalin's use of propaganda in Russia, or the way in which the USA, Europe, China and many other countries and regimes in Africa, South America and beyond have all used and continue to use propaganda in various ways at different points in history. He begins to critique the idea that any media are ever neutral: he questions such concepts as 'neutrality' and 'objectivity', wondering whether anyone can really ever be 'objective', as everyone has a personal history and a bias from that background. Expanding his reading, Themba begins to debate and analyse with his colleagues at work and in a variety of online discussion boards numerous theoretical issues about concepts such as race, class, gender, equity, ethics, praxis, habitus, social injustice, social capital and agency. He engages in electronic activism and begins to be increasingly fascinated by the process of political and social analysis: he enjoys reminiscing and debating with his colleagues in the diaspora about Zimbabwe, problem-solving debates about the oppression that people there have suffered throughout so many years, envisaging the ways in which things could improve.

Themba has begun to develop a critical awareness, a rounded view, of what occurred at the time of the farm takeover. He is now highly critical of the actions of the ZANU-PF government and the 'war vets' they employed to take over the farms. He is critical also of the colonial history of Zimbabwe, and realises that the dispossession of the white farmer – and hence of himself and his family – was in part linked to an earlier sense of oppression that the ZANU-PF government had themselves been affected by, initiated and perpetuated by colonialism. Themba realises that the two are possibly tied together into an ongoing reaction, and that the only way to break through that seemingly never-ending chain of action and reaction is to reclaim a sense of freedom, dignity and autonomy through mutual respect and tolerance of other people, so that one need no longer perpetuate either oppressive behaviours against others nor unthinking reactions to domination. From his readings on Paulo Freire, Themba recognises the wisdom of Freire's advice that to become fully human one must escape the duality of the oppressor-oppressed mentality:

... almost always, during the initial stage of the struggle, the oppressed, instead of striving for liberation, tend themselves to become oppressors, or "sub-oppressors." The very structure of their thought has been conditioned by the contradictions of the concrete, existential situation by which they were shaped. Their ideal is to be men; but for them, to be men is to be oppressors. This is their model of humanity.... At this level, their perception of themselves as opposites of the oppressor does not yet signify engagement in a struggle to overcome the contradiction;... the one pole aspires not to liberation, but to identification with its opposite pole..... It is not to become free that they want agrarian reform, but in order to acquire land and thus become landowners — or; more precisely, bosses over other workers. (Freire, 1970, pp. 27-8).

When Themba engages in his final course of study for his PhD focusing on e-learning in adult education in the field of social sciences, using educational research methodology, he selects to focus on the writing of Freire and the idea of 'critical pedagogy' (Freire, 1970). He realises that, without initially being conscious of this, he has gradually been undergoing similar processes of 'conscientization' to those that Freire advocated in terms of the realization of a more critical level of consciousness that has the power to begin to transform his understanding and his life. Themba begins to expand his reading to encompass the works of Foucault, Bourdieu, Habermas, Marx, Weber and Durkheim and a bookcase full of other social sciences theorists. He begins to reflect on issues relating to power and inequality, the means by which multiple inequalities in society are produced and reproduced, the colonial and post-colonial subjugation and social categorization of people according to class, gender, race, sexual preference and religion in southern Africa, and the perpetuation of continuing divisions between those 'with' authority and those 'without'. He engages in extensive dialogue with his teachers and fellow students about these issues and begins to achieve new levels of reflection and understanding.

Themba starts to question his own way of thinking, speaking and his previous failures to speak out against many things with which he instinctively disagreed. He challenges assumptions he previously held about the circumstances of his life, questioning his former presuppositions. He realises that he has in the past often just gone along with the ideas of the dominant social and political forces that have oppressed him and his fellow workers and taken away their freedom to be fully human. He analyses the southern African political and socio-economic situation, his former life in Zimbabwe and his current position in South Africa. Reading Freire again, he begins to observe that he is engaging in an 'authentic struggle' to transform the circumstances in which he is situated. He notices that this process has involved a critical recognition of the causes of his own political, social and economic oppression and understands that he is beginning to carry out 'transforming action' to create a new situation that will lead eventually to fulfilment and human

completion (Freire, 1970: 29). Considering the work of Bourdieu, Themba reflects on the extreme unfairness of the past apartheid discriminatory power relations in the colonial history of South Africa and the lack of freedom of black Zimbabwean Africans in the former colonial period. He also reflects on the present political situation in Zimbabwe and recognises that, despite the illusion of liberation from past colonial oppression, there are so many deficits and problems with the ruinous and corrupt policies of the current political regime that his previous fellow farm workers in Zimbabwe are now in a worse situation than they have ever been. He begins to notice more the ways in which some people without merit or conscience perpetuate their own power and wealth at the expense of their fellow human beings. At the same time, he sees that sometimes people who seem to be better, wiser, kinder and more intelligent human beings – or even just those who are 'different' - are excluded from many opportunities, oppressed and marginalized because of deeply ingrained prejudices about gender, race, ethnicity, class status, religious or political views which labelled and marginalized them as 'the other', as 'lesser human beings'. Themba's anger about and resistance to many forms of social injustice grows. Developing an awareness of the power of strategies of non-violent civil disobedience to protest against injustice, and facilitated by his university library physical and online collections, he includes amongst his readings the works of Dr Martin Luther King Jr., Mahatma Ghandi, Henry David Thoreau, Maya Angelou, Malcolm X, Alice Walker, Jackie Robinson, Frederick Douglass and Nelson Mandela, amongst others, though always, in focusing on his role as a teacher, he comes back to Freire's contributions.

Armed with his new understandings, Themba begins to gain new levels of self-confidence. He teaches e-learning in the local adult education college to his students with a new awareness of the writings of Freire. He develops a sense of praxis through new teaching methods and reflects on critical pedagogy. He encourages his students to utilise their educational resources and their own access to computers to engage in their studies critically and with meaningful attention to their own place in the world. He raises their awareness of the underlying mechanisms and complex operations of power and politics, race, gender and class in South Africa, reading with them some of the works of Nelson Mandela, Ghandi, Alice Walker and Martin Luther King. He encourages them to critique the curriculum, to consider the ways in which knowledge is produced and reproduced and to develop new insights from their own reflections and actions. Themba begins to write his own blog and some articles for a local newspaper, criticising various aspects of local life that he feels need to change. Following the completion of his PhD, he becomes active in various local and Zimbabwean campaigns on civic duties and social responsibility.

One day, some years later, he tells his students the long story of his journey, and in analysing that development, he relates how access to information and resources through education, books, journals and the use of computers has transformed his life and that of his family. Themba tells them that if there had been no farm take-over, his life would, in contrast to his current situation, have remained relatively impoverished, as there had been no access to education, to reading materials or to ICT at the farm or in his former life and he would never have become a teacher. He narrates the ways in which he has gained personal awareness and autonomy from a growing understanding of the ability that he has to play a useful role in transforming the world around him. He tells them his one remaining desire is to go to visit Zimbabwe to see if his mother is still alive. His students cheer at this touching story and wish Themba well on his forthcoming trip to Zimbabwe.

Feeling highly positive after the class, Dr Themba Gweru strolls into the local university library on his way home. Browsing through the academic books on the shelf, he comes across a slightly dusty edited volume entitled *International Explorations of Technology Equity and the Digital Divide*. Interested, he flips through the pages and sees a chapter about Zimbabwe. Embedded within the chapter is a story about a farm worker in Zimbabwe who escapes to South Africa following a raid on the commercial farm where he works. Freeing himself from poverty through his own efforts, the fictional farm worker in the story changes his life through study and hard work, learning ICT skills to become an adult education e-learning teacher. In the margin, someone has written a critical note in pencil, *"Ridiculous, naive and implausible: how could this character ever exist? Lacks evidence; provides little critical analysis!'* A hair-raising shiver runs up Themba's spine. The echoes of this coincidence are just too much! Fearful, he looks around the library suspiciously to see if anyone is watching him. Has someone had him under surveillance all these years? Is he being stalked because of the radical views expressed in his political blog? But there is no-one in the library who looks in the least suspect. Only a few young, anxious-looking students read quietly at the long wooden tables. An old woman in a nearby study booth taps noiselessly away on a computer, her head down. A fly buzzes at the sunny window. It must be a coincidence. Relieved, Themba laughs out loud at the irony and apparent unbelievability of this situation. Then, quickly, he swivels round, glancing behind to see who is reading over his shoulder. He looks at us directly. We see him, standing there by the books, the wiry neat figure of an older man, wearing glasses, a dignified expression on his lined face. Themba smiles at us: a warm, loving smile. For a moment, we glimpse through his eyes in golden shadows the streets of Harare, see its vivid Zimbabwean beauty, its suffering and sorrow. Then Themba turns away and quickly disappears into the Johannesburg sunshine to plan his journey home.

DISCUSSION

The fictional narrative of Dr Themba Gweru outlined above originates from Zimbabwean research reporting the difficult experiences of the many hundreds of thousands of farm workers in Zimbabwe who have experienced the negative effects of poor governmental management of the land distribution program during the years 2000-09 (Addison and Laakso, 2003; GAPWUZ, 2009, Isaacs, 2007; Zimbizi, 2001). Themba is fortunate. He is not one of those who was tortured or killed in the violent raids on commercial farms, nor one of those who lost all his family during the land displacements. By contrast, Themba escapes to discover a new world in South Africa.

Once there, he gradually transforms his life through the opportunities provided by adult education, achieving qualifications through dedicated study with books, computers, teachers and colleagues. He engages in numerous discussions with his teachers and fellow students, gradually changing his own mindset. He vastly expands his knowledge through self-generated learning, reading many works of inspiring literature which nourish him and provide both hope and new purpose. He is particularly influenced by Freire's dedication to the transformational potential of revolutionary critical pedagogy. He recognises in Freire's work a profoundly inspirational philosophy based on praxis, a combination of action and reflection informed by dialogue and active engagement in positive change (Freire: 1970, p.32). This is effected through education, which becomes for Themba an achievable way of overcoming the multiple oppressions of social injustice that in the previous colonial era and now in different ways in the post-independence era, have disenfranchised many people in southern Africa. He wants to achieve transforming action without propagating further bitterness in acts of revenge or hatred, however. Freire's work is infused with the kind of love for humanity, freedom and justice

(Freire, 1970, p. 26) that drives Themba forward as a teacher and social activist.

Themba's story is both motivational and highly optimistic, given his advanced age when he leaves Zimbabwe. A few like Themba have escaped from the multiple oppressions of past and present conditions in rural Zimbabwe to find and benefit from opportunities in the diaspora elsewhere. Yet, mostly, the experiences of Zimbabwean migrants have not been so fortunate. Many of the estimated three million migrants from Zimbabwe (Bloch, 2006; Meldrum, 2007) have been de-skilled, suffering serious economic, migration and asylum difficulties and barriers to education and skills rather than the reverse. They become 'displaced persons' amongst the estimated 2.9 per cent of the global population now living outside their home countries (Bloch, 2006, IOM, 2005).

This chapter is written in the context of desperate poverty, human rights abuses, inequality and oppression in education, in ICT access and in general living conditions in Zimbabwe. To locate this within the wider field of international technological inequity and the digital divide, the situation in Zimbabwe is summarised in the title of this chapter as a 'digital abyss'. This 'abyss' involves a profoundly dehumanizing lack of access to the information, social communication and knowledge resources enabled by ICT, notably for the rural poor and displaced farm workers in Zimbabwe. This void of extreme poverty has occurred as a result of a combination of multiple oppressions deriving from the former colonial era and the current post-independence political situation in Zimbabwe. In this context, the hopeful revolutionary critical pedagogy of Freire is selected from multiple potential interpretations of critical social theory because it demonstrates one potential optimistic way out of oppressive circumstances (Spener, 1990, 1992). The fact that the story of Themba is one of a migrant worker escaping to South Africa is an indication of the seriousness of the deficits affecting daily living and health conditions as well as in education and

ICT within the country itself. A recent WOZA report (2009, p. 2) states that in 2008, 70% of urban WOZA respondents 'said their children had been sent away from school because they had not paid fees on time, or paid the levy, or paid a fees top-up' and '99.6% of parents said there were no teachers in school in 2009 because of industrial action by teachers'. WOZA reported that the situation is worse in rural areas (WOZA, 2009), in which there may often be no electricity, no water, no books and run-down buildings with no chairs or toilet facilities, let alone adequately trained teachers or computers.

Although some small education and ICT projects have enabled positive community-based educational access to occur in some local urban Zimbabwean situations (ECOWISA, 2009) such actions are relatively rare and often time-limited. In this context, the application of critical social theory in the form of Freire's philosophy reified in the story of Themba provides an inspirational potential for local emancipation in Zimbabwe. While Freire's philosophy has been seen by some as relatively simplistic and tending towards an 'either-or' analysis of oppression (Smith, 1997; 2002), it is appropriate for the conceptualization of education – in this case, in the form of digital literacy – for the rural poor suffering from multiple socio-economic and educational disadvantages in Zimbabwe. If the fictional Themba – or rather, his real-life counterparts - were to return to Zimbabwe to seek out solutions to the challenge of working with the rural poor, such teachers might be able to utilise new-found knowledge, social capital and ICT skills to apply for external resources to create new community-based family educational centres. In this, the real-life Thembas – both male and female - of a new Zimbabwe could facilitate students to gain the skills and knowledge to become emancipated from poverty, gaining the qualifications to apply for jobs that would empower and liberate local people from their conditions of poverty and oppression. Freire's problem-solving approach to education involves challenging the

difficult circumstances of oppressed people and finding solutions which provide a potential way out from poverty and disadvantage. Hence as a revolutionary critical praxis it is deeply relevant to the Zimbabwean situation. Though educators such as Themba would be faced with the multiple challenges of Zimbabwe's poor educational and ICT infrastructure, as reported in the first part of this chapter, a problem-solving approach involving the concerted action of numerous partners in a community of e-learning practice might do much to provide new educational and technological opportunities for Zimbabweans desperate to receive such assistance (WOZA, 2009, 2010).

FUTURE RESEARCH DIRECTIONS

There is a need to carry out further research to examine issues relating to the digital divide affecting Africa as a whole and Zimbabwe in particular. These countries are being left behind as the world has rapidly developed further new technology take-up and ever higher levels of teledensity. A severe problem exists regarding the ways in which a lack of access to ICT affects the countries listed in Figure 2: there is a need to investigate this situation more and to recommend the implementation of large scale solutions to improve digital access, particularly to new 3G and fixed line broadband services in Africa. Concomitant with this, there is a need to investigate the extent to which the self-reported well-being of people is affected by poverty combined with lack of access to new technologies, and to run country-wide programs to improve this situation. New models for the development of management approaches to promote equity in e-learning including, for example, the CAMEL (Collaborative Approaches to the Management of e-Learning) model for communities of practice development (JISC infoNet, 2006; Jameson, 2008), are recommended. In particular, there is a need to carry out more in-depth local research on the educational and ICT situation affecting Zimbabwe, expanding the data collected on education in research recently conducted by WOZA (2009). A lack of up to date information on the precise numbers of teachers, schools, adult education and training facilities and their physical infrastructure, state of networking, state of e-learning development and ICT readiness in Zimbabwe suggests that a large scale country-wide research project in this field is both necessary and highly desirable to supplement the international data currently available in order that effective action to redress the 'digital abyss' can be carried out. Given the severity of multiple educational, socio-economic, health and welfare problems affecting Zimbabwe, it is appropriate that local action research projects grounded in or developing from the perspective of critical social theory and an emancipatory praxis such as that of Freire are applied to facilitate self-empowerment for Zimbabwean people through improved access to education and social enterprise.

CONCLUSION

This chapter is written in two parts. In the first part, issues relating to the digital divide affecting Africa as a whole and Zimbabwe in particular are analysed from a factual perspective, with reference to a number of different sources of global information about ICT access and resources from, for example, the ITU and the Afrobarometer surveys (Afrobarometer, 2002, 2006), considered alongside information on various aspects of demographic well-being. The dire situation affecting Zimbabwe in relation to the digital divide is briefly explored, with reference to the World Values Survey report and Happy Planet Index indicating that Zimbabwe is the unhappiest country in the world. In the second part of the chapter, the story of a hypothetical male Zimbabwean farm worker called Themba is narrated. Themba's gradual transition from unqualified farm worker to teacher of e-learning in an adult education college and PhD graduate is told, with reference to the writings of

Paulo Freire in *The Pedagogy of the Oppressed* and the process of 'conscientization' that Themba undergoes over a period of some years. The story of Themba highlights one hypothetical example of the isolating effect that the lack of ICT capability has on the population of Zimbabwe, notably the rural population, as the reader realises that none of Themba's newly found autonomy and rapid self-development would have been possible if he had stayed in Zimbabwe. Owing to the critical lack of ICT resources in the rural areas of Zimbabwe, people such as Themba would forever miss out on the opportunities he had in later life in Johannesburg. Unless real-life counterparts of Themba take concerted action to improve the situation locally in Zimbabwe, this depressing state of affairs is likely to continue.

REFERENCES

Addison, T., & Laakso, L. (2003). The Political Economy of Zimbabwe's Descent into Conflict. *Journal of International Development. Journal of International Development*, *15*, 457–470. doi:10.1002/jid.996

Africa Research Bulletin (2008). *Social and Cultural: Education Zimbabwe: Education has fallen victim as teachers flee or are killed.* Africa Research Bulletin – 17551, May 1st – 31st 2008: Blackwell Publishing Ltd.

Afrobarometer (2002). Key Findings About Public Opinion in Africa: What do Africans think about democracy and development? *Afrobarometer Briefing Paper No. 1: The Afrobarometer Network.* Accessed 2 May 2009: http://www.afrobarometer.org/papers/AfrobriefNo1.pdf

Afrobarometer (2006). The Status of Democracy, 2005-2006: Findings from Afrobarometer Round 3 for 18 Countries. *Briefing Paper No. 40: The Afrobarometer Network.* Accessed 2 May 2009: http://www.afrobarometer.org/papers/AfrobriefNo40_revised16nov06.pdf

Al-Samarrai, S., & Bennell, P. (2007). Where has all the education gone in sub-Saharan Africa? Employment and other outcomes among secondary school and university leavers. *The Journal of Development Studies*, *43*(7), 1270–1300. doi:10.1080/00220380701526592

BBC. (2002). *BBC News Talking Point: Can we narrow the digital divide?* Online discussion, accessed 15 January, 2010: http://news.bbc.co.uk/1/hi/talking_point/2369155.stm

Bloch, A. S. (2006). Emigration from Zimbabwe: Migrant Perspectives. *Social Policy and Administration*, *40*(1), 67–87. doi:10.1111/j.1467-9515.2006.00477.x

Brookfield, S. D. (2005). *The Power of Critical Theory for Adult Learning and Teaching*. Maidenhead, England: Open University Press.

Chan, S., & Primorac, R. (2004). The Imagination of Land and the Reality of Seizure: Zimbabwe's Complex Reinventions. [Columbia University School of International Public Affairs: Gale Group.]. *Journal of International Affairs*, *57*(2), 63.

Chikwanha, A. Sithole, T.& Bratton, M. (2004). *Working Paper No. 42: The Power of Propaganda: Public Opinion in Zimbabwe*. Afrobarometer Working Papers, accessed 2 May, 2009: http://www.afrobarometer.org/papers/AfropaperNo42.pdf

Dzidonu, C. K., Rodrigues, T., & Okot-Uma, R. (1989). The Emerging Global Electronic Messaging and Networking Technologies: An Analysis of their Potential Developmental Impact in Africa. *African Development Review*, *10*(1), 189–210. doi:10.1111/j.1467-8268.1998.tb00104.x

E-Knowledge for Women in Southern Africa (EKOWISA). (2009). *EKOWISA-OKN High Glen Community Centre Information Needs Analysis Report by Zunguze, M.* Harare, Zimbabwe: EKOWISA.

Freire, P. (1970). *Pedagogy of the Oppressed. Harmondsworth*. Middlesex, England: Penguin.

Fuchs, C. (2008a). *Internet and Society: Social Theory in the Information Age*. New York: Routledge, Routledge Research in Information Technology and Society Series, Number 8, Fuchs, C. (2008b). Critical Theory in Age of the Internet, presentation at the *1st World Forum of the International Sociological Association*. September 6, 2008, Barcelona, accessed 2 May, 2009: http://fuchs.icts.sbg.ac.at/i&s.html

GAPWUZ. (2009). *If something is wrong…The invisible suffering of commercial farm workers and their families due to "Land Reform"*. Report produced for the General Agricultural & Plantation. Workers Union of Zimbabwe [GAPWUZ] by the Research and Advocacy Unit [RAU] and the Justice For Agriculture [JAG] Trust: Report available on the kubatana.net website. Accessed 15 January, 2010: http://www.kubatana.net/docs/…/gapwuz_suffering_farm_workers_091111.pdf

Horkheimer, M. (1982). *Critical Theory*. New York: Seabury Press.

Inglehart, R., Foa, R., Peterson, C., & Welzel, C. (2008). Development, Freedom, and Rising Happiness: A Global Perspective (1981–2007). *Perspectives on Psychological Science*, *3*(4), 264–285. doi:10.1111/j.1745-6924.2008.00078.x

IOM (International Organization for Migration) (Ed.). (2005). *World Migration 2005: Costs and Benefits of International Migration*. Geneva: IOM.

Isaacs, S. (2007). *ICT in Education in Zimbabwe*, Survey of ICT and Education in Africa: Zimbabwe Country Report. Washington DC: InfoDev, World Bank, accessed 2 May, 2009: http://www.infodev.org/en/Publication.437.html

ITU. (2007a). ICT-D Statistics online, reported from the Information and Telecommunications Union (ITU) *World Telecommunication/ICT Indicators Database 2006.* Online, accessed 2 May 2009: http://www.itu.int/ITU-D/ict/statistics/ict/index.html

ITU. (2007b).World Information Society Report 2007: Beyond WSIS - Access to ICTs: Executive Summary. *International Telecommunications Union United Nations Conference on Trade and Development,* Geneva 2003 and Tunis, 2005, online, accessed 2 May, 2009: http://www.itu.int/osg/spu/publications/worldinformationsociety/2007/WISR07-summary.pdf

ITU. (2007c). *Telecommunication/ICT Markets and Trends in Africa,* ITU Statistical Report, Geneva, Switzerland: International Telecommunications Union. Accessed 2 May, 2009: http://www.itu.int/ITU-D/ict/statistics/material/af_report07.pdf

ITU. (2007d). *Connect Africa: Facts and Figures,* statistical report online at ITU website. Accessed 2 May, 2009: http://www.itu.int/ITU-D/connect/africa/2007/bgdmaterial/figures.html

ITU. (2009). *Measuring the Information Society: the ICT Development Index,* ITU Report, Geneva, Switzerland: International Telecommunications Union. Accessed 15 January, 2010: http://www.itu.int/ITU-D/ict/publications/idi/2009/index.html

Jameson, J. (2008). *Leadership: Professional Communities of Leadership Practice in Post-Compulsory Education:* Discussion in Education Series. Bristol: ESCalate, accessed 2 May, 2009: http://escalate.ac.uk/5130

JISC infoNet (2006) *The CAMEL project: collaborative approaches to the management of e-learning.* Online: retrieved 2 May, 2009: http://www.jiscinfonet.ac.uk/publications

Kipling, R. (1993). *The Elephant's Child, from the Just So Stories. (Republished from 1902 original),* Hanworth. London: Kenago Books.

Logan, B. I. (1999). The Reverse Transfer of Technology from Sub-Saharan Africa: The Case of Zimbabwe. *International Migration (Geneva, Switzerland), 37*(2), 437–463. doi:10.1111/1468-2435.00079

Masunungure, E., Ndapwadza, A., & Sibanda, N. (2006). Support for Democracy and Democratic Institutions in Zimbabwe. *Afrobarometer Briefing Paper No.27.* http://www.afrobarometer.org/papers/AfrobriefNo27.pdf

Matsika, K. (2007). Intellectual Property, Libraries and Access to Information in Zimbabwe. *IFLA Journal, 33*(2), 160–167. doi:10.1177/0340035207080556

Meisenhelder, T. (1994). The Decline of Socialism in Zimbabwe. [Crime and Social Justice Associates: Gale Group.]. *Social Justice (San Francisco, Calif.), 21*(4), 83.

Meldrum, A. (2007). Refugees flood from Zimbabwe, newspaper article. *The Observer,* 1 July 2007, accessed 2 May, 2009: http://www.guardian.co.uk/world/2007/jul/01/zimbabwe.southafrica

MISAZimbabwe. (2008). *Foreign currency billing system, deprivation of right to communicate.* Article on Media Institute of Southern Africa website, accessed 2 May 2009: http://www.misazim.co.zw/index.php?option=com_content&task=view&id=429&Itemid=5

MISAZimbabwe. (2009). *MISA-Zimbabwe statement on the State of the telecommunications sector in Zimbabwe.* Article on Media Institute of Southern Africa website, accessed 2 May 2009: http://www.misazim.co.zw/index.php?option=com_content&task=view&id=382&Itemid=5

New Economics Foundation (NEF). (2009) *The (un)Happy Index Planet 2.0*. Research report produced by the New Economics Foundation. Written by Saamah Abdallah, Sam Thompson, Juliet Michaelson, Nic Marks and Nicola Steuer. London, UK. Accessed 18 January, 2010: http://www.happyplanetindex.org/

Nherera, C. M. (2000). Globalisation, Qualifications and Livelihoods: the case of Zimbabwe. *Assessment in Education*, 7(3), 335–362. doi:10.1080/09695940050201343

OECD. (2004). *OECD African Economic Outlook 2003/2004: 22 Country Studies: Zimbabwe*. Paris: Organisation for Economic Co-operation and Development, accessed 2 May, 2009: http://www.oecd.org/document/44/0,3343, en_2649_15162846_32404716_1_1_1_1,00.html

Peresuh, M., & Ndawi, O. P. (1998). Education for All—the challenges for a developing country: the Zimbabwe experience'. *International Journal of Inclusive Education*, 2(3), 209–224. doi:10.1080/1360311980020302

Power, S. (2003). How to Kill a Country: Turning a breadbasket into a basket case in ten easy steps—the Robert Mugabe way. *The Atlantic Magazine*, December, 2003, accessed 2 May 2009: http://www.theatlantic.com/doc/200312/power

Roller, L.-H., & Waverman, L. (2001). Telecommunications Infrastructure and Economic Development: A Simultaneous Approach, *American Economic Review, American Economic Association*, 91(4):909-23, September. Accessed 14 January, 2010: http://www.e-aer.org/archive/9104/91040909.pdf

Smith, M. K. (1997, 2002) Paulo Freire and informal education, *The Encyclopaedia of Informal Education*. Accessed 15 January, 2010: http://www.infed.org/thinkers/et-freir.htm

Spener, D. (1990, 1992) *The Freirean Approach to Adult Literacy Education*, ESL Resources: Digests, National Center for ESL Literacy Education. April 1990, Revised November 1992. Accessed 16 January, 2010: http://www.cal.org/caela/esl_resources/digests/FREIREQA.html

Times, Z. (2009). *Zimbabwe scores badly on ICT*. Newspaper article published in Harare: March 26, 2009, accessed 2 May 2009: http://www.thezimbabwetimes.com/?p=14058

WEF. (2009). *The Global Information Technology Report 2008-2009*. INSEAD Business School: World Economic Forum, available online at http://www.weforum.org accessed 15 January 2010: http://www.weforum.org/en/initiatives/gcp/Global%20Information%20Technology%20Report/index.htm

Williams, M. (2008). *Broadband for Africa: Policy for Promoting the Development of Backbone Networks*. Washington, DC: World Bank Report funded by InfoDev, accessed 2 May 2009: http://www.infodev.org/en/Publication.526.html

Women and Men of Zimbabwe Arise (WOZA). (2009). *The State of Education in Zimbabwe - a Dream Shattered: A Women and Men of Zimbabwe Arise (WOZA) perspective, February 2009*. Famona, Bulawayo, Zimbabwe, accessed 15 January, 2010: http://wozazimbabwe.org/?p=315

WOZA. (2010) *Looking back to look forward - education in Zimbabwe: a WOZA perspective – January 2010*. WOZA Report on Education: Bulawayo, Zimbabwe, accessed 15 January, 2010: http://wozazimbabwe.org/?p=607

Zimbizi, G. (2000). *Scenario Planning for Farm Worker Displacement*. Harare: Zimbabwe Network for Informal Settlement Action (ZINISA).

Zimbizi, G. (2001). *Study on Socio-Economic Status of Children and Women on Commercial Farms, Mining and Peri-Urban Areas.* Report prepared for UNICEF, accessed 15 January, 2010: http://www.unicef.org/evaldatabase/files/ZIM_01-800_Part1.pdf -

ADDITIONAL READING

Alexander, J. (2006). *The Unsettled Land: State-Making & the Politics of Land in Zimbabwe, 1893-2003.* James Currey.

Bond., P. (1996). *Uneven Zimbabwe: A Study of Finance, Development, and Underdevelopment.* Africa World Press.

Cheney, P. (1990). *The Land and People of Zimbabwe.* New York: Lippincott.

Chikuhwa, J. (2004). *A Crisis of Governance.* Zimbabwe: Algora Publishing.

Dashwood, H. S. (2000). *The Political Economy of Transformation.* Zimbabwe: University of Toronto Press.

Dzimba, J. (1998). *South Africa's Destabilization of Zimbabwe,* 1980-89. Macmillan. 1998.

Economist Intelligence Unit. (2001). *Country Profile: Zimbabwe.* London: Economist Intelligence Unit.

Freire, P. (1985). *The politics of education.* New York: Bergin and Garvey.

Harmon, D. E. (2002). *Southeast Africa: 1880 to the Present: Reclaiming a Region of Natural Wealth.* Philadelphia, Penn.: Chelsea House Publishers.

Harold-Barry, D. (2004). (editor). *Zimbabwe: The Past is the Future: Rethinking Land, State, and Nation in the Context of Crisis.* Weaver Press.

Herbst, J. I. (1990). *State Politics in Zimbabwe.* Berkeley: University of California Press.

Hill, G. (2003). *The Battle for Zimbabwe: The Final Countdown.* Zebra.

Kriger, N. J. (1992). *Zimbabwe's Guerrilla War: Peasant Voices.* New York: Cambridge University Press.

Lan, D. (1985). *Guns & Rain: Guerrillas and Spirit Mediums in Zimbabwe.* University of California Press.

Makoni, B., Makoni, S., & Mashiri, P. (2007). Naming Practices and Language Planning in Zimbabwe. *Current Issues in Language Planning, 8*(3), 437–467. doi:10.2167/cilp126.0

Masters, W. A. (1994). *Government and Agriculture in Zimbabwe.* Westport, Conn.: Praeger.

Munro, W. A. (1998). *The Moral Economy of the State: Conservation, Community Development, and State-making in Zimbabwe.* Athens: Ohio University Center for International Studies.

Owomoyela, O. (2002). *Culture and Customs of Zimbabwe.* Westport, Conn.: Greenwood Press.

Potts, D. (1993). *Zimbabwe.* Santa Barbara, Calif.: Clio Press.

Raftopoulos, B. (2004). In Tyrone Savage (Ed.). *Zimbabwe: Injustice and Political Reconciliation.* African Minds.

Renwick, R. (1997). *Unconventional Diplomacy in Southern Africa.* New York: St. Martin's.

Robinson, J., & Duckett, A. (1995). *I Never Had It Made: An Autobiography of Jackie Robinson, Hopewell, N.J.: Ecco Press. Rotberg, R. I. (2002). Ending Autocracy, Enabling Democracy: The Tribulations of Southern Africa, 1960–2000.* Cambridge, Mass.: World Peace Foundation.

Rubert, S. C., & Rasmussen, R. K. (2001). *Historical Dictionary of Zimbabwe*. Lanham, Md.: Scarecrow.

Skelnes, T. (1995). *The Politics of Economic Reform in Zimbabwe: Continuity and Change in Development*. New York: St. Martin's Press.

Sylvester, C. (1992). *The Terrain of Contradictory Development*. Zimbabwe: Westview Press.

Tamarkin, M. (1990). *The Making of Zimbabwe: Decolonization in Regional and International Politics*. Savage, Md.: F. Cass.

Thomas, D., & Muvandi, I. (1995). *The Demographic Transition in Southern Africa: Reviewing the Evidence from Botswana and Zimbabwe*. Santa Monica, Calif.: Rand.

Weiss, R. (1995). *Zimbabwe and the New Elite*. New York: British Academic Press.

West, M. O. (2002). *The Rise of an African Middle Class: Colonial Zimbabwe, 1898-1965*. Indiana University Press.

Chapter 7
Paulo Freire's Liberatory Pedagogy:
Rethinking Issues of Technology Access and Use in Education

James C. McShay
University of Maryland, USA

ABSTRACT

This chapter explores why there is a need for scholars to not only systematically couple discussions about technology use along with technology access, but ground their inquiries in a theory of critical multiculturalism as they seek to fully understand ways for minimizing the digital divide. In order to help explain why using this critical framework is important, this discussion is set against the historical backdrop of the country of Brazil whose past in many ways parallels the United States with regard to its history of oppression and servitude of people based upon their racial heritage. Moreover, this work provides a brief discussion of Paulo Freire's work with African Brazilians and how he helped them to develop critical understandings about how hegemonic structures limited the extent to which they were able to experience their own humanity. This chapter draws from the historical experiences of African Brazilians as a way to deconstruct how issues of technology and educational inequalities are examined in the U.S. The author of this chapter claims that if U.S. educators are to help prepare students to become productive and reflective decision-makers, they must first acquire tools for understanding their own social realities and learn ways for re-creating them to reflect the ideals of democracy and social justice. Furthermore, the author made calls for educational scholars develop a new language that captures the spectrum of questions at the center of the digital divide debate concerning access and use, but also foregrounds issues of liberation, agency and social change.

DOI: 10.4018/978-1-61520-793-0.ch007

Copyright © 2011, IGI Global. Copying or distributing in print or electronic forms without written permission of IGI Global is prohibited.

INTRODUCTION

Within the scholarly universe of literature about technology and its so called revolutionary role in reforming K-12 education, there has been ongoing debate about what constitutes the digital divide and how can it be ameliorated (Atwell, 2001; Warschauer, Knobel & Stone, 2004; DiMaggio, Hargittai, Neuman, & Robinson; 2001). The scope of this debate in U.S. schools is often shaped by a binary structured argument of technology *access* vs. *use*. More specifically, one part of this dialectical is that students' access to computer-related technologies is a predominant factor in creating equal educational outcomes. The assumption that serves as the basis for this argument is that if a student has physical access to the technology both in and outside of school, then problems and issues associated with the digital divide will be alleviated. The oppositional construct, *technology use*, is formulated by the notion that the way in which technology is employed can have a significant impact on the production of unequal learning outcomes due to students' different race, class and gender identities.

Whereas scholars (Atwell, 2001; Pomerantz; 2001) have made claims about the plausibility of both sides of the binary and have drawn conclusions such as "minorities and the poor are less likely to own computers and have Internet access at home compared to their middle-class, White counterparts" (p. 253) or "increasing student computer literacy skills would make them more competitive and functional in their world" (p. 512). What is most apparent is that the education outcomes reflected in both arguments are grounded within a discourse of liberal multiculturalism, a very popularly held philosophical position in a majority of schooling institutions throughout the United States (Kincheloe and Steinberg, 2002; McLaren cited in Duarte & Smith, 2000; Sleeter & Grant, 2007). Within this particular discourse, an educator's orientation towards teaching and learning tends to be focused on removing per-

ceived barriers that work to limit the ability of a student who is typically labeled as at-risk, poor, or minority so that they achieve the same learning outcomes as their mainstream, high-SES, and White counterparts. McLaren (cited in Duarte & Smith, 2000) elaborates on how liberal multicultural perspectives are reflected in notions about social equality:

From the point of view of liberal multiculturalism, equality is absent in the U.S. society not because of Black or Latino cultural deprivation but because social and educational opportunities do not exist that permit everyone to compete equally in the capitalist market place. Unlike their critical counterparts, they believe that existing cultural, social and economic constraints can be modified or "reformed" in order for relative equality to be realized. (p. 219)

Dominant epistemologies espoused by schools determine what is considered official school knowledge and how should it be measured. Therefore, in the context of technology reform, both arguments in the debate, presuppose that by providing students with access to technology or ensuring that it is used in ways to effectively promote their achievement are the only acceptable ways of removing those barriers. I argue that we must go beyond solely using liberal multicultural oriented optics of technology *access* and *use* to examine questions about how to bridge the digital divide. Educators must search for critical social theories that explain how technology can be used to help prepare students to become productive and reflective decision-makers who can participate effectively in an increasingly interdependent world—a critical goal for K-12 education in the 21st century.

In this chapter, I will explore why there is a need for scholars to not only systematically couple discussions about *technology use* along with *technology access*, but also ground their inquiries in a theory of critical multiculturalism as they seek to fully understand ways for minimizing the divide. In order to help explain why using this critical framework is important, I will set this work against the historical backdrop of the country of Brazil whose past in many ways parallels the United States with regard to its history of oppression and servitude of people based upon their racial heritage. I argue that issues of educational equality and opportunity as well as other social issues can be better understood if they are examined in another context. Drawing from the influential work of Paulo Freire (Freire, 1972; 1974; Shor & Freire, 1987), I will use his concept of critical pedagogy to help provide an illumination of our current social reality in the U.S. schools which allow for the doling out of technology resources and opportunity based upon a student's social location. More specifically, this analysis of technology *access* and *use* within the historical context of Brazil will create a generative space for constructing critical questions and learning new insights which help to identify problems, issues, and concerns related to the digital divide and reframe questions about technology *access* and *use*, in ways that point to how dominant political, social and economic systems work to define goals and priorities for technology and schooling in the U.S.

Towards this end, the first section of this chapter will provide a brief discussion of the theory of liberal multiculturalism and how since the end of the civil rights movement, it has become a predominant discourse in K-12 schooling. Within this section, I will also explore how this discourse has either advanced or impeded current technology reform efforts that address issues of equity in schooling. The second section provides a historical overview of Paulo Freire's widely known work using Culture Circles as a method to help African Brazilian farmers and laborers develop literacy skills and simultaneously construct a critical awareness of the ways in which the governing elite used technology or limited its access as a political and economic tool for domi-

nation (Freire & Macedo, 1987; Kahn & Kellner, 2007). It was through learning how to read and write that many African Brazilians learned how to unlock the psychological and material chains that held them in an economic and social purgatory. Furthermore, I will also highlight Freire's actual use of technology in these Cultural Circles to support the African Brazilian's acquisition of a new critical social consciousness. It was in this context that Freire referred to technology *use* as liberatory media (Kahn & Kellner, 2007). Finally, the last section of this chapter explore the benefits for educational scholars using a Freireian theory grounded in critical multiculturalism to help illuminate challenges and possibilities for addressing questions surrounding technology *use* and *access* and their implications for the educational reform movement in the U.S.

BACKGROUND

Liberal Multiculturalism and Schooling in the U.S.

Briefly examining the historical development of liberal multiculturalism within U.S. schools can help provide a context for understanding the ways in which it has shaped scholars current questions about the educational equality and the digital divide. As mentioned earlier, liberal multicultural conceptions about schooling grew out of the civil rights era of the 1960's. The impassioned early supporters of this movement were driven by the realization that schools enacted K-12 curricula that valorized the histories, experiences, and values of mainly European, middle-class, and suburban identities (Pang, 2005; Spring, 2007). The systemic historical and social exclusion of most non-White groups from the school curriculum raised serious questions about the purpose of education for students of color, particularly African American youth. Proponents of this movement charged that schools' failure to provide equitable learning environments for students of color would have damning implications for their academic, social, and psychological development. Overtime, there were calls for schools to be deliberate in their efforts to serve the needs of all students regardless of their social identity group membership. These calls resulted in the push for greater inclusion and the utilization of human relations and compensatory approaches to address inequalities in schools. This push led to the incorporation of cross-cultural learning, self-esteem development, and social skills building activities into the standard curriculum. Moreover, liberal multicultural schooling reform efforts attempted to improve educational outcomes for students of color by implementing programs that provided additional academic support to help them acquire the skills necessary for overcoming educational disadvantages (Sleeter & Grant, 2007).

Even though some of these efforts received some degree of acceptance, emerging multicultural critics (Giroux cited in Duarte & Smith, 2000; Grant & Sleeter, 2007) maintained that the purpose and scope of this conception of schooling was limited, and as a result, unable to reach its overarching goal. They claimed that these conceptual approaches did not do enough to challenge the ways in which monocultural power structures that school institutions were based upon produce and perpetuate school inequalities. The key argument made by subscribers to this particular orientation is that inequality is created by the lack of access to social and economic opportunities and that by increasing access, inequality can be eradicated Kincheloe and Steinberg (2002). This view is reflected in the commonly used practice of adding multicultural content to the curriculum as a way to attain the ultimate goal of increasing access to societal opportunities for all students. Part of the conventional thinking was that if students developed an appreciation, understanding and respect across lines of cultural difference, they would be less likely to discriminate against the culturally, and social economically different other. These

educational experiences would have important implications for future professionals who might have the authority to make decisions in the work place that affect the life chances of the culturally different.

Curriculum decisions can also be driven by the concern about how so called at-risk students can be perceived as not having the academic skills necessary to academically successful. Therefore, remedial or accelerated learning programs are used to help fill in gaps in students' knowledge as a way to help mainstream them into general education. In this example, educators perceive students' low performance in a particular academic area as a barrier to their achievement, and consider intervention a natural response to improve the students' chances of accessing future educational and professional opportunities.

I would agree that the aforementioned efforts could potentially be very important steps that can help realize a vision of educational equality. However, I contend that only embracing a formulaic solution of increasing *access* as a way to eliminate educational inequality reflects an underdeveloped theoretical orientation. This particular stance renders one incapable of challenging dominant epistemologies espoused within schools, which produce and reproduce unequal educational outcomes for students of racially and economically marginalized groups. Kincheloe and Steinberg (2002) comment further about the limitations of liberal conceptions of multiculturalism:

[Such a decontextualized] perspective insists that multicultural educators can bring about an unspecified positive change without either clarifying the nature of change or understanding the historical, social and epistemological dimensions of all educational metamorphosis. Viewing liberal multicultural programs and cultural production, is struck by the fact that oppression and inequality are virtually invisible, that the assimilationist goal is unchallenged. (p.13)

Liberal Multiculturalism and Technology Access and Use

Kincheloe and Steinberg's remarks raise important questions about the ways in which current technology reform efforts focused on eliminating the digital divide in U.S. schools have been shaped by liberal conceptions of multiculturalism. After a careful review of the literature, it is increasingly clear that many scholars (Atwell, 2001; Warschauer, Knobel & Stone; 2004; DiMaggio, Hargittai, Neuman, & Robinson; 2001) are concerned with how differential levels of access and use of digital technologies in U.S. schools will work to perpetuate already existing societal disparities based on race, class, and gender. For instance, Atwell (2001) asserts that low SES and minority families are less likely to have access to digital technologies and the Internet, resulting in what he calls a gap between "the information haves" and "information have-nots" (p. 252). Warschauer, Knobel & Stone (2004) make a similar claim:

[On one hand], if computers and the internet are distributed equally and used well, they are viewed as powerful tools to increase learning among marginalized students and provide greater access to a broader information society. One the other hand, many fear that unequal access to new technologies, both at school and at home, will serve to heighten educational and social stratification, thereby creating a new divide. (p. 563)

Hargittai's (2002) research findings support these claims regarding concerns surrounding technology access and its impact on the digital divide. She uses the terminology of *technical access* and *effective access* to make distinctions between how individuals are impacted differently due to the way in which they use the technology. For example, Hargittai refers to *technical access* as having equal access to the Internet and *effective access*, which refers to how one goes about extrapolating the information they need from the

content made available to them on the Internet. There is no doubt that examining metrics such as these help scholars understand how these different forms of access work to inform the development of strategies for closing the divide, which is an essential goal of current technology reform efforts in education. However, I maintain that we must go beyond solely looking at issues of technology *access* and *use* through the lens of liberal multiculturalism to examine questions about how to bridge the digital divide. Furthermore, I argue that educational scholars must be intentional in their efforts to explore how various historical, ideological and institutional forces impact not only who has access to computer related technology and how is it being used, but to also answer: Whose interests are most served? Whose forms of knowledge are most validated? And, who should be involved in shaping the goals of current technology reform efforts and what should be the ultimate outcome (Gale & Densmore, 2003).

In order to explore these questions, I will turn my focus on the influential work of Paulo Freire (Freire, 1972; 1974; Shor & Freire, 1987). His seminal works help us to better understand how pedagogies rooted in a critical multicultural theory might help to transform convention tools used to challenge disparities associated with the digital divide. Freire, a critical pedagogue, focused on helping members of historically oppressed groups acquire tools for developing a critical social consciousness that would lead to individual and community empowerment. I contend that this exploration will help elucidate questions that can help broaden the field of analysis for scholars who are concerned with helping to create new discourses in K-12 technology school reform that reflect an equitable and liberatory vision for all students.

THE OPPRESSION OF AFRICAN BRAZILIANS

Many historical analysts of the colonization of Brazil have examined the legacy of oppression of

African Brazilians after the abolition of slavery in 1888 (Lovell, 1994; Andrews, 1992). It is within this context that I examine how monocultural education was used as a way to economically and socially exclude African Brazilians from full participation in their society. Moreover, I explore in this section how these historical practices continue to have profound present-day implications for how educational scholars in the U.S. should think about issues of technology, education and equity.

Similar to other societies that were multiracialized through colonization, schools often functioned as tools to establish and maintain cultural and economic hegemony. In this sense, either having access to schooling or not, both worked effectively to achieve the intended goal of colonization. Prior to World War II, schools as either a nationally or locally government-sanctioned institution were not promoted as an affordance of Brazilian citizenship (Andrew, 1992). Since the creation of the Brazilian state in 1822, receiving a formal education was a privilege and was mostly available to wealthy White Brazilians. Even with local Brazilian governments assuming the responsibility to make education a funded mandate after World War II, the educational divides between the poor and wealthy, light skinned and dark, continued to be strikingly wide.

As late as 1950, almost sixty percent of White Brazilians were labeled as literate, compared to thirty-one and twenty-six percent, Mulatto and Black respectively (Andrew, 1992). For a majority of African Brazilians, access to formal education continued to be both limited and of poor quality. This circumstance helped to maintain their status in bottom rungs of Brazil's economic class structure (Lovell, 1994). When formal education was made accessible to Blacks, it had tied to it a hidden deculturalizing agenda. According to Spring (2007), deculturalization is a form of cultural genocide, it is the educational process of eliminating the pre-existing culture of a subjugated group and substituting it with the culture of the dominant group. Freire and Macedo (1987) argued that the

colonial structure of schools served to inculcate the African natives with myths and beliefs that belittled their culture, history, and language (p. 143). Education as a means to expeditiously "colonize the minds" of Blacks had extremely negative impact on future generations of this population. Accepting this colonial mentality meant that African Brazilians would need to view that their place within the social and economic class structure created by Brazil's Portuguese colonizers as normal and just.

Paulo Freire and a Pedagogy of Liberation

Paulo Freire, a renowned scholar, activist and native of Brazil, has produced many scholarly works which have had a significant impact on the body of literature which examines the role of education in challenging historically-rooted, political, economic and social practices that limit full democratic societal participation. Freire's work as a critical educator is best understood within the historical context of Brazil and his role in helping African descendants acquire tools that could be used for their social and political liberation (Freire, 1972; 1974; Shore & Freire, 1987). He viewed critical education as a means to promote social advocacy and self-empowerment. And it was in this regard that he helped Blacks develop both literacy skills and a critical social consciousness with the belief that these ways of thinking and acting in the world were essential for dismantling oppressive societal structures and practices.

Freire (1972; 1974) and Giroux (cited in Duarte and Smith, 2000) have referred to this attempt to help members of oppressed groups and their political allies rescue history, experience, and vision within an educational context as a form of critical pedagogy. Giroux (1992) elaborates further on the concept of critical pedagogy:

Essential to a critical pedagogy is the need to affirm the lived reality of difference as the ground on which to pose questions of theory and practice. Moreover, a critical pedagogy needs to function as a social practice that claims the experience of lived difference as an agenda for discussion and a central resource for a project of possibility. (p. 102)

Kincheloe and Steinberg (2002) expand upon the goals of critical pedagogy within a context of schooling. They suggest that teachers should be able to help students overcome social barriers by engaging the students in the exploration of different ways of reading the work, methods of resisting oppression and vision of progressive democratic communities. The goal of a critical educational approach is to prepare students to see themselves as sites of political struggle in relation to oppressive and democratic forces, and move them to a recognition of the forces that shape their identity, the various stages of reflective self-awareness and the strategies their personal empowerment demand (Kincheloe and Steinberg, 2002).

Freire's use of critical pedagogy was best exemplified by his work leading an Adult Education Project of the Movement of Popular Culture in the 1950's. It was through this project that he introduced the use of Culture Circles to help African Brazilian workers and other poor working class adults acquire literacy skills and tools for identifying, critiquing and changing their own social realities. Understanding the meanings the dominant culture often associated with the language of schooling, Freire referred to this project as Culture Circles because it was a way to challenge the conceptions of authority and dominant epistemologies often espoused by schools. Through expanding the lexicon of schooling, Freire used new language that contested the ways traditional education paradigms that have historically shaped meanings about the educational process. This was reflected in how the roles of teacher, students and their practices were identified. For example, teachers were referred to as coordinators, students were called group participants and the method by which they interacted was called dialogue. This

method Freire termed dialogic inquiry created a context for group participants who could meet to reflect on their reality as they make and remake it (Shor, 1987, p. 98) It was through this process of dialogue that Freire sought to help the Black laborer participants interrogate the own social and political realities and develop a sense of individual and collective agency which could be used to bring about positive social change.

The question remains, how does the history of African Brazilians and Freire's use of Culture Circles help to offer new perspectives about the digital divide and its ill effects on humans' ability to fully experience their own humanity? Moreover, Freire's illuminating critique of education and its ability to subjugate or liberate the minds and bodies of Black Brazilians necessitates further inquiry to help identify implications for technology and schooling. I contend that examining these questions concerning equity and the digital divide become even more essential when considering the strong push over the last two decades for U.S. schools to expedite the process infusing technology into the K–12 curricula.

Freire's aforementioned dialogical inquiry process provides a framework for understanding how technology can play a role in supporting pedagogies that create opportunities for liberatory learning. According to Sleeter and Delgado (2004), there are four major components of Freire's dialogical process:

1. Supports a pedagogy of empowerment in which the teacher acts as a partner with the students
2. Uses problem posing as a way to help students critically examine their own experience and historical location
3. Students are creators of knowledge and their classroom practices reflect democratic ideals
4. Class materials are used as tools for expanding students' analyses

Critical Pedagogy and Technology

All four of these components were reflected in the pedagogies Freire used in his literacy project in Brazil not only as a means to help African Brazilians develop literacy skills but to also develop a critical understanding about their own social reality. As a way to enhance the dialogical process, Freire introduced analog technology in the form of a slide projector to support learning in these Culture Circles. These slides depicted authentic representations Brazilian culture and society, thereby creating a shared visual context for analysis and critique carried out by the African Brazilians. Establishing this collective learning environment allowed for what Kahan and Kellner (2007) call reflective distancing to occur. Furthermore, tied to this analysis, were discussions that helped form the foundation for learning about phonemics, grammar, and syntax.

Another major goal of technology use within the Culture Circles was to build capacity for political agency within African Brazilian communities. Kahn and Kellner elaborated the following regarding Freire's vision for technology use as a way to support this form of critical learning: He argued for the importance of teaching media literacy to empower individuals against manipulation and oppression, and of using the most appropriate media to help teach the subject matter in question (p. 435). Freire had a clear sense as to how technology could play a powerful role in emancipating the oppressed, however, he also knew that technology use, if left unexamined, could further support their marginalization. This concern was driven by the understanding that governments will valorize its own social progress and prosperity by promoting its use of modern and innovative tools to develop its infrastructure. Claims made by the Brazilian government in this regard made it far more challenging to initiate rigorous national debates about the social consequences of technology for fear that this critique would be viewed as an anti-government political stance (Freire, 1972).

This view provides a new context to examine Hargittai's (2002) conceptualizations about the technology constructs of *technical access* and *effective access* previously discussed earlier on in this chapter. It is clear that Freire considered it important to provide Black laborers with access to appropriate technologies and to use them in ways that helped them learn how to read and write. However, Freire also showed a very deep concern for understanding how the technology could be used as a tool to promote liberation or subordination. It is in this vein, I argue that language espoused by Hargittai to describe the pedagogical and practical significance of technology *access* and *use* severely limits the ability of leaders in educational technology reform to reframe this dialectical in ways that could be supported within a critical multicultural framework. Questions posed earlier by Gale & Densmore (2003) such as whose interests are most served and whose forms of knowledge are most validated, will continue to be obfuscated when attempts are made to understand underlying causes that create the digital divide. It is with regard to this notion that Giroux (1992) provides further elucidation about how language can shape discourses surrounding schooling and society:

[Within critical pedagogy] we are talking about how educational paradigms begin to generate new language and raise new questions which affects how we view the very realities we engage. When people say we write in a language that isn't clear as it could be, while that might be true they're also responding to the unfamiliarity of a paradigm that generates questions suppressed in the dominant culture. (p. 151)

Drawing from Giroux's remarks, it is imperative that educational researchers should not only systematically couple discussions about *technology use* along with *technology access*, but also ground their inquiries in a theory of critical multiculturalism as they seek to fully understand ways for minimizing the divide. My contention is that educational scholars need new language that captures the spectrum of questions that relate to *access* and *use* but also brings to the foreground a concern about liberation, agency and social change. This new language would help to shape not only the questions they pose as researchers, but also transform understandings by first, identifying what are considered problems associated to the digital divide, and second, recognizing ways in which all of society will benefit by working to bridge the gulf. It is clear that the causes of the digital divide and identifying ways of using technology to help bridge the chasm between those who have access and those who do not, become central questions to understanding the ways to support marginalized groups' efforts to gain full access to societal participation.

Gomez (2006) explained how the possibilities of full societal participation for the poor and marginalized could be realized for all those who became technically, digitally and critically literate. She articulates the following:

And in the Internet sphere, this [embodiment of characteristics of inquiry creativeness, and agency] could be possible if citizens were able to use and understand the process of creating messages and forwarding them, in other words, to declare his word, "to write to the world". Once this can be achieved, the practices of digital literacy would provide maximum benefit to the individual and the community (p. 55).

Interpreting Gomez' (2006) remarks within the context of technology reform efforts in K-12 education and the digital divide, I contend that students need to demonstrate competencies in both academic content areas and the use of digital learning technologies. I also maintain that students must understand how knowledge within those academic content areas is constructed and the ways is it used to shape meanings, frame worldviews, and influence decision-making. Furthermore, students

would need to be able to extend this understanding to how digital technologies can be used as tools for domination as well as for liberation. Educational scholars who work to reform technology efforts in schools with the goal of bridging the digital divide must heed the words of Kincheloe and Steinberg (2002) who remind us that we must not fail to understand that power wielders-especially from the corporate world-have gained unprecedented access to the construction of individual consciousness and identity (p. 12). Therefore, as scholars investigate research questions surrounding the digital divide, they must be mindful of the ways in which the U.S system of capitalism has shaped their epistemological orientation.

FUTURE RESEARCH DIRECTIONS

Understanding the ways in which liberal conceptions of multiculturalism have shaped current technology reform efforts that have sought to bridge the digital divide in U.S. schools should be a key goal of educational scholars. It is essential that as researchers examine problems, issues, and concerns related to the digital divide and interrogate questions about technology *access* and *use*, there is special attention to how dominant political, social and economic systems work to define education goals, and hence, shape the types of research questions that are posed. Giroux (1979) supports Freire's view that in a democracy there must be a concern for empowering people in ways that allow for the creation of their own meanings and be able to draw from that new knowledge a process for acting in their social world. With this in mind, and using students of color as an example, these learners who like their White counterparts must be adept in using digital technologies to demonstrate competencies in academic areas such as literacy, mathematics, and science in order to gain access to future educational and professional opportunities. However, these students should also learn the ways in which these same technologies are used as tools to promote and perpetuate narratives that reflect a world view constituted by mainly White, male, middle-class, Christian, heterosexual and able bodied identities.

Students should also understand the ways technology can be used to represent their own histories, experiences, and forms of knowledge with the goal of creating awareness about ways to challenge dominant systems of oppression. One such example to help elucidate how educators might provide learning experiences for students that focus on increasing proficiency in technology use and simultaneously developing an awareness about issues related to critical multiculturalism are digital stories. Many scholars in the field of education have begun to examine the learning benefits associated with digital story development (Lambert, 2007; Ohler, 2008; Robin, 2008). According to Ohler (2008) digital stories rely on relatively inexpensive personal forms of digital technology that use an array of media such as video, music, computer-generated graphics, and narration to construct a coherent three to five minute narrative. Typical uses of digital stories in academic classroom settings include the development of personal narratives, historical documentaries, or instructional presentations (Robin, 2008). Creating digital stories requires that the student use various tools that convey the authenticity and diversity of the human experience, and can in fact become a form of liberatory media. Similar to Freire's use of the slide projectors to help African Brazilians learn to read, write and become socially aware, digital stories can depict the lived experiences of students in ways that allow for acquisition of academic knowledge, critical reflection, analysis, and dialogue about issues of diversity, equity and social justice.

CONCLUSION

This chapter explored the potential benefits for educational scholars using a Freireian theory grounded in critical multiculturalism to help illuminate challenges and possibilities for addressing questions surrounding technology *use* and *access*. More specifically, it argued that we must go beyond solely using a lens of liberal multiculturalism to examine questions about technology *access* and *use* and how they are implicated in debates about how to bridge the digital divide. This work provided a brief discussion of Paulo Freire's work with African Brazilians and how helped them to develop critical understandings for how hegemonic structures limited the extent to which they were able to experience their own humanity. Furthermore, it explored how Freire's version of critical pedagogy offered them a language of empowerment and self-determination. This chapter drew from the historical experiences of African Brazilians as a way to deconstruct how societal and educational inequalities are examined in the U.S. I contended that if U.S. educators are to help prepare students to become productive and reflective decision-makers, they must first acquire tools for understanding their own social realities and learn ways for re-creating them to reflect the ideals of democracy and social justice. Towards this end, it is essential that researchers must heighten their awareness about how current discourses in technology education reform frame the scope and purpose of scholarly work in this area and also search for a new language of critical inquiry. I maintained that doing this would bring a paradigmatic shift in how issues of technology and inequality are addressed in the current educational reform movement.

REFERENCES

Andersen, D. A. (2001). *The internet and web design for teachers: A step by step guide to creating a virtual classroom*. New York: Longman.

Andrews, G. (1992). Racial inequality in Brazil and the United States: A statistical comparison. *Journal of Social History, 26*(2), 229–263.

Apple, M. W. (2003). Freire and the politics of race in education. *International Journal of Leadership in Education, 6*(2), 107–108. doi:10.1080/13603120304821

Damarin, S. K. (2000). The 'digital divide' versus differences: principles of equitable use of technology in education. *Educational Technology, 40*(4), 17–22.

DiMaggio, P., Hargittai, E., Neuman, W. R., & Robinson, J. P. (2001). Social implications of the internet. *Annual Review of Sociology, 27*, 307–336. doi:10.1146/annurev.soc.27.1.307

Freire, P. (1972). *Pedagogy of the oppressed.* New York: Continuum.

Freire, P. (1974). *Education for critical consciousness.* London: Continuum.

Freire, P., & Macedo, D. (1987). *Literacy: reading the word and the world.* Massachusetts: Bergin & Garvey.

Gale, T., & Densmore, K. (2003). *Engaging teachers: Towards a radical democratic agenda for schooling.* Philadelphia, PA: Open University Press.

Giroux, H. (1979). Review: Paulo Freire's approach to radical educational reform. *Curriculum Inquiry, 9*(3), 257–272. doi:10.2307/3202124

Giroux, H. (1992). *Border crossings: Cultural workers and the politics of education.* New York: Routlege.

Giroux, H. (2000). Insurgent multiculturalism and the promise of pedagogy. In Duarte, E., & Smith, S. (Eds.), *Foundational Perspectives in Multicultural Education* (pp. 195–212). New York: Longman.

Gomez, M. (2006). Contemporary spheres for the teaching education: Freire's principles. *Turkish Online Journal of Distance Education, 7*(2), 52–65.

Gorski, P. (2005). Education and the digital divide. *Association for the advancement of computing in education journal, 13*(1), 3-45

Hargittai, E. (2002). Second level digital divide: Differences in people's online skills. *FirstMonday,* http://www.firstmonday.org/issues/issue7_4/hargittai/

Kahn, R., & Kellner, D. (2007). Paulo Freire and Ivan Illich: technology politics and the reconstruction of education. *Policy Futures in Education, 5*(4), 431–448. doi:10.2304/pfie.2007.5.4.431

Kincheloe, J. K., & Steinberg, S. R. (2002). *Changing Multiculturalism.* Buckingham, UK: Open University Press.

Lambert, J. (2007). *Digital Storytelling: Capturing lives, Creating Community.* Berkeley, CA: Digital Diner Press.

Lovell, P. (1994). Race, gender, and development in Brazil. *Latin American Research Review, 29*(3), 7–35.

McLaren, P. (2000). White terror and oppositional agency: Towards a critical multiculturalism. In Duarte, E., & Smith, S. (Eds.), *Foundational Perspectives in Multicultural Education* (pp. 195–212). New York: Longman.

Ohler, J. (2008). *Digital story telling in the classroom: New media pathways to literacy, learning, and creativity.* Thousand Oaks, CA: Corwin Press.

Pang, V. (2001). *Multicultural Education: A Caring-centered, Reflective Approach.* Boston: McGraw Hill.

Pomerantz, L. (2001). Bridging the digital divide: Reflections on "Teaching and learning in the digital age". *The History Teacher, 34*(4), 509–522. doi:10.2307/3054203

Robin, B. (2008). Digital storytelling: A powerful technology tool for the 21st century classroom. *Theory into Practice, 47*, 220–228. doi:10.1080/00405840802153916

Shor, I. (1987). *Pedagogy for liberation*. MA: Bergin & Garvey.

Sleeter, C., & Bernal, D. (2004). Critical pedagogy, critical race theory, and antiracist education: Implications for multicultural education. In Banks, J. A., & Banks, C. (Eds.), *Handbook of Research on Multicultural Education*. San Francisco, CA: Jossey-Bass.

Sleeter, C., & Grant, C. (2003). *Making choices for multicultural education: Five approaches to race, class, and gender*. New York: Merrill.

Spring, J. (2007). *Deculturalization and the struggle for equality: A brief history of the education of dominated cultures in the United States* (5th ed.). Boston: McGraw Hill.

Warshauer, M., Knobel, M., & Stone, L. (2004). Technology and equity in schooling: Deconstructing the digital divide. *Educational Policy, 18*(4), 562–588. doi:10.1177/0895904804266469

ADDITIONAL READING

Apple, M. W. (2003). Freire and the politics of race in education. *International Journal of Leadership in Education, 6*(2), 107–108. doi:10.1080/13603120304821

Barzilai-Nahon, K. (2006). Gaps and bits: Conceptualizing Measurements for digital divides. *The Information Society, 22*, 269–278. doi:10.1080/01972240600903953

Becker, H. (2000). Who's wired and who's not? *The Future of Children, 10*(2), 44–75. doi:10.2307/1602689

Bee, B. (1981). The politics of literacy. In Mackie, R. (Ed.), *Literacy and Revolution: the pedagogy of Paulo Freire*. New York: Continuum.

Best, S. Kellner, D. (2001). *The postmodern adventure: science, technology, and cultural studies at the third millennium*. New York: Guiford Press.

Bolt, D., & Crawford, R. (2000). *The digital divide: Computers and our children's future*. New York: TV Books.

Driedger, D. (2004). Pedagogy of the oppressed: Flights in the field. *Convergence, 37*(2), 27–37.

Feenberg, A. (1991). *Critical theory of technology*. New York: Oxford University Press.

Freire, P. (1997). *Pedagogy of the Heart*. New York: Continuum.

Lankshear, C., & Snyder, (2000). *Teachers and technoliteracy*. Sydney, Australia: Allen & Unwin.

Lindsay, B., & Poindexter, M. (2003). The internet: Creating equity through continuous education or perpetuating a digital divide? *Comparative Education Review, 47*(1), 112–122. doi:10.1086/373959

Robinson, J. P., Dimaggio, P., & Hargittai, E. (2003). New social survey perspectives on the digital divide. *IT & Society, 1*(5), 1–22.

Sayers, D., & Brown, K. (1993). Freire, Freinet and 'distancing': Forerunners of technology-mediated critical pedagogy. *NABE News, 17*(3), 32–33.

Schofield, J. W., & Davidson, A. L. (2004). Achieving equality of student Internet access within schools: Theory, application, and practice. In Eagly, A. H., Baron, R. M., & Hamilton, V. L. (Eds.), *The social psychology of group identity and social conflict* (pp. 97–109). Washington D.C. APA Books. doi:10.1037/10683-006

Spring, J. (2005). *Conflict of interest: The politics of American education* (5th ed.). Boston: McGraw Hill.

Warschauer, M. (2002). Reconceptualizing the digital divide. *First Monday*, 7(7). Http://www.firstmonday.dk/issues/issue7_7/ warschauer

Warschauer, M. (2003). *Technology and social inclusion: Rethinking the digital divide.* Cambridge, MA: MIT Press.

Watkins, W. H., & Lewis, J. (2001). *Race and education: The Roles of history and society in educating African American students. Boston: Allyn Bacon.* Victoria: Chou.

Wilhelm, A. (2000). *Democracy in the digital age.* New York: Routledge.

Wilson, E. J. (2000). Closing the digital divide: An initial review. Briefing the president. Washington, D.C.: Internet Policy Inst. http:///www.internetpolicy.org/briefing/ErnestWilson0700.html

Chapter 8
Digital Equity and Black Brazilians:
Honoring History and Culture

Patricia Randolph Leigh
Iowa State University, USA

ABSTRACT

In this chapter, the author examines the history of the colonization of Brazil through the transatlantic Black slave trade and the effects this history had upon digital equity experienced by Black Brazilians in the information age. This examination is conducted using the philosophical lenses of critical theory and critical race theory (CRT). Coming from these perspectives, the author joins other scholars in the belief that racism does, in fact, exist in Brazilian societies and joins with those who aim to dispel 'the myth of racial democracy' and the myth of racial harmony in a country with roots in a race-based system of slavery and peonage. The author contends that issues of digital equity and equality of opportunity can only be effectively addressed if one has a deep understanding of the factors that led to inequities that preceded the information age. With this in mind, the author shares various culturally-based grass-roots efforts along with government initiatives she observed during a preliminary investigation of digital equity in this segment of the African Diaspora.

CRITICAL AND HISTORICAL PERSPECTIVES ON DIGITAL EQUITY

In examining the issue of digital equity today, many researchers have gone beyond exploring mere access to computer-related technologies by individuals or groups in society (e.g. Mack, 2001; van Dijk, 2005). Consequently, in this age of rapid change, technological development, and the widespread availability of information communication technologies (ICTs), the broader definition of technology and digital equity has evolved which includes differential access to and use of computer-related resources by various groups in society. Employing social justice perspectives to examine these issues, the author focuses upon the effects that the information and digital age has upon

DOI: 10.4018/978-1-61520-793-0.ch008

Copyright © 2011, IGI Global. Copying or distributing in print or electronic forms without written permission of IGI Global is prohibited.

specific social identity groups who have histori-cally been oppressed and discriminated against. The use of historical methodologies and critical social theories aid in this analysis as attention is centered upon those within the African Diaspora.

Critical Social Theories

A critical social theory is one that examines power relationships and addresses issues of oppression and domination. Typically, such a theory addresses issues of racism, classism, sexism, and/or other forms of discriminatory practices, behaviors, and policies aimed at specific social identity groups that have been historically underserved. Social identity grouping involves how individuals align themselves or are assigned by others according to race/ethnicity, religion, gender, language, and other aspects of culture. The shared and most salient characteristics of those within the African Diaspora include those subjugated as a result of colonization due to their Black African lineage and heritage and associated racial and ethnic identities. The colonization of North and South America, particularly the United States and Brazil, was facilitated through the racist transatlantic Black slave trade. As a result, the overwhelming major-ity of Blacks living in these countries today align themselves within or are assigned to this social identity grouping, despite differences among them concerning other cultural traditions and cultural features (i.e. language).

Critical Theory and Critical Pedagogy

The author believes that an appropriate critical so-cial theory to use in examining digital equity within and among marginalized social identity groups is critical theory along with critical pedagogy, which attempts to implement the tenets of critical theory within educational settings. The tenets of critical theory deal with power and dominance and the oppressed and the oppressors within a society. It is

heavily influenced by Marxist philosophy, which historically has focused upon society's economic base and means of production. As such, Marxists believe that institutions and systems support the economic base and the hierarchies of workers and owners in production (Block, 1994). They are concerned with justice in the general society as well as in schools. Critical theorists and scholars aim to expose power relationships, dominance, oppression, and injustices.

Reproduction Theory, Hegemony, and Resistance Theory

Many critical theorists also believe in reproduc-tion theory. Reproduction theorists contend that many institutions, including educational institu-tions, serve to maintain or reproduce the system of government, economy, and the power hierarchies that exist in a society. For example, the U.S. and Brazilian educational systems and pedagogical practices carried out within them will reproduce existing unjust systems, inequities, and power re-lationships. They also believe that, in many cases, the dominated or oppressed become complicit or contribute to their domination or oppression by internalizing the perspectives of those in power. The term referred to as hegemony speaks to the widespread adoption of dominant views within a society even when the adoption of such views contributes to the demise of the individual or group who adopts such views or perspectives. In addition, acceptance of dominant views and nar-ratives often defies common sense, as explained in the following:

Hegemony supposes the existence of something which is truly total, which is not merely second-ary or superstructural...but which is lived at such a depth, which saturates the society to such an extent, and which...even constitutes the substance and limit of common sense for most people under its sway, that it corresponds to the reality of social experience... (Williams, 1980, p. 37 as cited in Block, 1994, p. 71)

On the other hand, many critical theorists also espouse resistance theory. Resistance theories claim that in many cases the dominated and oppressed, although they are sometimes virtually powerless to control or change their fate, do not accept oppression nor do they accept negative views of themselves. Furthermore, these theorists contend that the oppressed offer considerable resistance to powerful forces. (Martusewicz and Reynolds, 1994, pp. 6-9)

The author of this chapter believes that we can see examples of reproduction theory, hegemony, and resistance theory operating within marginalized groups. As a result of media representations, school curricula and curricular materials, and how they have been treated historically, marginalized social identity groups may internalize and take on the negative views that their oppressors hold and, consequently, view themselves as inferior. Whereas, others in these same groups, even though they may have no power to affect change, refuse to accept those negative views of themselves.

In summary, critical theory and critical pedagogy, along with reproduction and resistance theories, can potentially shed light upon the experiences of those in the African Diaspora, particularly concerning technology and digital equity in various social institutions and settings.

Critical Race Theory

Other critical social theorists would agree that critical theory does an effectual job of examining and analyzing injustices, oppression, and inequities that affect the rich and poor in societies but they may believe that the social construct of race is not adequately addressed. Consequently, critical race theory (CRT), which has roots in critical theory, critical legal studies, and the law, was developed in the 1970s by Harvard law professors in the U.S. Of course, race is the salient feature for defining the social identity grouping that the critical race theorists focused upon and African American experiences, within the context of the U.S. legal system, were initially the target of their analysis.

Critical race theorists have often called upon its major tenets when examining Black and White racial relationships within the U.S. The first is the claim that racism is pervasive in American society and is hegemonic in nature. Further, racism has taken on the face of normalcy. Consequently, racist attitudes, views, policies, and behaviors are so ingrained in the society, that only the most egregious, explicit acts are identified as racist. (see Bell, 1980, 1987; Delgado, 1995)

Other scholars have applied the tenets of CRT in their analysis of the treatment of other marginalized social identity groups within the U.S. (i.e. Solorzano & Bernal, 2001). Because of shared history and the salience of the social construct of race within this theoretical perspective, the author of this chapter contends that CRT can shed light upon the past and current experiences of Blacks within the African Diaspora. As such, the sometimes invisible yet endemic nature of racism in Brazilian society and its institutions are examined. These analyses are important if one is to understand issues of digital equity in Brazilian society, particularly in educational settings.

PRELIMINARY RESEARCH PROCESS

In an effort to begin the initial stages of an investigation of digital equity and inclusion for Black Brazilians, the author spent eight weeks in Brazil during the summer of 2009. Six of those weeks were spent in the city of Salvador, which is in the state of Bahia. The purpose was to locate and analyze as much information as possible from discussions and meetings, observations, reports, and other primary and secondary historical resources. In this chapter, the author presents the exploratory process, which sets the stage for future ethnographic inquiries aimed at uncovering ways in which digital equity is engaged African Brazilian cultures. The author will hereafter be referred to in the first person voice for the remainder of the chapter.

I began the first week of the trip by presenting a conceptual paper at an international conference in Salvador, Brazil. During this conference I had planned to develop networks with universities and academic scholars in the area who were studying topics related to social justice, education, and digital equity for African Brazilians. With the help of colleagues in the U.S., my visit was preceded with email exchanges with individuals who were very instrumental in making my time in Bahia productive. At their advice, I sent ahead and arrived with a letter of introduction and support for my investigations from the chair of my academic department. Although I was unable to connect formally with any universities in the area or their faculty, the contacts I ultimately made and the experiences that ensued far exceeded my expectations. All of my experiences will not be directly reflected in this chapter, but I believe it is important to note key activities that informed and provided a background, as well as colored my perspectives as I accomplished my work during this time in Salvador, Bahia. I believe it is also important that I unravel for the reader, the connections and processes by which experiences and opportunities were afforded me.

The overall aim of making these connections and forming these networks was multifaceted: (1) Facilitate language usage and communication; (2) Secure knowledge of the economic, political, and social (racial) climate in the state and country; (3) Gain entrance into African Brazilian religious and secular communities; (4) Expand knowledge of African Brazilian culture; (5) Explore the state of education for African Brazilians; and (6) Learn of initiatives to address digital inequities for African Brazilians. Since any single contact provided opportunities to fulfill several facets of the overall aim, I will describe how I came to know an individual or group and what activities transpired.

After the first week of conference meetings and initial contacts, I was alone in Salvador. However, an important Salvadorian contact, which I will discuss later, was key in seeing to my well-being and was in contact to assure that my situation was conductive to a productive stay. I should mention that I am a monolingual English speaker. I found that in order to accomplish my goals, I needed a full-time, bilingual interpreter and personal assistant to arrange my appointments and meetings, provide transportation around the city, and, in cases where I was interacting with non-English speakers, interpret and translate from Portuguese to English, and the reverse. Although now working for the state government of Bahia, my Salvadorian contact had previously taught economics and tourism for one of the state universities in Salvador. He arranged for me to meet one of his former students, who agreed to take on this position of interpreter and personal assistant for the remainder of my stay. This interpreter and assistant was invaluable to this investigative process and, with the exception of the few visits arranged by the professional guide discussed below, she accompanied me to most of the meetings and activities described in this chapter.

Early on, my university colleague, who also was my co-presenter at the Salvador conference mentioned above, introduced me to an individual who would, on occasion, act as a professional guide. In past years this colleague had coordinated study-abroad trips to Salvador and the professional guide was the primary Brazilian contact, organizer, and tour guide for the students. She is knowledgeable of Bahian history and has contacts within the African Brazilian communities. This professional guide took me on a day tour of the city at the end of my first week in Brazil and later provided me entrance into the Ilê Axé Opô Afonjá terreiro (Candomblé religious community), the Ilê Axé Oyá terreiro and carnival *bloco* (drumming corps that performs in the annual carnival), and the Calabar community and school. (I discuss the Candomblé religion and the Opô Afonjá terreiro in greater detail in subsequent sections of this chapter.) I learned about the history of the Calabar favela (slum-like shanty town) and its school from three teachers who were present during my visit. During these outings and visits, the guide also provided translation and interpretation between Portuguese and English.

As mentioned previously, the professional guide I secured has ties in the African Brazilian community and is, therefore, familiar with the work of the Steve Biko Cultural Institute (this institute is discussed in greater detail later in the chapter). Consequently, she arranged the first meeting with the Executive Director, in which I expressed my research interests and he explained the work of Steve Biko. Soon after this meeting, the Director of Projects and Communication for Steve Biko, contacted me and provided transportation to the institute on several occasions for meetings with him and the Executive Director. Both directors were very interested in my work and teaching at my home university and in aiding me in successfully completing my preliminary investigation in Salvador. Further, the Director of Projects and Communications at Steve Biko was responsible for other important contacts. He arranged for my invitation to sit in on a meeting with two Bahian government officials: the Director of Technology for Socio-environmental Development and the Superintendent of Scientific and Technological Development. This meeting concerned the state of Bahia's economic base and infrastructure and its public education system.

During the course of my visit, the Director of Projects and Communications at Steve Biko also made me aware of the role of the Secretary to Promote Equality state governmental office. As a result, and with the help of my interpreter and personal assistant, I was able to, secure a meeting with a key official and learn of community initiatives her office had undertaken. This office was created two years prior at the request of the

community an its selected priorities include issues concerning violence against women, sustaining the Quilombos (communities developed by escaped slaves), and increasing the power of Black institutions. I was also invited by the Secretary to Promote Equality official to attend a public meeting of the Council for the Development of the Black Community (CDCN) that followed. The CDCN is under the auspices of the Secretary to Promote Equality, therefore some attendees were affiliated with the government and others were from the local community. A representative from the office of the President of Uruguay addressed the group concerning a project aimed at strengthening technology access to those of African descent while looking at how culture contributes to inequality in Latin America.

Two additional individuals assisted me greatly in my endeavors in various ways. The first, although a native of Bahia, now works in Rio de Janeiro with the Foundation Center of Science and Distance Education of Rio do Janeiro State (CECIERJ). We had met electronically prior to my trip because I was planning to join five of my U.S. colleagues in Rio to collaborate with CICIERJ during my last week in Brazil. This individual from CECIERJ then put me in touch with a contact in Salvador and we all were in email communication before I arrived in Brazil. They were both keenly aware of my research interests, therefore, soon after my arrival, the contact in Salvador arranged for me to meet two people who proved very instrumental in the achievement of my goals. As I described earlier, the first was the individual who became my interpreter and personal assistant. Additionally, at his invitation, I met the second individual at his home. This individual has prime responsibility for the Coordination of Social Articulation in the state and reports to the governor of Bahia. It was during this meeting that she introduced me to the various social programs sponsored by the government and explained her role in them.

Subsequently, this Coordinator of Social Articulation became a point of contact for me within the government and we met many times thereafter. She was responsible for providing contacts with other key individuals working on state-funded projects. In addition, she also arranged for my invitation to visit a school in the Cajazerias community, a favela on the outskirts of Salvador, for a presentation from the Bradesco foundation on their digital inclusion projects and to observe the technology center in the school, which was privately funded by the Bradesco foundation. During that same occasion, the Coordinator of Social Articulation also made a presentation on a grassroots project aimed at the betterment of Cajazeiras and communities like it. Later that day, she escorted me to her capoeira class, a form of martial arts central to the African Brazilian culture, which was also held in the Cajazerias community.

Towards the end of my visit, the Coordinator arranged for me to visit a Jejê (a form/house of the Candomblé religion) museum and to met the president the Association of the Bantu Cultural Heritage Preservation (ACBANTU), who was holding a meeting of the group on the grounds of the museum. After my museum visit and meeting with the president of ACBANTU, I remained to observe the ACBANTU meeting and to hear the African songs of blessing for the food collected for donations to the surrounding community.

Through the Coordinator of Social Articulation, I met the Executive Coordinator of Cidadania Digital, a digital inclusion project for the state of Bahia, who subsequently introduced me to numerous staff and key individuals in the technology centers I would visit. She also provided transportation with a government car and driver each day I reported for a site visit. The first day of our meeting, the Executive Coordinator accompanied me to the training center for Cidadania Digital, which is located on the Pólo Universitário Santo Amaro de Ipitanga (PUSAI) campus in the Lauro de Freitas community. There I was introduced to the class of managers and monitors in training for the technology centers (CDCs) and learned the intricacies of the training program for this digital inclusion project. The Executive Coordinator of Cidadania Digital also arranged for my meeting with the priest and the director of the technology center at the Ilê Asé Opô Ajagunã Candomblé religious community and my visits to the Omi Dúdú cultural center and to Ilê Aiyê, the African Brazilian carnival bloco facility. I also met with numerous other staff working on aspects of the Cidadania Digital project or other governmental projects dealing with technology initiatives and/or those targeting underserved groups.

Again the purpose of this preliminary investigation was to begin to put the question of digital equity within historical, political and social contexts. The following sections will provide a history of the colonization of Brazil, descriptions of important aspects of the African Brazilian culture (the Candomblé religion and capoeira martial arts), more detailed descriptions of the Cidadania Digital project, and examples of the environments or communities in which its technology centers are housed (Steve Biko, Opô Ajagunã, Omí Dúdú, and Ilé Aiyé).

HISTORY OF BLACKS IN THE COLONIZATION OF BRAZIL

In previous explorations of digital equity, I used the term 'analog divide' in reference to inequities in non computer related resources that existed for Blacks in the U. S. prior to the information age and pointed out,

...the racist attitudes and policies that gave rise to the analog divide have been woven into the fabric of our society and have resulted in structural inequalities that serve to maintain these divides. Again, it is these previous injustices and resulting structural inequalities that we must address and find solutions to before we can adequately address the issue of the digital divide. (Leigh, 2008, p. 3)

I now further contend that explorations of the histories of Blacks in the African Diaspora, which fostered analog divides, will shed light upon current digital inequities and illuminate ways in which we resist and create just solutions. In this instance, I examine the history of the colonization of Brazil through the transatlantic Black slave trade and its impact upon Black Brazilians in their past and present societies.

The history of the Brazilian sugar plantation complex, which used the coerced labor of Africans stolen from their homes, was preceded by Portugal's early involvement in slave trading of Black Africans on the European continent and within its country's borders. This Brazilian history also followed the enslavement by the Portuguese of indigenous residents and Blacks Africans in the colonization of the Atlantic islands, which were in convenient proximity to the western coast of the African continent (Schwartz, 1986, p. 7). This plantation complex, created on the Atlantic Islands, and the use of a "controlled and usually ethnically distinct labor force associated with sugar agriculture" (p. 7) served as a model and reflected what the Portuguese would accomplish in Brazil after their attempts to enslave indigenous American Indians failed. Therefore, by the time the transatlantic Black slave trade was well underway and Africans were being stolen and transported directly from Africa in alarming numbers and disturbing conditions to the New World, the Portuguese in Brazil had much experience to draw on. By this time, the Portuguese colonists had already dealt with the Black slave runaways and rebellions on the Atlantic islands, the Indian wars and Indian slave runaways and rebellions in Brazil, and, in many instances, the failure to assimilate and deculturalize either—all of which contributed to the demise of some sugar plantations and the loss of coveted profits for the Portuguese crown and its nobility (Schwartz, 1986). It is reasonable to conclude that the Portuguese had much to teach its neighbors to the north about the treatment and control of an enslaved Black African labor force.

There has been an ongoing debate concerning the nature of the system of Black African slavery used in the colonization of Brazil and the presence or absence of racism in post nineteenth century Brazilian societies that resulted. Scholars on one side of the issue, contend that the enslavement of Indians and later Black Africans on the sugar plantations of Bahia and other Brazilian states was far less severe in comparison to the conditions and treatment typical of those found in the southern plantations of the United States. In addition, many claim that the prevalence of racial mixing and the creation of a multiracial Brazilian society, even during the colonial period, mitigated racist attitudes and behaviors. Giberto Freyre is often cited as one most influential in propagating this line of reasoning and thought (see Schwartz, 1986, p. xv; Schwartz' preface in Mattoso, 1986, pp. viii-ix). On the other hand, scholars, including a group centered in São Paulo, contradict what they describe as a rosy or idyllic view of Brazilian history and current racial relationships. Consequently, those that oppose Gilberto Freyre's idea of post nineteenth-century Brazilian society characterize it as a 'myth of racial democracy' (see Apple, 2003, p. 108; Goncalves e Silva, 2005, p. 303; Machado da Silva, 2005 p. 297). This latter view and critique of Freyre is more congruent with the perspectives of critical theorists and critical race theorists concerning what transpired during the enslavement of Africans in Brazil and their interpretations concerning the resulting effects upon Black Brazilians in modern-day societies.

From Past to Present: Salvador, Bahia

Salvador, Bahia proved to be an appropriate site for this investigation due to its large population of Brazilians of African descent. Jones de Almeida (2003) states, "…many people are not aware that Brazil has the largest population of African descendants in the world outside of Nigeria" (p. 41). Of Salvador, particularly, Selka (2007) contends,

Hailed as the 'most African city outside of Africa,' Salvador is the cradle of Afro-Brazilian culture and the homeland of Candomblé. In colonial times, Salvador served as Brazil's major slave port.... It was here that Afro-Brazilian cuisine and traditional practices such as capoeira (an Afro-Brazilian martial art) emerged.... Salvador has done much to define culturally speaking, what it means to be Afro-Brazilian. (p. 10)

Coming to this preliminary investigation from the perspective that a history of racism and discrimination would certainly, in large part, determine the current experiences of African Brazilians in this digital age, I, nevertheless, was unaware of the path that this initial inquiry would take. As the process began to unfold, it became clear early on that in order to understand 'what it means to be Afro-Brazilian,' as Selka states above, one would need to understand the nature and history of those traditions, practices and beliefs that contributed to that identity in past ages. It is then that one might begin to unravel what this means for the Black Brazilian in terms of digital equity and inclusion in this information age. Thus, I draw heavily on personal observations, field notes, and research literature in order to reach these understandings. Reflecting on my many cultural experiences as well as visits to technology centers within cultural centers, it became apparent that the various meanings attached to the practice of Candomblé and capoeira inform 'what it means to be Afro-Brazilian' from historical, political, and social perspectives.

Religion and Martial Arts

Candomblé

It is difficult to separate the Candomblé religion and the capoeira dance/martial arts form when considering their historical contributions to the Black Brazilian identity. So in looking at Bahia from past to present in light of the African influ-

ence, it is important to have a basic understanding of this religion.

It wasn't long after I arrived in Salvador, Brazil in June 2009 that I heard the word, Candomblé, and the African-derived religion associated with this word, which up to that point was foreign to me. A trip to the historic Pelourinho district as a tourist and researcher led me to a small but pristine museum featuring the traditional dress of Candomblé women. It was there I learned, from the museum displays and the help of my professional guide, that during enslavement Africans in Brazil, as in other colonized countries, were not allowed to openly practice their religions and would therefore hide their beliefs and practices under the cloak of Catholicism—the religion of their masters (Professional Guide, personal communication, June, 19, 2009; also see Selka, 2007). As such, enslaved women disguised their levels as Candomblé priestesses by the fabric and styles of their traditional dresses (see Figures 1 and 2). I learned that the beautiful white fabric used for headdress, dresses, shirts or blouses is traditionally handcrafted by Candomblé women who provide the crochet-like nettle work in the cut-out portions of the material. The powerful visuals of black 'female' mannequins dressed in traditional garments spoke to the central roles that women play in Candomblé religious life in Bahia.

The knowledge I gained from this initial museum experience was strengthened considerably from subsequent discussions and meetings with individuals knowledgeable of and/or active in Candomblé communities and religious activities. These discussions and meetings took place both outside of as well as during my visits to and observations of two *terreiros* or Candomblé religious communities. Selka (2007) reports that Salvador, Bahia has some of the oldest and most respected terreiros. Ilé Axé Opô Afonjá is among those he describes as quite large and expansive (p. 73). I visited Opô Afonjá, which houses up to 70 families on land purchased by its founder in the late 1800s (Professional Guide, personal com-

Figure 1. Candomblé traditional dress portrayed in Pelourinho museum

Figure 2. Candomblé traditional dress portrayed in Pelourinho museum

munication, July 15, 2009) as well as the much smaller Ilê Asé Opô Ajagunã terreiro.

Candomblé devotees give homage to spirits known as *orixás*. Individual orixás have distinctive personalities as well as gifts. Individuals who go through the somewhat intensive initiation process into the religion to become Candomblécistas (Selka, 2007) typically adopt an orixá. While devotion to these orixás is common in all terreiros, there are some differences in traditions or practices depending on the African region or tribal culture that exerts the primary influence. As such, there are three Candomblé houses or nations from which different African-derived traditions are drawn. There are various descriptors for these three Candomblé houses or nations, depending on the region, sub-region or collective cultural traditions represented, and they include: 1) Ketu/Yoruba/Nâgo, 2) Bantu/Angola, and 3) Jejê/Fon (Coordinator Social Articulation, personal communication, July 7, 2009; Selka, 2007, p. 73). There are also differences in terms of present-day connections to Catholicism as well attitudes towards the religion's presence in the public sphere and as part of the Brazilian identity.

In today's Bahia, a visitor will be immediately exposed to Candomblé symbols and images as this once illegal and outlawed religion of enslaved Africans has been adopted into the dominant culture and seen as a part of the state, if not national, identity. However, Candomblé terreiros have different opinions concerning these phenomena. As stated earlier, enslaved Africans in Brazil had to practice their religions under the cloak of Catholicism. As a result, there was a mutual influence whereby Catholicism in Brazil ultimately adopted some Africanisms and, more importantly, Candomblé devotees adopted holy days and Christian saints that more closely matched their celebrations and the personalities or characteristics of Candomblé orixás or santos. These adaptations made it easier to disguise their own practices and, at the same time, these Catholic influences became part of this African-derived religion and remained after emancipation. Selka (2007) points out that Opô Afonjá is an example of a terreiro that has moved to rid itself of all Catholic influence in an effort to return to what they perceive as pure and authentic Candomblé practices. Similarly, such terreiros view the public use of Candomblé symbols and icons as a way of trivializing this religion and transforming it into folklore for tourist consumption. From this perspective, there is no honor in being considered part of a Bahian or Brazilian identity because the true and authentic religion is not being represented. However, other terreiros continue to maintain Catholic-influenced practices and consider them authentic to the religion passed down through slavery. In addition, there are terreiros who maintain good relationships with Catholic clergy, such as the Pastorial Afro, who promote issues of social justice for Black Brazilians, particularly within the church.

The Opô Afonjâ terreiro is located in a very poor surrounding community or favela. Favelas are shanty towns or slums made up of haphazard buildings occupied almost exclusively by African Brazilians at the lowest levels of Brazilian society. Favelas are typically on the outskirts of Brazilian cities and on its hillsides. Jones de Almeida (2003) further explains, "Favelas are generally regarded by officials as informal areas not fully incorporated into civil society. As such, those living in the favelas have virtually no access to services provided to middle- and upper-class neighborhoods" (p. 11). From a distance, favelas have an artistic quality and are often the subject of local artwork sold in galleries as well as on the streets and in open markets. However, visitors like myself as well as local outsiders are warned about entering these slum-like communities without the company of an 'insider', whose presence can somewhat guarantee ones safety. Therefore, just traveling to the security of the enclosed terreiro compound becomes an issue and, in addition, getting inside the terreiro compound requires invitation or permission. All of my visits to terreiros were pre-arranged either through a professional guide with ties to the terreiro or by a government official overseeing collaborative projects with the terreiro. I also traveled to and from the terreiros with knowledgeable guides and either by government car and driver or by private car with a hired driver.

Actually, riding through the streets of the favela is much what you would expect to see or find in any extremely impoverished area. However, the environment inside the terreiro is opposite from the disorganization and unsanitary conditions found outside the walls of the compound. The Opô Alfonjá community is well ordered and all buildings and grounds are well maintained. In ad-

dition, there are well-constructed shrines for the orixás and an elementary school on the premises.

The children were in classes during the time I visited Apô Alfonjá and learned that the school served approximately 350 children (Professional Guide, personal communication, July 15, 2009). The halls and classrooms were brightly decorated and adorned with pictures of the children at various celebrations. The school embraces what is referred to as Nâgo Pedagogy, which has African roots and supports the teaching of Black Brazilian history and identity. As such, elders in this Candomblé community share stories with children in the classrooms and the curriculum also includes capoeira, a type of martial arts that can be traced back to African slave communities in Brazil (Professional Guide, personal communication, July 15, 2009).

Capoeira

In the critical social theory literature, when scholars describe the assimilation process that generally occurs when dominant and subordinate groups clash, usually it is the subordinate group that is forced to relinquish its cultural values or volunteers to do so through the process of hegemony (e.g. Spring, 2004). As a result, the subordinate group begins to look more like their 'superiors' as they take on dominant traits and values to replace those that have been relinquished. From a critical multicultural perspective, forced or hegemonous assimilation is negative in that it involves the deculturalization of the group considered subordinate. Additionally, I submit that the adoption and even claiming of capoeira in the dominant Brazilian society is another example where a different kind of assimilation takes place for the Black Brazilian. It seems that the adoption of African-based practices does not indicate that they have value in the dominant society because they are African-based but rather because their African origins are hidden and/or forgotten as they are claimed by the Brazilian society as a whole. It might then be said that Black Brazilians are assimilated as their cultural practices are viewed as no longer belonging to them but to the larger society and they, therefore, look more like the dominant group in those aspects. They do not give up those practices but are forced to relinquish claim to those practices that are important to African Brazilian identity. In the context of 'racial democracy', Talmon-Chvaicer (2008) also discusses Gilberto Freyre's espousal of a unified homogenous Brazilian society and national identity that benefits from the influence of African and Indian cultures (p. 2). Talmon-Chvaicer gives a detailed historical analysis of "the three major cultures that inspired capoeira—Kongolese, Yoruban, and Catholic Portuguese" (p. 3). Of course, the first two represent African-based cultures and the latter European.

However, Talmon-Chvaicer (2008) provides a vivid picture of capoeira practice and attitudes towards it in colonial Brazil, focusing primarily on Rio de Janeiro. Initially, enslaved Africans were allowed to 'play' capoeira during festivals and celebrations. Slave owners and other White outsiders believed that Africans happily engaging in such activities proved that slavery was not a hardship and justified the institution. It is reported that early on, knives and other small weapons were used and violence would break out among capoeiristas. In addition, capoeira would attract large crowds of Blacks and that, along with the violence that often ensued, also caused the power elite to fear for slave rebellions. Consequently, the practice of capoeira by Blacks, slave or free, was outlawed and those in violation were punished and/or arrested. In terms of how capoeira has moved from this marginalized and illegal practice tied to the Black slave community to the prominence of a national sport and part of the national identity is addressed in Talmon-Chvaicer's comments:

The authorities…have tried to characterize capoeira as a national activity originating in Brazil, the country's national sport, and part of Brazilian folklore…. They have tried to force their convictions on the rest of the population and have *redefined capoeira according to their needs and interests…. In other words, most of the written sources reflect the convictions of the elite and ruling circles. Consequently, due to the marginal social status of slaves and former slaves in Brazil, the importance and influence of the Kongolese and Yoruban cultures have not found expression in these sources. (p. 2-3)*

Debates over the origins of capoeira center around the two styles present today in Brazil. Capoeira Angola claims to be of African origin brought to Brazil through the system of slavery whereas Capoeira Regional claims to be a Brazilian product of a multiracial (African-Brazilian and mixed race) nonracist society (pp. 151-155).

My exposure to the word, concept, and sport of capoeira was similar to my experience with Candomblé—it was early and often. I do not believe one could easily spend 24 hours visiting Brazil, without at least hearing of this national art form with African origins that was eventually made popular by the Black slave communities. Today capoeira seems omnipresent. It is masterfully performed by the *Balé Folclórico da Bahia,* a folk dance company, which also includes in its repertoire dances of the Candomblé and other dances brought to and performed by enslaved Africans in Brazil. At the end of my first week in Salvador

I attended their performance in the Teatro Miguel Santana in the popular Pelourinho district. In this venue I could appreciate the beauty and artistry of the dance without explanation. But capoeira is practiced not only on stage but in the daily lives of many Brazilians. When touring historic sites with a guide from the Steve Biko Institute, we came upon a capoeira classroom situated in an old fort area. The floors were beautifully tiled and the walls decorated with pictures of masters and students. An elderly master was present and welcomed us in. There was no one there to perform capoeira with the master, however, he explained the instruments used, particularly the musical instrument called the berimbau. The berimbau, believed to be of African origin, is a wooden bow with a single string and a gourd attached to the end of the bow. The berimbau is played using a small coin. The capoeira master demonstrated how the instrument is played and sang a song to accompany his music. Typically, the berimbau is played during the performance of capoeira and is considered an integral part of the practice of this martial arts form.

While visiting the Omí Dúdú cultural center in the Liberdade community, I met yet another master of capoeira whose class was recording their berimbau music in the government-funded technology center. We spoke briefly and he too was eager to share his expert knowledge of capoeira and invite me to attend and observe one of his advanced classes in the city. Due to limited time I was unable to do so.

Perhaps my most memorable experience with capoeira occurred deep into the community of Cajazerias, a favela lying outside Salvador. After visiting a school and technology center in the community, I accompanied the Coordinator of Social Articulation for the Bahian government, who is also a member of the community, to observe a class in which she participated. We traveled to and around Cajazerias in the private car of my interpreter and translator. The car was parked between some buildings in the favela and along the side of a clearing, the edge of which was a deep chasm of vegetation. At first, I believed the capoeira class was being held in the building at the end of the clearing but soon discovered that we were to descend into the area below using rock steps built into the side of the cliff-like drop off. As we descended I saw a clearing and a structure where the class would take place. The open-air structure consisted of a cement-like floor with four stucco or cement like-walls about five feet tall. There is a diagram of some type of vegetation with roots painted on one of the walls—above the diagram are the words, *grupo ango – regional* and under the diagram are the words *raizes no Brasil.*

Translated it seems to imply that both styles of capoeira—Angola and Regional —have roots in Brazil and are honored there. I was introduced to the master, who was very welcoming. The class I observed was for beginning females. The capoeira master played the berimbau, two young boys and an adult man played drums, and an advanced male capoeirista, who appeared to be an apprentice, aided the women as they went through their routines. Between exercises the master would explain the philosophy and history of capoeira and he later demonstrated more intricate moves with his student, the Coordinator that accompanied me, and his 'apprentice'. An interesting aspect that the master emphasized was that engaging in capoeira involved defensive movements on both participants and no one needed to be hurt. This was in sharp contrast to the early practice of capoeira as described in the literature (e.g. Talmon-Chvaicer, 2008). He spoke of the action and the reaction—when one capoeirista kicks, for example, the other would avoid the contact from the kick by squatting low and sharply bending the knees (Capoeira Master, personal communication, July, 2009). After this communication and as I watched the master demonstrate capoeira, I could see how the artful kicks and arm movements were action and reaction patterns. The class ended with a symbol of unity as all the students carried out synchronized movements to the music of the berimbau and drum beats. The master was also eager for me to learn more about capoeira and invited me to return but unfortunately again my schedule would not allow.

Capoeira, particularly Capoeira Angola, speaks to the history of the struggle for freedom, equality, and social justice for the Black Brazilian. As Black Brazilians began to form carnival groups in the 1970s, capoeira was included in processions and skits "to tell the history of Brazil and its blacks [using] such themes as slave revolts and opposition to slavery…" (Talmon-Chvaicer, 2008, p. 156). This is relevant in that those who want to maintain and promote African Brazilian culture and identities do not ignore the importance of the contributions capoeira makes to both.

GRASSROOTS EFFORTS WITH GOVERNMENT INITIATIVES

At the same time that I was exploring and experiencing African Brazilian culture I also was investigating how the digital divide was being experienced and addressed for this same group of Bahians. Below I describe a major government project aimed at serving identified Brazilian ethnic groups, including those of African descent.

Cidadania Digital

The State of Bahia government established a program in the year 2000 called Cidadania Digital in efforts to address issues of what it refers to as digital inclusion. Cidadania Digital translates as Digital Citizenship in English but seems to demand more explanation than the translation provides. Cidadania implies a sense of community building, community involvement, and individual responsibility to the other members and to the community to which one belongs. The government's use of this program title implies that communities will drive the efforts to promote digital inclusion. Although the government has changed hands in terms of political party affiliations of those in power, commitment to this project has remained constant and, according to the current administrators and staff involved, the number of communities involved and level of community involvement continues to increase since this change of governmental political leadership (Executive Coordinator, personal communication, July, 2009).

Cidadania Digital provides technology centers, referred to as CDCs, to poor communities and particularly targets Blacks, traditional communities, Indigenous, and the rural poor. The traditional configuration provided by this program includes ten desktop computers and a networked printer.

This hardware is maintained by the state, which assumes responsibility for repairs and updates. The operating system, which is 'Windows-like' software called 'Berimbau Linux', and a suite of software similar to MS Office, are written and maintained by state-employed programmers and analysts, and provided without cost to the centers. In addition, the Cidadania Digital project provides training for the managers and monitors of the centers.

The communities must make application to receive the computers, peripherals, and software that supply the centers along with training for monitors and managers, who oversee the centers and provide support for those who visit the centers. Therefore, the CDCs must have community support and are in no way forced upon the community. However, there are stipulations that the applicants must adhere to in order to receive initial and continued support. The CDCs must be open to the community for 8 hours per day during the week and the monitors must be on the job during those hours. Community members are not to be charged for use of the centers during those hours. The project also requires that the centers be used for educational purposes only and not for private business ventures. The community can use the CDCs for classes and other revenue generation after the mandatory 8 hours have passed but all activities but be pre-approved by the Cidadania Digital management team. The salary for the monitors does not come from the government—the community is responsible. Because Cidadania Digital targets economically poor and often historically marginalized groups, these corresponding communities most often lack the financial means to support the monitors. As such, Cidadania Digital recommends that communities seek out corporate and business sponsors to cover these costs as well as other expenses such as Internet access and electrical utility services. The current executive coordinator of Cidadania Digital reported this as a problem for some communities who are approved recipients of CDCs and she used an indigenous (Indian) community as an example of one which could not afford to support the monitor's salary and unable to secure corporate or business sponsorship.

With the support of the Cidadania Digital management team and staff, I was able to visit CDCs established in Salvador. Below are descriptions of a representative few.

Cidadania Digital Centers (CDCs)

Steve Biko Cultural Institute

Steve Biko, a cultural institute, was established since 1992 to promote affirmative actions. Directors and staff work diligently on projects aimed to prepare Black Brazilian youth for university entrance. Steve Biko is located in the historic district of Pelourinho. From the window of Steve Biko one sees an area that is picturesque yet, at the same time, tells a story of poverty and oppression (see Figures 3 and 4).

Steve Biko's mission is to eliminate racial discrimination and is primarily concerned with the education of Brazilian youth of African descent. As such, many of its projects reflect anti-racist and culturally relevant pedagogies. Steve Biko has a limited budget and is largely dependent upon the commitment of its administrative directors and other staff in carrying out its goals and objectives. However, the successful completion or continuation of some projects has been due to collaborations with private businesses or grant support from foundations. Steve Biko is relatively independent and free to carry out its own agenda, absent the pressures from those who may not share their philosophical ideology, which often comes with more lucrative partnerships and financial backings. Despite these economic challenges, Steve Biko boasts of significant successes in the lives of those who have benefited from its programs (Executive Director, personal communication, June, 2009).

Essentially, Steve Biko attempts to address issues concerning the deplorable state of public education experienced by these same students, which leaves the few who do manage to finish secondary school ill-prepared for the free state and federal universities. This complex situation is somewhat revealed in the fact that, although the level of education has increased for both Black and White Brazilians, the gap has remained the same throughout the years. Blacks consistently fail to complete the educational levels of their White Brazilian counterparts (Director of Projects and Communication, personal communication, June, 2009). This indicates that parallel lines will not

Figure 3. View of the Pelourinho historic district from the window of the Steve Biko Institute

Figure 4. View of the Pelourinho historic district from the window of the Steve Biko Institute

cross unless there is an intervention to dismantle the not only the racist systems and institutions that set this situation in motion but there is a need to uproot the institutionally embedded racist, yet sometimes invisible, policies that serve to maintain this trajectory and keep this gap in place.

Steve Biko is also aware of a very serious underlying problem affecting the education of Black youth in Brazil. That is, well over half of Black and Mixed children ages 5-9 years old work and do not attend school. The same is said for Black and Mixed Children ages 10-14 years (Director of Projects and Communication, personal communication, June, 2009). The problem then becomes how to motivate children and their families to relinquish these incomes for an education that currently reaps few rewards. This problem is obviously circular in nature and there is evidence that the economic and educational systems are serving to reproduce themselves in terms of power and socioeconomic hierarchies. In working with groups of students to alleviate these problems, Steve Biko engages in establishing democratic environments for the development of educational projects and aims to resist these powerful forces that continue to profoundly compromise and undermine the quality of life for Black Brazilians (Executive Director, personal communication, July, 2009).

The Steve Biko Institute applied for a CDC with the intent of developing technology skills in its students, which the directors believe are vital to their educational preparation. Once the equipment for the traditional CDC set-up was received, the directors began to face challenges in meeting the criteria outlined in the Cidadania Digital agreement. One of the challenges is the need to use space for classroom seating that was previously designated for computers. Potential loss of space for computer workstations would compromise the ability to meet the mandatory eight public access hours. Steve Biko also has financial challenges concerning the increased costs of electrical utility services and the monitor's salary because the institute has no corporate sponsors to absorb these

costs. Cidadania Digital management is aware of these challenges and is working with Steve Biko directors to address them.

Ilê Asé Opô Ajagunã

Opô Ajagunã is a traditional community built on the African-based Candomblé religion. This terreiro is significantly smaller than the Opô Afonjá terreiro discussed earlier, and, as such, has no residential housing within the compound. However, Opô Ajagunã, like the larger terreiro, has organized and well-maintained buildings and grounds within the compound, which are not reflective of the conditions found in the poor surrounding favela in which it is situated. The Opô Ajagunã community had applied and received a CDC and the priest underwent the training to carry out the duties of manager and monitor in his role as director of the center. I met and was welcomed by the priest and director of the Opô Ajagunã CDC upon my first visit to the terreiro. A tour of the facilities revealed well-constructed shrines (small houses) for the orixás with free ranging chickens and a rooster on the grounds, elementary school classrooms, a newly established technology center, and a newly constructed temple building for religious ceremonies and rituals.

The priest and I discussed the relationship of the terreiro to the outside community and related issues. While Selka (2007) gives a detailed description of Opô Afonjá's stances related to involvement with Catholicism and the Catholic Church, my discussions with the priest at Opô Ajagunã did not cover these topics directly. However, he did indicate that the Opô Ajagunã community was on good terms with Catholics who owned property in close proximity to the terreiro and they had a history of collaborating and cooperating in terms of the buying or trading of land (Director Ajagunã, personal communication, July, 6, 2009).

The Opô Ajagunã school serves children of the Candomblé devotees and children of the surrounding communities as well. Classes were not in session during my two visits to the Opô Ajagunã terreiro but children were present in preparation for the inaugural ceremonies for the technology center. A significant number of the children are from the community surrounding Opô Ajagunã and are part of the PETI (translated, Program to Eradicate Child Labor) government initiative. Through PETI, monetary incentives are provided to families of school-aged children as an effort to remove the children from the labor market. In essence, the incentives are to release children from the obligation of working and contributing to the family income and allowing them to attend school with the support of the families. The Opô Ajagunã school provides the education for PETI children and integrates them into the terreiro life (Director Ajagunã, personal communication, July, 6, 2009). The children were friendly and seemed well cared for. They are quick to smile and pose for pictures and seem to enjoy the presence of visitors. The children played on and around the swing set on the dirt-covered playground, as there are no paved areas, sidewalks, roads or pathways within the compound.

A new Opô Ajagunã temple building was constructed to replace the original, which was transformed into a historical museum. In the new temple, orixás are depicted on the walls by using draped fabric that alludes to their clothing and artifacts that represent their personalities and/or gifts (see Figure 5). Much of the dedication services for the new CDC took place in this building the day of my second visit.

In the technology room, there are approximately ten computer stations, with several stations that can accommodate individuals in wheelchairs if necessary. This is a traditional CDC set-up with computers running free software that resembles the Microsoft Office suite in terms of functionality. The long range objective of the CDC is to aid in maintaining or rescuing the African Brazilian culture by promoting digital literacy and providing access to information technologies, such as the Internet, to the poor. However, the priest/

Figure 5. Depiction of an Orixá on interior wall of the Opô Ajagunã terreiro building

director indicated that students would begin by learning how to operate the computers and how to use the basic software. When I asked about security around this technology, he noted that Opô Ajagunã is a tight knit community where everyone is working together. Therefore, security is not seen as a problem within the community. He also expressed pride that although his was a poor community with unpaved dirt grounds, a modern technology center is being provided for the benefit of its members (Director Ajagunã, personal communication, July, 9, 2009).

Upon invitation, I returned to the community, three days after my initial visit, for the dedication of this new CDC. Since the dedication was an auspicious occasion, many of the Candomblé women wore their traditional dress. After all officials had arrived, including the religious father of the Opô Ajagunã terreiro, along with members of other Candomblé communities, the program began in the temple building. Afterwards, the celebration moved outside the technology room. The priest led the group in a song of dedication, which included African handclaps and rhythms. The plaque outside the door of the CDC was unveiled to tell all visitors that this center was donated by the state government of Bahia as part of the Cidadania Digital project.

A reception was also part of the day of celebration. Food is an important aspect of culture and the African-born, deep-fried food known as Acarajé, served at this reception, is central to the African Brazilian culture. In addition to Acarajé, a traditional sweet made from tapioca was prepared for these CDC dedication festivities. This ended the day-time celebrations. I was also invited to religious rituals and blessings for the center that were to take place later that evening in the temple. However, I was unable to attend because of lack of a suitable guide and escort for the confusing ride through the evening streets of the surrounding favela.

Omí Dúdú

Omí Dúdú is a self-proclaimed product of a national Black movement in Brazil aimed at reclaiming, rescuing and preserving the Black Brazilian culture (http://nucleoomidudu.org.br/home/). The people that comprise Omí Dúdú share the view that culture is more than a collection of artifacts but rather a human right. Further, although they claim no allegiance to any political party, they believe that their aim to restore the Black Brazilian identity and culture should be viewed as a political stance (http://nucleoomidudu.org.br/a-ong/organizacao.html).

Omí Dúdú was founded in 1988 in the community of Bairro da Liberdade and its primary audience is the Black youth, ages 16-24 years, who are typically from families suffering from extreme poverty and who have poor educational backgrounds. These youth also come from schools and communities that foster violent environments. By rescuing the Black aesthetic and identity through projects and programs that prioritize and educate around culture, it is believed that the destiny of Black youth can be significantly and positively affected (http://nucleoomidudu.org.br/a-ong/interacao-com-a-comunidade.html).

I visited Omí Dúdú with great interest and anticipation because I had been told of its non-traditional CDC.

I found the Omí Dúdú center physically housed in the larger Viva Nordeste center and building. The Bairro da Liberdade community (or Freedom Neighborhood) is a well-known favela that Omí Dúdú serves. The irregular and seemingly sporadic and often colorful structures that house the desperately poor in the favelas of Brazil seems to be a taken for granted feature of the landscape. I myself am conflicted by the beauty from a distance that is often captured by the local artists and the invisibility of the close-up and not so beautiful lives to which many residents are relegated (see figures 6 and 7). Although I visited this facility by government car with my interpreter and government staff, I had informed another professional guide of my visit the previous day. She responded by email a day later and stated, "Today that is one of the neighborhoods with the highest levels of violence, so the groups working there are really essential to the population, particularly to the youth whose lives are being threatened by that (Professional Guide, personal communication, July 9, 2009). Courses to promote self-esteem and the knowledge of history and culture are taught along with other technical offerings. However, the emphasis on Black Brazilian identity and pride are central to the curriculum. For example, Black women are taught how to care for their hair in natural, traditional ways and are discouraged from using straightening techniques and products (Omí Dúdú Guide, personal communication, July 8, 2009).

With the needs of the community in mind, Omí Dúdú applied for and received a CDC with a non-traditional configuration. The ten-computer and printer configuration was replaced by digital recording equipment that allowed the designated technology room to be transformed into a recording studio. Some of the technical classes taught at Omí Dúdú were related to how to operate, program, and maintain the equipment. This popular music

CDC allows groups to schedule this recording studio to create and produce their works. At the time of my visit, a singing group was critiquing the last tracks they had laid down and a capoeira class had scheduled this CDC for a recording session only. In the latter case, all of the capoeiristas had gathered with their master and their individual berimbau instruments. Many forms of music are central to the African Brazilian culture and identity—for example, the samba, the drums of the Black carnival blocos, and the music the Candomblé rituals and of the capoeira (see Selka, 2007, p. 55; Talmon-Chvaicer, 2008, p. 152). It seems fitting to me that a center, whose mission is to preserve this aspect of culture and pride, would do so through the technology available in this digital age.

Ilê Aiyê

Ilê Aiyê is an African Brazilian-centered institution that grew out of the carnival tradition and has strong ties to the Candomblé religion and community (Jones de Almeida, 2003). In terms of what some scholars refer to as the 'myth of racial democracy' in Brazil (see earlier discussion), Selka (2007) states, "What is truly revolutionary about Ilê Aiyê is that it emerged to confront racial inequalities during the military dictatorship, a time when anything that challenged the image of Brazil as a racial democracy was seen as subversive" (p. 88). Like other African-based institutes and communities discussed earlier, "Today Ilê Aiyê states that its objective is to preserve, valorize, and promote Afro-Brazilian culture" (p. 89). Going further, its own promotional material/brochure states, "Ilê Aiyê has established itself as one of the most distinguished agents in recovering the self-esteem and the self-consciousness of the black community in the city capital of Bahia, by strengthening Brazilian culture by keeping and propagating Afro-Bahian roots" (see also http://www.ileaiye.org.br/).

The Ilê Aiyê facility, also located in the Liberdade community, is vast and houses an elementary school and facilities for evening and vocational classes. It is known especially for teaching drumming to Black youth, its carnival band, and its place in the annual carnival. Ilê Aiyê was established in 1974 (Jones de Almeida, 2003) and first appeared in the carnival of 1975 (Jones de Almeida, 2003, p. 49; Selka, 2007, p. 88). In the past Blacks were not allowed to participate in the White carnival blocos so this Black carnival bloco was established. Now, and since its inception, Ilê Aiyê only allows the participation of Blacks or those of African descent. There is criticism of this racial policy and "the leadership has been under increasing pressure to open up participation to non-Blacks" (Jones de Almeida, 2003, p. 51). However, others believe that a change of policy would compromise the beauty and integrity of the carnival bloco (p. 51).

The Ilê Aiyê community school, Escola Mãe Hilda, "was founded by, and is named after, the mother of Ilê Aiyê's founder and president…. His mother is also a high priestess of a [Candomblé] terreiro" (Jones de Almeida, 2003, p. 49). The school, founded in 1988, focuses on young Black children and is a place where the history of Blacks is told. This facility houses the classrooms for elementary-aged children and I briefly observed academic classes of various levels in session. According to the director, Ilé Aiyé leaders did not want to 'lose' the children once they finish their final grade-level, so they began to offer classes of interest for returning older students.

The Ilê Aiyê facility is largely decorated with printed fabric made with the ink stamps that are produced in-house. The fabric is actually produced elsewhere but returned and used to cover tables, make banners, cover chairs and other furniture. I also was shown a room where students were learning how to sew purses and another where they were taught how to make shoes/slippers from the fabric. These items are not sold at the school itself but at other market venues.

Figure 6. View of the Liberdade favela from the Omí Dúdú center

Another part of the facility contains a very large room where practices and rehearsals for carnival take place. This room is also used for other performances and events. Each year a banner is made that depicts the theme of carnival for this bloco. These banners are made from the specially designed fabric that decorates the walls of the building. One year the theme centered on civil rights figures and activists from the U.S. (e.g. Malcolm X, Martin Luther King). This banner is also displayed prominently on one of the walls in the facility (see Figure 8).

Ilê Aiyê also received a fully-equipped CDC, with a traditional ten-computer configuration, through the Cidadania Digital project. During the tour of the facility, particularly when viewing the CDC classroom space, the director stated that Ilê Aiyê has no business partners and receives no financial support from any corporate sponsor. He also had stated that there are no fees attached to any services, including courses and classes in their general facilities or in the CDC space. When asked about the source of the funds needed to support the running of this large endeavor, the director indicated that financial support comes, in large part, from the carnival bands and from clothes that are made and sold through Ilê Aiyê.

Next door to the CDC is another technology room used for teaching classes related to information technology. The room was full at the time we observed it and there was a class in section. The CDC itself appeared to have all the equipment in place and running. We were introduced to the monitor on duty but at that time there were only two users in the room. When asked if there were classes offered after the 8-hour minimum for free community access had been met, the director's response was negative. He further stated that all services were free and that the reason that no classes were offered in the CDC and the reason the room was empty at the time had to do with their lack of sponsorship and their inability to afford to pay the monitor. How such challenges potentially and likely compromise the success of an inclusion project is among the issues that can be explored in future research

FUTURE RESEARCH DIRECTIONS

This preliminary investigation was carried for the express purpose of informing future ethnographic research. What has become clearer is the fact that there are various paths this future research could take. The relatively short time I spent in Salvador, Bahia allowed me to experience a sampling of African Brazilian culture and to get a sense of how issues of digital equity and inclusion are being addressed. However, the information that was not readily available concerned how successful these initiatives, such as the Cidandia Digital project, are in sustaining support for the centers they have established. There is a need to explore this issue for government-sponsored programs as well as for privately-funded technology centers, such as the one funded by the Bradesco Foundation in the Cajazerias community. It is also important to examine the extent of success the various communities have had in meeting their goals through the use of technology and the technology centers. Since the centers I visited had as their goal the

Figure 7. View of the Liberdade favela from the Omí Dúdú center

maintenance of the African Brazilian culture, an investigation of if and how this was accomplished would inform similar initiatives.

There is also potential for researching digital equity beyond Salvador and Bahia. During the last week of my eight-week visit in Brazil, I joined five of my U.S. colleagues in Rio de Janeiro. It was there that we partnered with the Foundation Center of Science and Distance Education of Rio do Janeiro State (CECIERJ) on a symposium on distance education and discussed opportunities for future research collaborations. CECIERJ is currently involved in developing and disseminating distance education materials and experiences to impoverished individuals and communities within its reach. A future ethnographic study might look more broadly at the state of digital equity for African Brazilians and examine the successes of this and various other solutions.

Regardless of how this future ethnographic work ultimately takes shape, I believe it has potential for not only informing what interventions work in Brazil but what might be successful in other countries, such as in U. S. urban areas of concentrated poverty.

CONCLUSION

I began this chapter by providing a summary of the theoretical frameworks I use in examining issues of digital equity and ones I would use particularly in this preliminary investigation in Brazil. When exploring issues of social justice and equity, issues of power and dominance, oppression and subjugation, and hegemony and resistance typically rise to the surface. Critical social theories, such as critical theory, provide the lenses through which to view these phenomena. When the social construct of race is believed to contribute significantly to the experiences of those central to the investigation then critical race theory sheds much needed additional light and understanding. By situating the exploration of digital equity for Black Brazilians against the background of the debate around the existence of a racial democracy in Brazil's multiracial country, I allowed for its deconstruction through a CRT perspective. Those arguing that the presence of a racial democracy in Brazil is a myth would gain support from one of CRT's central tenets. Critical race theorists contend that racism is so ingrained in societies that it is taken for granted and appears normal and only the most egregious acts are considered racist. I came to this preliminary investigation with these perspectives and biases. I also acknowledge that as an African American female, socialized in the U.S., my experiences with issues regarding race construction in my home country likely colored how I made sense of how race is constructed in Brazil. Nevertheless, grounded by the historical context for the Brazilian experience provided in this chapter, I would argue that the racist laws, policies, and practices that allowed for the colonization of Brazil with enslaved Africans became invisible and institutionalized after emancipation. The favelas of Salvador, filled with Brazilians of African descent living at the lowest levels of poverty, are taken for granted, normal parts of the landscape. These Brazilians, who are in the numeric majority, suffer on quality of life issues such as education, housing, employment and overall standard of living. The racist policies that denied access to education during slavery, for example, can be said to be institutionalized in a public education system that is by all accounts deplorable and one that all but the poor and Black manage to avoid. In addition, the need for school-age children from economically impoverished families to work causes them to reject even these poor educational opportunities. These and other factors lead to very high rates of illiteracy among the occupants of Salvador's favelas, which potentially compromises their ability to participate in democratic processes. As such, to support Freyre's claim of a racial democracy (as cited by Apple, 2003, p. 108; Goncalves e Silva, 2005, p. 303; Machado da Silva, 2005 p. 297), would imply that all vestiges of historical racism have been eliminated in Brazilian society and its institutions and that at least the large majority of its citizens are able to participate in its democratic processes. Critical multiculturalists contend that in order to achieve social and global justice, it is paramount that the differing cultures and traditions within and among our societies are honored and embraced (e.g. Nieto & Bode, 2008). While the acceptance and inclusion of African Brazilian culture and traditions into the Brazilian identity can be viewed as a positive step towards a multicultural and egalitarian society, it should not be mistaken as a sign that racism has been eliminated and quality of life and democratic participation issues have been addressed. Moreover, CRT offers another way of looking at the move for a national identity that includes African Brazilian culture and traditions.

In reiterating Talmon-Chvaicer's (2008) statement, "The authorities…have tried to force their convictions on the rest of the population and have redefined capoeira according to their needs and interests" (p.2), it seems the CRT tenet referred to as interest convergence can explain why the previously outlawed and denigrated beliefs and practices of Candomblé and capoeira are now venerated as part of the national identity. What is more, arguments against these cultural aspects being subsumed into a national identity along with efforts to maintain the integrity of African Brazilian culture are often viewed as subversive by the dominate society (Talmon-Chvaicer, 2008). Interest convergence occurs, according to critical race theorists, when the needs of the oppressed are in alignment with the needs of the oppressor and they further argue it is only then that actions will be taken to benefit the oppressed. In other words, social justice will not be achieved for the subjugated race unless benefit also flows to the dominant group and does not threaten the latter's superior status in society. The benefit to Brazil in embracing African Brazilian culture in its national identity is that it can appear as a multiracial, non racist, racial democracy without having to address quality of life issues that obviously fall along racial lines. Bell (1980), for example, believes that the U. S. benefited in similar ways by outlawing racist segregation in the 1954 *Brown v. Board of Education* Supreme Court case and thus repaired its world image after fighting in WWII against a racist Nazi regime.

In terms of digital equity, because I also came to this exploratory process believing history impacts a group's present experiences and opportunities, I was not surprised to find that Black Brazilians in Salvador are by and large left behind in this digital age. However, I had no preconceived notions about how issues of digital equity were viewed and/ or being addressed by and for them. I did not learn of the Cidadania Digital until after my arrival in Brazil, therefore I did not anticipate which groups would be involved in such a digital inclusion project nor did I anticipate the importance of maintaining the African Brazilian culture to the groups and individuals to whom I would speak. Despite, the hegemonic messages claiming that Brazil, in general, and in Bahia, in particular, are absent racial issues, practically all African Brazilian groups and institutions I encountered are engaged in some form of resistance to these messages and most have as their ultimate mission

Figure 8. Ilê Aiyê carnival banner honoring U.S Black civil rights leaders

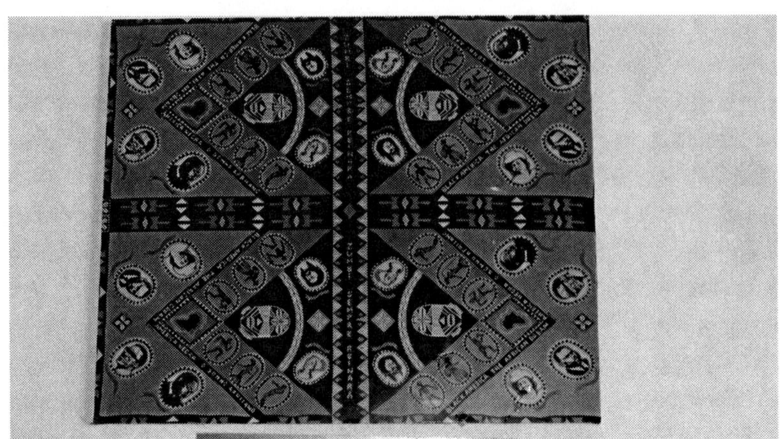

and goal to recapture, rescue, maintain, and/or valorize African Brazilian culture and traditions. Those involved in this government-sponsored Cidadania Digital initiative are willingly involved and with the purpose of meeting those ends. As indicated above, both time and future research will speak to the success of Cidadania Digital but this preliminary investigation leads me to believe it has great potential for addressing issues of equity while honoring histories and cultures of underserved peoples.

REFERENCES

Apple, M. W. (2003). Freire and the politics of race in education. *International Journal of Leadership in Education, 6*(2), 107–108. doi:10.1080/13603120304821

Bell, D. (1987). *And we are not saved: The elusive quest for racial justice.* New York, NY: BasicBooks.

Bell, D. A. (1980). Brown and the interest-convergence dilemma. In Bell, D. (Ed.), *Shades of Brown: New perspectives on school desegregation* (pp. 90–106). New York: Teachers College Press.

Block, A. A. (1994). Marxism and education. In Martusewicz, R. A., & Reynolds, W. M. (Eds.), *Inside out: Contemporary Critical perspectives in education* (pp. 61–78). Mahwah, NJ: Lawrence Erlbaum Associates, Publishers.

de Almeida, J. (2003). Unveiling the mirror: Afro-Brazilian identity and the emergence of a community school movement. *Comparative Education, 1*(47), 41–63.

Delgado, R. (Ed.). (1995). *Critical race theory: The cutting edge.* Philadelphia, PA: Temple University Press.

Goncalves e Silva, P. B. (2005). A new millennium research agenda in Black education: Some points to be considered for discussion and decisions. In King, J. E. (Ed.), *Black education: A transformative research and action agenda for the new century* (pp. 301–308). Mahwah, NJ: Lawrence Erlbaum Associates.

Leigh, P. R. (2008). Historical perspectives on analog and digital equity: A critical race theory approach. In Kidd, T., & Chen, I. L. (Eds.), *Social information technology: Connecting society and cultural Issues* (pp. 1–11). Hershey, PA: IGI Global.

Machado da Silva, T. J. (2005). Black people and Brazilian education. In King, J. E. (Ed.), *Black education: A transformative research and action agenda for the new century* (pp. 297–300). Mahwah, NJ: Lawrence Erlbaum Associates, Publishers.

Mack, R. L. (2001). *The digital divide: Standing at the intersection of race & technology.* Durham, NC: Carolina Academic Press.

Martusewicz, R. A., & Reynolds, W. M. (1994). *Inside Out: Contemporary critical perspectives in education.* Mahwah, NJ: Lawrence Erlbaum Associates, Publishers.

Mattoso, K. M. de Queiros (1986). *To be a slave in Brazil.* New Brunswick, NJ: Rutgers University Press.

Nieto, S., & Bode, P. (2008). *Affirming diversity: The sociopolitical context of multicultural education.* Boston: Pearson Education.

Schwartz, S. B. (1985). *Sugar plantations in the formation of Brazilian society: Bahia, 1550-1835.* Cambridge, MA: Cambridge University Press.

Selka, S. (2007). *Religion and the politics of ethnic identity in Bahia, Brazil.* Gainesville, FL: University Press of Florida.

Solorzano, D. G., & Bernal, D. D. (2001). Examining transformational resistance through a Critical Race and LatCrit Theory framework: Chicana and Chicano students in an urban context. *Urban Education, 36*(3), 308–342. doi:10.1177/0042085901363002

Spring, J. (2004). *Deculturalization and the struggle for equality: A brief history of the education of dominated cultures in the United States.* New York, NY: McGraw Hill.

Talmon-Chvaicer, M. (2008). *The hidden history of capoeira: A collision of cultures in the Brazilian battle dance.* Austin, TX: University of Texas Press.

van Dijk, J. A. (2005). *The deepening divide: Inequality in the information society.* Thousand Oaks, CA: Sage Publications, Inc.

ADDITIONAL READING

Andrews, G. R. (2004). *Afro-Latin America, 1800-2000.* New York: Oxford University Press.

Assunção, M. R. (2004). *Capoeira: A history of an Afro-Brazilian martial art.* New York, NY: Routledge.

Baskaran, A., & Muchie, M. (2007). *Bridging the Digital Divide: Innovation systems for ICT in Brazil, China, India, Thailand, and Southern Africa.* London: Adonis & Abbey Publishers Limited.

Butler, K. D. (1998). *Freedoms given, freedom won: Afro-Brazilians in Post-Aboliition São Paolo and Salvador.* Piscataway, NJ: Rutgers University Press.

Capone, S. (2010). *Searching for Africa in Brazil: Power and tradition in Candomblé.* Durham, NC: Duke University Press.

Curto, J. C., & Lovejoy, P. E. (2004). (Eds.). *Enslaving connections: Changing cultures of Africa and Brazil during the era of slavery.* Amherst, NY: Humanity Books.

Curto, J. C., & Soulodre-LaFrance, R. (2005). *Africa and the Americas: Interconnections during the slave trade.* Trenton, NJ: Africa World Press.

Dantas, B. G. (2009). *Nago grandma and White papa: Candomblé and the creation of Afro-Brazilian identity.* Chapel Hill, NC: University of North Carolina Press.

Davis, D. J. (1999). *Afro-Brazilians: Time for recognition.* London: Minority Rights Group.

Davis, D. J. (2009). *White face, black mask: Africaneity and the early social history of popular music in Brazil.* East Lansing, MI: Michigan State University Press.

Despland, M. (2008). *Bastide on religion: The invention of Candomblé.* London: Equino Publishers.

Eakin, M. C. (2007). *The history of Latin America: Collision of cultures.* New York, NY: Palgrave Macmillan.

Freyre, G. (1986). *The masters and the slaves: A study in the development of Brazilian Civilization.* Berkeley, CA: University of California Press.

Hanchard, M. (1999). *Racial Politics in Contemporary Brazil.* Durham, NC: Duke University Press.

Harris, M. (1964). *Patterns of race in the Americas.* New York, NY: Walker and Company.

Klein, H. (1999). *African slavery in Latin America and the Caribbean.* New York, NY: Oxford University.

Kraay, H. (1998). (Ed.). *Afro-Brazilian culture and politics: Bahia, 1790s to 1990s.* Armonk: M. E. Sharpe Incorporated.

Matory, J. L. (2005). *Black Atlantic religion: Tradition, transnationalism, and matriarchy in the Afro-Brazilian Candomblé.* Princeton, NJ: Princeton University Press.

Merrell, F. (2005). *Capoeira and Candomblé: Conformity and resistance in Brazil.* Princeton, NJ: Markus Wiener Publishers.

Schwartz, S. B. (1992). *Slaves, peasants, and rebels: Reconsidering Brazilian slavery.*

Warschauer, M. (2003). *Technology and social inclusion: Rethinking the digital divide.* Cambridge, MA: MIT Press.

Chapter 9
Digital Equity in a Traditional Culture:
Gullah Communities in South Carolina

Patricia Randolph Leigh
Iowa State University, USA

J. Herman Blake
Medical University of South Carolina, USA

Emily L. Moore
Medical University of South Carolina, USA

ABSTRACT

In this chapter, the authors explore the history of the Gullah people of the Sea Islands of South Carolina. In examining the history of oppression and isolation of Black Americans of Gullah descent, the authors look at how a history of racism and inequity set the stage for the digital inequities experienced by Gullah communities since the onset of the information age. They find that despite the Gullahs' tenacious struggles for education and literacy during enslavement, many are left behind in this age of digital technology. The authors examine the effects that the isolated and closed Gullah communities, which were forced conditions during slavery, had upon many Gullahs' reluctance and resistance to engagement in information communication technologies (ICTs) centuries later. They contend that this continued isolation inadvertently contributed to the loss of Gullah land as well as a pattern of gentrification that severely compromises Gullah traditions and values.

DOI: 10.4018/978-1-61520-793-0.ch009

Copyright © 2011, IGI Global. Copying or distributing in print or electronic forms without written permission of IGI Global is prohibited.

INTRODUCTION

In many parts of South Carolina, rural as well as urban, the traditional culture of Gullah people is confronted by rapidly changing circumstances. As neighborhoods are transformed by gentrification and entire communities displaced by development, Gullah peoples still utilize deeply-rooted values and social institutions to hold on to their unique culture (National Park Service, 2005; Housing Authority Council, 2005). Information technology has impacted Gullah communities in distinctive ways—while the digital divide so prominent in minority communities has become an agent of change the traditional culture of Gullah peoples is both a challenge to information technology as well as challenged by information technology.

This chapter uses the analytical framework of critical race theory (CRT) to develop insights into digital equity and a traditional culture. One of the central features of critical race theory is the use of the human narrative to critique as well as illustrate key issues and increase our insights and under-standings. Furthermore, the authors draw on this major tenet of CRT, which is the belief that the use of narrative can serve to liberate and emancipate historically underserved and marginalized indi-viduals and groups. Consequently, the storytelling and counter-storytelling found in CRT writings contradict and interrupt the dominant narratives and hegemonous perspectives surrounding the histories and experiences of these same social iden-tity groups. Critical race theorists rely heavily on narrative to 'unsilence' and make heard the voices from the bottom rungs of societies' hierarchies in their efforts to dispel misconceptions, myths, and untruths. While Derrick Bell (1987, 1992) and Richard Delgado (1995), for example, employ allegory and fictional storytelling to achieve these purposes, the authors of this chapter believe that other forms of narrative resulting from methods traditional to qualitative research, such as the use of interview and 'oral' data, as well as community engagement can also be effective in lifting up the unheard voices of Gullah people. In this chapter we use compelling narratives to show the intricacies of Gullah culture and its relation to the broader social structure. In doing so, we also discuss and critique, from CRT perspectives, the importance of property and property rights in U.S. society in determining how Gullah communities are viewed by the dominant and powerful elite and how they have come to be, once again, disempowered and threatened, through gentrification, with the loss of the very culture and traditional ways honored and voiced in their narratives.

The particular narratives used in this chapter are condensed from biographies/autobiographies as well as accounts from our more recent research to present eloquent narratives that speak to the issues of CRT and the maintenance of Gullah/Geechee culture. These narratives emanate from the lives of the following:

- **Charlotte Forten** (1837-1914): An African-American woman from the North who taught at Penn School on St. Helena Island (Beaufort County) in the years after the Civil War. Her journal speaks power-fully to the social constraints as well as the intense desire of Gullah people for education.
- **Robert Smalls** (1839-1915): Born a slave in Beaufort County, SC; a hero in the Civil war; a delegate to the 1868 South Carolina Constitutional Convention; an ardent ad-vocate for mandatory public education; a Reconstruction-era legislator in South Carolina. A recent biography, *Yearning to Breathe Free* has been published by Andrew Billingsley (2007).
- **Septima Poinsette Clark** (1898-1987): A Charleston native, taught for sev-eral years on Johns Island, (Charleston County); went on to community organiz-ing as well as establishing the *Citizenship Education Schools* that provided rudimen-tary literacy skills to thousands of African

American adults. Her two autobiographies (1962, 1986) *Echo in My Soul* and *Ready from Within* speak to the early to mid-20th Century conditions.

- **Ervin R. Simmons**: (1957-): Born on the still-isolated Daufuskie Island, SC, recounts his experiences learning in a two-room school that did not go beyond the eighth grade.

In this chapter, the authors also explore how Gullah experiences, histories of oppression and subjugation, and their own resulting attitudes, all which have created and reinforced isolated communities, have also contributed to present-day digital inequities and to the creation of what is known as a digital divide in the Low Country of South Carolina. The digital divide refers to differential access among groups of individuals to computer-related resources and includes differential uses of these same resources. The authors purport that one cannot understand or begin to rectify the digital divide unless one is willing to fully confront and attack the problem of non computer-based inequities that preceded it and continue to persist. From this perspective, this digital chasm of 'haves and have-nots', known as the digital divide, came about as a result of many inequities that began and were nurtured long before the information or digital age and is indicative of the enormous inequalities in access to information technologies that exists between racial, ethnic and social-economic groups. "[T]he digital divide is conceived as," in van Dijk's (2005) work, "a social and political problem, not a technical one. Physical access is portrayed as only one kind of (material) access among at least four: motivational, material, skills, and usage" (p. 3). Historically disadvantaged racial and ethnic groups, such as the Gullah people of the Sea Islands, have long been denied equal access to economic and educational opportunities and these inequities manifested themselves in very specific political, social, and cultural contexts.

Consequently, a history of the Gullah people provides the appropriate context for analyzing how digital equity plays out for Gullah descendents in the 21st century.

HISTORY OF THE GULLAH PEOPLE

Who are the Gullah?

Pollitzer (2005) reports: "The homeland of the Gullah people is a coastal strip 250 miles long and 40 miles wide where low, flat islands, separated from the mainland by salt water rivulets, feel the tides twice each day" (p. 4). These are the Sea Islands of South Carolina and Georgia. The people of African descent that inhabit this strip are commonly called Gullahs, particularly by South Carolinans, but sometimes referred to as Geechees especially in Georgia (p. 6). The ancestors of the residents of the Sea Islands were Africans brought to the U. S. on slave ships after being forcibly taken from their West or Central African homes, perhaps with intermediate stops of forced labor in the West Indies. What is unique about the Gullahs has been their uncanny ability to preserve many aspects of African culture into the 21st century when it is a widely accepted view that slaveholders went to great lengths to deculturalize and strip the slave of all vestiges and remembrances of a past life and culture (e.g. Davis,1983; Lerner, 1992; Spring, 2004). The ability to retain and preserve African traditions, African syntax and language within an English/Gullah dialect, and African ways of working and producing goods is often explained by the relative isolation of the Gullah people and communities during and after slavery (see Toepke & Serrano, 1998). This isolation perhaps removed some pressure on the African slaves to deculturalize and prevented the process of assimilation into the dominant culture to fully take place for both the slaves and their Gullah descendents. The history of centuries of enslavement followed by more than a century of oppression and relative isolation influenced the education of Gullahs from the antebellum years into contemporary America.

Education and the Gullah Culture

From their earliest arrival in the Low Country, education was a very high priority for Gullah people. Finding they were living among and subjected to literate peoples, literacy became an urgent priority. (Williams, 2005) Evidence that slaves sought and obtained literacy is found in the determination of legislators in South Carolina to *prevent* literacy among enslaved peoples. During the antebellum era—in less than 100 years—South Carolina passed three laws prohibiting literacy among slaves in 1740, 1800, and 1834. The details spelled out in each act showed that Blacks were able to circumvent the laws because of their recognition of the importance of literacy.

The act of 1740 was simple and straightforward, enacting a fine for any person teaching or enabling any slave to be taught to write or employing any slave as a scribe (see Williams, 2005, p. 207). This act was most likely encouraged by fears that the slaves who killed more than twenty Whites the previous year in the Stono rebellion of 1739 had communicated their plans in writing. The act of 1800 acknowledged that the previous law was "found insufficient for keeping them in due subordination" (Williams, 2005, p. 13, 207). Therefore the law was extended to prohibit "all assemblies and congregations of slaves, free Negroes, mulattoes and mestizos…assembled or met together for the purpose of mental instruction" (p. 13, 207). The act went further and prohibited an assembly for the purpose of mental instruction or religious worship even in company with White persons "either before the rising of the sun, or after the going down of the same" (p. 13, 207). Punishment for this offense was not to exceed twenty lashes. Thus the prohibition was extended to free persons of color as well as White persons and included harsher punishment.

Evidently this did not prevent slaves and free persons of color from becoming literate however, and in 1834 the previous act was amended to add both a fine and prison for any White person teaching or assisting in teaching any slave to read and write; for a free person of color the punishment included a whipping up to fifty lashes as well as a fine not exceeding fifty dollars. If a slave, the punishment was a whipping up to fifty lashes but there was no fine. What is more, the 1834 legislation prohibited slaves and free persons of color from keeping "any school or other place of instruction for teaching any slave or free person of color to read or write" (Williams, 2005, p.16, 208). The continued expansion of the prohibition on literacy among persons of African descent in South Carolina indicates that, in spite of the yoke of slavery, people were seeking and acquiring literacy, and, what is more, they were establishing organizations to spread literacy more widely.

This brief description of the efforts of Sea Island Blacks to obtain an education and literacy in the face of a system determined to keep them ignorant shows that Gullah people would not passively succumb to their enslavement. Untaught and unknowing, they went to extraordinary lengths to overcome their constraints. It is also clear that the perpetrators of slavery could not suppress the quest for knowledge any more than they could suppress the quest for freedom. Punishment, no matter how harsh, could not suppress the desire to learn (Blake, 2007; Williams, 2005).

Education and the Gullah/ Geechee Narrative

The experience of Gullah peoples is unique in the history of African Americans in particular and America in general. From African roots the people persisted through the crucible of capture, the middle passage and eight generations of enslavement followed by a century of oppression. That the culture and the people survived at all is an eloquent testimony to their character. Within the survival patterns the quest for education has always been paramount.

While we select three icons and a contemporary resident of the Gullah experience to exemplify the significance of education, one could select almost any individual and find a similar narrative. Here we particularly speak very briefly about Charlotte Forten, Robert Smalls, Septima Poinsette Clark and Ervin R. Simmons. In their lives they covered the antebellum period for Gullah people as well as the century from slavery to freedom.

Charlotte Forten

Born in Philadelphia, Charlotte Forten came from a family with strong abolitionist fervor. When she learned there was a need for teachers in the Port Royal area in the early days of the Civil War she applied and, after considerable persistence, was accepted (Taylor, 2005). At age 25, Charlotte Forten was one of the first African American teachers among Gullah people. She taught on St. Helena Island in Beaufort County.

Her experiences testify to the extraordinary desire of Gullah people for education. In her first experience she taught in a small room in a church on St. Helena Island, where there were 140 freed children crowded into one room. She had great difficulty understanding their Gullah language, but she had no difficulty understanding their desire to learn. In her letters she wrote:

I never before saw children so eager to learn, although I had had several years' experience in New England schools. Coming to school is a constant delight and recreation to them. They come here as other children go to play. The older ones, during the summer, work in the fields from early morning until eleven or twelve o'clock, and then come into school after their hard toil in the hot sun, as bright and as anxious to learn as ever.

Of course there are some stupid ones, but these are the minority. The majority learn with wonderful

rapidity. Many of the grown people are desirous of learning to read. It is wonderful how a people who have been so long crushed to the earth, so imbruted as these have been—and they are said to be among the most degraded negroes of the South—can have so great a desire for knowledge, and such a capability for attaining it. (Forten, 1864, p. 591)

Taylor (2005) concludes her biographical sketch of Forten in stating,

[She] led an active and challenging life in the spirit of her family's legacy. She was a spiritual woman, as well as one who was widely read and well written. As a schoolteacher, she sought out opportunities to contribute to the uplift of her race through education and demonstrated her self-determination by not taking "no" for an answer when she decided to participate in the Port Royal Commission undertaking in South Carolina. She pursued her intellectual activities through reading and writing throughout her life, as she faced ongoing bouts with her health. (p. 134)

Robert Smalls

Born in Beaufort, SC, Robert Smalls built on his life as a slave to enrich the education of Gullah people in particular, but eventually all residents of South Carolina. Robert Smalls' appreciation and support of education began with himself, extended to his own children, and beyond to the newly freed children of Beaufort, Black and White children of the entire state of South Carolina and to higher education in general. Personally, Smalls achieved national fame for his heroic exploits in the early days of the Civil War. On May 13, 1862 he delivered the armed Confederate steamer *Planter* to the Union navy blockading the Charleston Harbor. Only 23 years old, Smalls had to struggle to achieve a level of literacy to match his acclaim, a struggle that lasted throughout his life and expressed in his commitment to education for all. As his prestige increased, Smalls hired professional educators and tutors to teach him to read and write. Miller (2008) stated that Smalls' son born after the Civil War stated that in 1865 his father could not spell his name correctly and that his children tutored him, acted as his secretaries, and corrected his letters for all his life. Although he personally struggled to acquire literacy, he made sure his children were educated, and established a school for "the Negro children of Beaufort" (p. 22).

His major contributions came as a delegate to the South Carolina Constitutional Convention in 1868. There he introduced a resolution "providing for a system of common schools, of different grades, to be open without charge to all classes of persons" and another resolution that "required that all parents and guardians send their children between the ages of seven and fourteen to some school, at least six months for each year, under penalties for non-compliance" (South Carolina, 1868, p. 100). It was the vision and leadership of Robert Smalls that resulted in legislation that created the first free and compulsory universal public school system in America, provisions that were continued in South Carolina in later years when educational opportunity was denied to Blacks.

Septima Poinsette Clark

Septima Clark was born in Charleston, SC in 1898. Her entire life was dedicated to education, literacy, and learning as keys to community uplift, social change and personal satisfaction. She began her teaching career in 1916 at 18 years of age when she graduated from Charleston's Avery Institute. At that time the City of Charleston did not hire "colored people" so Septima Clark accepted a position on Johns Island—a long boat ride from her Charleston home. Residing and teaching on Johns Island marked the beginning of a lifetime of service to poor and uneducated Blacks in vulnerable communities.

Her career was guided by the values and principles she learned from her parents. Her father, Peter Porcher Poinsette, was born a slave on the plantation of Joel Poinsett near Charleston. As a slave boy her father went to school every day with the son of his master. They rode together on the horse with the young master holding the reins, and the young slave seated behind him holding the books. The slave remained outside with the horse while the master sat in the classroom and learned. Peter Porcher Poinsette never learned to read or write, and was a very old man when his daughter, Septima, taught him how to "sign" his name.

Her mother, Victoria Warren Anderson, was born in Charleston but as a young child was taken to Haiti where she was raised by her uncle. There she also received a very good education and learned to read and write well. Septima was the second child of Victoria and Peter Poinsette. The parents were committed to education and under very difficult circumstances struggled to ensure that each of their eight children completed elementary and secondary schools, and two of them (Septima included) completed college. The contrasting educational experiences of her parents, and her young years as a teacher on a remote island greatly influenced Septima Clark's career of service through teaching.

She had a passion for learning that came out in all of her teaching. Living and teaching for two years on Johns Island, Septima Clark gained deep insights and understandings into motivations of adults to overcome their illiteracy as well as the desires of children to learn. Fundamentally, education and literacy were the foundation of leadership, self-confidence, pride and community uplift.[1]

Four decades after her youthful experiences as a teacher, Septima Clark returned to Johns Island to establish a program of adult education and literacy that became a model for transforming Black communities throughout the South. In January 1957, working with Esau Jenkins a Johns Island native, and her niece, Bernice Robinson, a beautician from Charleston, Septima Clark established the first Citizenship Education School. This was a learning experience for adults on Johns Island, providing them with the literacy skills they needed to become registered voters. Her pedagogical strategy was unique. The first session they asked the participants what they wanted to learn. Initially the people expressed three desires: (a) How to write their names, (b) How to read the election laws sufficiently well to qualify for registering to vote, and (c) How to fill in money orders and other forms to purchase merchandise from mail order houses (Clark, 1962, p. 147). With this "curriculum" Septima Clark and Bernice Robinson began an adult education program that rapidly grew to include schools on other sea islands and ultimately throughout the South. Later the curriculum was expanded to include the Bible, church hymnals, and life insurance policies (personal interviews of Septima Clark).[2]

Some of her examples reveal the challenges in bringing literacy to Gullah people who had been denied an education all their lives. They include the difficulty of a gentleman learning to hold a pencil—rather than a hoe. It took two hours for him to master the task. In other examples men who had signed their names with an X for many years, objected to "changing their names" when they saw how it should actually be written. Yet all learned and almost all of them became voters (Clark, 1962, p. 147-155).

Ervin R. Simmons

Ervin Simmons was born January 21, 1957 on Daufuskie Island, South Carolina, still one of the most remote Gullah communities in the United States.[3] In excerpts from a recent essay he describes his family's commitment to education and how they guided him.

My formative years took place in an isolated island community of African Americans who had relatively little interaction with the larger American society except through radio, television and occasional trips to the mainland. Although we lacked a full understanding, we were an intimate part of the Gullah traditions of the descendants of Africans enslaved in the region who maintained many of their unique cultural features.

Although there was a small community of whites on the island our interaction with them was extremely limited—sometimes friendly between the women while often adversarial between the men.

My family consisted of my parents, four brothers and one sister. My maternal and paternal grandparents as well as other relatives lived nearby and were intimately involved in my formative experiences. I learned my childhood lessons well....

My mother and my four grandparents gave me my earliest foundations in education. The grandparents followed our progress in school very carefully and insisted we do well in our studies. My paternal grandfather—Jake Simmons—had to leave school as a child to work and support his parents. He could neither read nor write. Of the many lessons he taught me one of the most important was the significance of education and learning. He monitored our school attendance very carefully and one of his mandates was that we should never skip school—absence was never acceptable for any reason.

My mother insisted that we do our homework every day as well as attend all our school sessions. Like Grandpa Jake she would never let us skip school. In the early 1960s it was common for the lights to go out for several days at a time in the winter months. Electricity was still a new phenomenon with us, the island was almost all black, and when a tree fell on the lines the light company did not rush to make repairs. My mother never allowed any excuses for failure to complete our homework. She had lots of kerosene lamps in the house and it was not unusual for us to sit under lamps at night reading and spelling. Momma was serious about us doing our homework under the lamps.

I learned valuable survival skills from my father and grandfathers. While they stressed education they also taught me the skills that would allow one to take care of your family. By the time I was an adolescent I was skilled in hunting, trapping, fishing, crabbing and casting the shrimp net. I was equally comfortable in my little boat (bateau) in the waters around the island, and in the woods hunting deer and raccoons, or trapping otters and minks. I also learned how to maintain and operate a still making moonshine whiskey that we sold to selected customers. My daddy's largest still "ran" about 25 barrels and produced up to 250 gallons of moonshine.

I did excellent work in school as well as in the natural environment. I even learned how to combine them to ensure greater success. Reading and spelling were dominant in school. Every day we had to learn ten new words and spell them correctly. I always wrote my ten words on a paper and took them hunting with me. In the woods I would sing out the words to myself until they came naturally when I had to stand up in class and spell.

Our school went from grades 1-8 and only had two rooms. We had two teachers—one of them, Miss Frances Jones, was a Daufuskie Island native. She was very demanding and strict but I quickly learned that once a student established he was serious and made high grades the teacher gave that person breaks. My parents and grandparents pushed me and I made very high grades—in fact I was the top student in every class. I was particularly good in arithmetic. I had come across something on Benjamin Banneker and felt I wanted to go further in the study of mathematics.

In my last year of elementary school—8th grade—we had our first male teacher. He was also white. He was an energetic and enthusiastic person as well as an excellent teacher. I continued to follow the guidance of my parents and grandparents and graduated as the valedictorian of the class.

To me, life was good. I had excelled in the classroom and I was doing well outside of school. In the early years sitting in school around the wood stove in the winter months, tightly bunched into a circle, we felt warmth and a sense of family among ourselves even though it was extremely cold outside. Many of us were from the same family or related in one way or another. The island was a safe nest for me; I did not feel like I was poor or educationally deprived.

On Daufuskie Island when you finish the 8th grade you had to move to the mainland to attend high school. You had to leave your family and live with strangers in a new environment. My safe nest was destroyed and high school was a very different life. I went to three different high schools and my world was more uncertain, however my educational aspirations remained high in spite of low performance. I graduated from high school toward the bottom of the class. (Blake & Simmons, 2008, p. 3-5)

Pollitzer (2005), in speaking of the changes that occurred in Gullah country between the time of emancipation and the turn of the twentieth century, states:

The succeeding fifty years have seen even more rapid change…. The public schools discourage the use of the Gullah language and teach the goals of a society geared to production and progress rather than the ideal of equilibrium that has long characterized this culture unconcerned with time. (p. 4)

This may have been the beginning of the disconnection between Gullah communities and the schools that are to serve them. How their unrelenting struggles for education through enslavement and even soon after emancipation might have derailed because of this disconnect, may be more easily understood by looking at the nature of modern day Gullah communities.

THE GULLAH/GEECHEE NARRATIVE: GAINING TRUST

Community: The Gullah/Geechee narrative is not easily accessible to outsiders. Individuals are very private about their lives, and communities are closed. One of the most difficult challenges to those seeking to learn and understand the narratives of the people is gaining their trust—an indomitable task. The third author—Dr. Emily L. Moore—during her tenure at Iowa State University became a highly respected and trusted scholar of Gullah culture through her interest in HIV/AIDS prevention in vulnerable populations. Her first-person account of gaining the confidence and trust of Gullah communities is instructive for those seeking to bridge the digital divide in similar communities.

A major interest in Gullah communities was spurred by the dramatic increase in HIV/AIDS among African American women, particularly in the childbearing ages. This scourge has impacted urban and rural African Americans throughout the United States. The research shows that the pandemic is fueled by: denial about the disease; stigma and fear of its victims; silence; poverty; and profound distrust of intervention efforts by the government. To gain insight and understanding, I concentrated on rural communities. Since 1996 I have been actively conducting research on HIV/AIDS prevention in Gullah communities in coastal South Carolina.

The first challenge was gaining the trust of the community. I stayed in the home of a community member and actively participated in all programs, including church services, family gatherings, and community meetings. Clearly I was an outsider. However, I did not attend these activities as a researcher; but, as a welcomed visitor. The common saying is that "trust is earned". I realized that I had gained trust in the community when I received a call in my office at Iowa State University asking why I missed the community meeting.

It was at this point that I was able to begin my formal research.

I spent a lot of time sitting on front porches, drinking sweet tea and listening. This interaction and communication process brought me to the realization that effective communication was the foundation of trust. In order to reach my interests, I had to be engaged in their interests. If they wanted to discuss diabetes and hypertension, that is what we discussed.

In retrospect I realized that some of the key issues emerged from community isolation. Even when located physically close to a road, a town, or an attraction, the community was isolated with a sense of invisibility. This increased the sense of a divide between the Gullah peoples and the larger society. I learned there were values rooted in their history. The primary value was a focus on the family. The families were close knit and rarely accepted outsiders. However, in my case, they understood my respect for the community and their values. A second value relates to religion and spirituality. In the churches and in general communication the sense of faith is profound. Outsiders who do not reflect an appreciation of their faith will not be accepted.

There is also suspicion and mistrust. The history of the people is fraught with exploitation of their land and labor causing a sense of mistrust for those outside the community. After more than a decade of interaction, communication and building trust, I am able to engage communities in the discussion of very sensitive issues. My experience speaks to the challenge of developing culturally sensitive and competent research among Gullah communities in relation to digital equity.[4]

Land and Property Ownership

We began this chapter by referring to the gentrification of Gullah communities through the encroachment of land development companies. Before making the connections to technology and digital equity in this information age, it seems appropriate to look briefly at the historical role that land has played in U. S. society. Since the colonization of the U. S., land ownership has been a key factor in determining the power elite. Those who came with charters from European nobility and/or the economic means to secure and cultivate parcels of land, wielded much power and influence (Pollitzer, 2005; Spring, 2004). This was particularly true in slaveholding states and for the plantation owners in those regions. The Sea Islands of South Carolina and Georgia offer examples of how such land ownership, power, and influence played out during the colonial period. Land use on these plantations and the technological skills needed to ensure successful enterprises determined the course of the enslaved people we now refer to as Gullahs and Geechees. Although these were non computer-related skills, they were nevertheless supportive of the technologies of the day.

The topography of the Sea Islands was very similar to that of coastal areas of western African countries including those lying along what is referred to as the Rice Coast or Grain Coast. Slave trading plantation owners preferred Africans from that area of the continent because of the skills they possessed in rice production. It should be pointed out that statistics show that many Africans came to South Carolina and Georgia first via enslavement in the West Indies (Pollitzer, 2005, p. 47). Creel (1988) states, "Their value was a long familiarity with planting and cultivation of rice and indigo, the quality of which was said to have surpassed that grown in Carolina" (p. 36). Keep in mind that these technological skills concerning rice production were 'owned' by the Africans prior to their enslavement. However, because of the nature of the system of slavery and its totality, which grew out of the Black transatlantic slave trade, these skills became the property of the plantation owner much like the land and the African. In essence, the slave owner also had ownership of the rice producing technologies and technological skills

by virtue of owning the African as property. All property rights, including technologies and technological skills, accrued to the property owner. It is also important to note that those technologies and technological skills possessed by Africans prior to enslavement, which did not benefit the White slave owners, did not survive to the same degree into later centuries (Pollitzer, 2005, p. 12).

Critical race theorists often speak of property rights when referring to the plight of African Americans in the past and in modern day U.S. society (i.e. Tate, 1996; Harris, 1995). Ladson Billings and Tate (1995) contend that property rights take precedence over other individual or human rights and further state:

The grand narrative of U.S. history is replete with tensions and struggles over property—in its various forms. From the removal of Indians (and later Japanese Americans) from the land, to military conquest of the Mexican, to the construction of Africans as property, the ability to define, possess, and own property has been a central feature of power in America.... However, the contradiction of a reified symbolic individual juxtaposed [to] the reality of "real estate" means that emphasis on the centrality of property can be disguised. Thus, we talk about the importance of the individual, individual rights, and civil rights while social benefits accrue largely to property owners. (p. 53)

While property is a construct typically associated with ownership of tangibles, including land and real estate, and property rights are tied to ownership of those tangibles, critical race theorist, Cheryl Harris (1995) introduced the notion of the property and property rights being associated with the intangible construct of Whiteness. This notion suggests that being White and claiming Whiteness as a social identity brings with it certain benefits as a result of owning that particular identity by membership in that particular socially constructed racial category. Thus, Whiteness, in and of itself, is property and those owning such property have rights that include the right to enjoy one's property and "the absolute right to exclude" (p. 282-283). Harris further explains, "The possessors of whiteness were granted the legal right to exclude others from the privileges inhering in whiteness; whiteness became an exclusive club whose membership was closely and grudgingly guarded" (p. 283).

Property rights in Whiteness had severe implications for Black enslaved Africans and their descendents in terms of how these rights impacted and superseded 'real' property rights tied to land owned by Gullahs and their descendents after emancipation and into the 21st century. This impact takes the form of gentrification in modern day Gullah communities.

Gentrification and Gullah Land

Pollitzer (2005) states, "The Gullah people face a crisis today as the demand for their land and marsh encroaches upon home and farm and threatens their way of life. They are ill-equipped to meet the challenges of a modern era" (p. 4). This demand for Gullah land that Pollitzer speaks of is in the form of gentrification.

The Housing Assistance Council (HAC) in its 2005 report defines gentrification as "the process by which higher-income households displace lower-income residents of a community, changing the essential character and flavor of that community" (p. 1). This report is reflective of case studies of various rural areas including Beaufort, SC, which, in this instance, contains many Gullah communities and where the gentrification is caused by in-migration of retirees. The variables used by HAC in each case study were population growth, percent of new residents in the past five years, median income growth, growth in new housing units, housing value growth, and housing cost burden. The HAC (2005) report states, "Increases in all or many of these categories within a county were taken to be indicative of gentrification, whereby new residents with higher incomes move into a

community and purchase homes at increasing prices, forcing low-income residents to face higher costs" (p. 6). The result is that poor families are forced off their land and/or out of their homes and, in most cases, there is no longer any affordable housing available for those who were displaced. But beyond these economic concerns related to affordability of housing for low-income families are concerns for erasure of local cultural values and traditions brought on by gentrification. The HAC reports that gentrified rural communities studied "have also lost long-held traditions and elements of local culture in exchange for a homogenous, suburbanized new identity" (p. 3).

The description of Gullah communities provided in the above sections, gives some insight on how "essential character and flavor" of those communities might be affected by gentrification. Concerning rural communities in general, the HAC (2005) report offers,

In the literature of rural sociology, it has long been maintained that rural or frontier areas have distinct cultures and sets of values that revolve around self-reliance, conservatism, a distrust of outsiders, the centrality of churches, a strong work ethic, and social structures that emphasize the family (particularly extended family). (p. 5)

The narrative offered by Moore in this chapter verifies that this distrust of outsiders is particularly present in Gullah communities. Jones-Jackson (1987), also an African American female, concurs and explains how she encountered challenges in her efforts to carry out ethnographic research in Gullah communities. She observed, "I soon realized that I had the status of 'outsider' and no outsiders, black or white, are welcomed warmly into Sea Island activities or viewed without suspicion until they have spent considerable time among the people" (p. 2). Pollitzer (2005) adds, "Certain cultural traits are common, with variations, throughout Africa. A distinctive feature in most of the continent, the extended family has been traditionally the primary economic as well as social unit" (p. 29). Concerning the Gullah specifically, most of whom have their origins in West Africa, Pollitzer goes on to say,

Throughout the sea islands the extended family has been the all-important but flexible social unit, tying together a network of relatives in different homes. Related people in clusters of houses close together promote a web of kinship, an economic and social unit, often buying land and attending social functions together. (p. 8)

All of these descriptors are important in examining how gentrification affects rural Gullah communities.

This practice of buying land together has particular implications. The HAC (2005) report further explains:

The Gullahs have lived on the land for hundreds of years and passed it down through successive generations. This type of property is referred to as "heirs' property." All members of a family own parcels of land as "tenants in common" and when a family member dies, the land is divided among the next generation. After several generations, one plot of land can be owned by literally dozens of people without a clear legal definition of who is actually responsible for the land.... Also, most of the native islanders failed to write wills or use other formal means to pass on their land. This now occasionally prevents owners from being able to develop their property and at other times the owners may lose their land in tax sales if the land has outstanding tax debt. (p. 37)

So as developers come into the area eager to purchase properties that will yield high profits after attracting a high income populace, they find they are hindered somewhat by this 'heirs property' tradition. Smith (1991) notes, "The land ownership issue was ambiguous because of the norm of family inheritance: everyone received a small

portion of land whether they farmed or not. This represented a symbolic inclusion in the extended family" (p. 294). If developers are determined, however, they use this long-held tradition to their advantage and to the disadvantage of the Gullah families still practicing their traditional ways. Those Gullah who do want to give up their title to the land, particularly those who have moved away and will be unaffected by the loss of the land, will pressure others to follow suit. In some cases, if one person relinquishes through sale, his/her rights to a portion of the group-owned property, the sale of the entire tract of land will be forced in order to divide up the profits to the remaining owners. Some speculate that illiteracy and the lack of knowledge concerning business transactions make the historically uneducated Gullah very vulnerable to profit seeking developers (Smith, 1991). Jones-Jackson (1987) contends, "The population that remains behind on the islands is having to adjust to the incoming population of resort residents, as well as resort development for tourism. Developers have not been responsive to local needs" (p. 165). She further explains:

Land is being taken out of production. Acreage once used to grow food is now restricted and controlled by high taxes and county regulations implemented in response to development. Small livestock farmers located next to tracts under new ownership may be pressured to sell out if their activities seem incompatible with the projects of the developers. Lands are lost, too, when people unused to record keeping lose tax receipts and forget to pay yearly bills, or when land descends to heirs who have moved away and have no interest in it. (p. 166)

The HAC (2005) report provides further insight:

African Americans in Beaufort County are part of the regional Gullah culture, which has held fast to traditions from their African past. As the Gullah are displaced to make way for new residents, the vibrancy of their culture is threatened. The practice of "heirs property" is an informal and vernacular form of land tenure that has served the Gullahs' needs since the end of the Civil War. The pressures from land developers and tax assessors may lead to the end of this practice. (p. 46)

Thus, it is evident that this form of gentrification caused by in-migration of high-income tourists and retirees affect and threaten Gullah families and communities in various ways. Not only are they displaced from their land and homes and, like other victims of gentrification, find that they can longer afford to even live in the area they and their ancestors have occupied for centuries, but their very way of life has been threatened. Smith (1991) states, "…the land upon which fishing and subsistence farming relied was being forced out of traditional land-use patterns and into retirement villages and seasonal plantation. Many traditional cultures that have lost control over their land base have been swept away by modernization" (p. 294). In examining these dynamics it becomes apparent that land use is tied to culture. By the same token a community that has a land-based culture also has an economic base tied to land. Therefore, it is very difficult, if not impossible, to separate land use, economics, and culture in a traditional Gullah community. Gentrification that affects land use also affects the economic base and culture of the community.

Grabbatin (2008) exposes the connection between gentrification and the loss of rural land that produces the grasses used for the time-honored cultural tradition of sweetgrass basketry. Sweetgrass basketry making, along with its roadside stands, is central to Gullah culture and its artistry and craftsmanship, which have been passed down for generations. Redefining land use that jeopardizes the resources needed to produce these cultural artifacts as well as altering roadway infrastructures that also alter the ability to market these artifacts at familiar roadside stands, put at

risk the possibility that this craft will continue to pass down through generations and will be central to Gullah culture and its economic base in future generations (Grabbatin (2008).

On the other hand, Thomas (1980) examines the impact of tourism, resulting from gentrification of Gullah land and communities, on the employment of Black Gullah. She notes that as "more and more whites began to move into the islands…. more and more blacks have turned away from the agricultural sector and begun to take other jobs" (p. 2). As pointed out earlier (Hargrove, 2009; Pollitzer, 2005; Smith, 1991), farming and agriculture are central to the cultural and economic base, so changes in employment patterns impact both.

The connection between culture and land use is also exemplified by how the cultural practice of passing ownership of property down to families and not individuals was undermined and destroyed by gentrification. Extended families and households find they are no longer living in close proximity as closed communities are scattered and their non computer related social communication networks are jeopardized. The historically proud and self-reliant Gullah, who continue to practice traditional ways and speak the Gullah language passed down from slavery, themselves become objects of either disdain, pity, or at best curiosity as they are forced to mingle and live among these predominately White residents of European ancestry. Their long history of suffering with formidable inequities has determined their paths in many ways and these inequities have followed them into the twenty-first century—inequities that made possible this gentrification and, at the same time, this gentrification has exacerbated the inequities. Moreover, the presence of new high-income residents in previously low-income areas has implications for digital equity among the Gullah people and their new neighbors. Again, the reverse may also be true in that isolation and the lack of educational opportunities in the past, as outlined in previous sections, have determined the Gullah's lack of participation in information age communication technologies. This lack of engagement actually facilitated the process of gentrification that began several decades prior to this digital age.

EXPLORATIONS OF DIGITAL EQUITY FOR GULLAH PEOPLE

As the U.S. moved into the age of the industrial complex and later into the information age of the twentieth and twenty-first centuries, the emancipated Africans and their descendents in Gullah communities were typically left behind in terms of ownership and even use of new technologies. Though many Gullah families now own the land on which they or their ancestors were once enslaved, this ownership is vulnerable. It can be argued that this vulnerability, which began well before the onset of the information age with the encroachments of land developers concomitant with the influx of wealthier new residents (Hargrove, 2009; Smith, 1991; Thomas, 1980), nevertheless, now lies in this technology gap. One of the most prevalent technologies of this age is the computer, which is central to the knowledge industry that has emerged from it. Likewise, information communication technologies (ICTs) are computer-based and allow for the dissemination of information and knowledge via various electronic networking capabilities (e.g. Internet, Web, etc.). This type of communication is contrary to Gullah culture. As we discussed earlier, the isolation of Gullah ancestors during slavery, and years after, makes most communities closed to outsiders. This isolation and hostility to outsiders is exacerbated by the desire to maintain cultural values, traditions, and ways of life. In addition, these electronic networking tools are threats to the culture. These attitudes are similar to those held in some developing countries. Hellsten (2007) explains, "there has been some local resistance to modernization in a sense that it is seen to be a sign of further cultural

colonization of developing countries" (p. 19). Other aspects of African-derived cultures, such as self reliance mentioned earlier (HAC, 2005), may also contribute to this reluctance to engage with new technologies. Hellsten (2007) further notes,

for many African communities, development does not often mean merely the acquisition of sophisticated technology and material benefits; it also means searching for the intellectual and social conditions that will permit internal, positive freedom for human beings in the form of self-realization. In search for self-development many African peoples simply do not care about new technology. When development is seen as self-development, learning about mechanical and technological details loses its importance. (p. 18)

Technology and Knowledge

However, scholars such as Fallis (2007) claim that no matter how low the income, no one can really afford the luxury of not being connected through computer-related informational networks. He claims, "access to information technology is a necessity in modern life" (p. 29) and further points out that "it has even been suggested that access to information technology is now a basic human right" (p. 29). To Fallis, "the principal value of access to information technology… is that it leads to *knowledge*" [emphasis in the original] (p. 29). Consequently, it is this access to information and knowledge that is seen as a major factor in measuring the extent of the digital divide. As discussed earlier, digital divides can be characterized as differential ownership, access, and uses of computer-related resources. This differential ownership, access, and uses of ICTs by Gullahs, compared to dominant White communities and counterparts, and thereby differential access to important information and knowledge, has great potential for disadvantaging Gullah communities in their efforts to maintain their lands and culture. Denied access to knowledge concerning property rights, legitimate claims to property, market values, and accepted business procedures, practices, and processes is brought about in large part by prevention from and resistance to engagement in informational communication networks. Consequently, those who are engaged in such networks, such as land developers, not only have access to this knowledge but also benefit from a potential increase in the influence and social capital that is derived from ever expanding social connections. Hacker, Mason, and Morgan (2007) report, "research shows that those who combine online communication with offline social interaction expand their social networks and increase their social capital" (p. 121). Thus knowledge, power, and influence are more likely to rest in the hands of those who want to facilitate gentrification than in the hands of those who want to resist it.

We have discussed here how the digital divide, particularly in terms of access to information and knowledge via ICTs, has facilitated gentrification in Gullah communities but we should also point out how this gentrification, in turn, widens this same divide. It is widely accepted by scholars studying this phenomenon that those who have more material resources have greater access to and significantly different uses of ICTs. Gentrification by definition brings with it an influx of high-income families into typically low-income communities (HAC, 2005). Before this influx, the divide existed between the concentrated, coherent Gullah/Geechee communities of the Sea Islands of South Carolina and the dominant 'outsiders' in U.S. society. As these 'outsiders' become 'neighbors' inhabiting their expansive houses and gated and/or fenced resorts and housing developments, they bring with them greater ownership and access to advancing technologies and greater uses of and engagement in ICT networks. Thus the digital divide widens between the Gullahs and these 'others/outsiders'–giving new meaning to Collins' (2009) term, 'outsider within' (p. 13-15)[5].

FUTURE RESEARCH DIRECTIONS

In this chapter, we have demonstrated how the unique history and culture of this population of African Americans, known as Gullah, have shaped how digital equity has played out for them and continues to affect their lives today. Problems related to technology equity for Gullah communities of rural South Carolina are decidedly different from those experienced by African Americans in our densely populated urban/industrial cities. A result of knowing these varying histories and the digital consequences that emerged leads us to look for differing solutions. In this chapter, we demonstrated how lack of access to modern digital technologies and networks have jeopardized Gullah way of life, yet other scholars may argue that these same value-laden, assimilation-rendering technologies have potential to destroy cultures. Hellsten (2007) weighs in by claiming that helping poor people "get connected with the outside world and become involved in matters concerning their own lives is not an attempt to rob cultural traditions; it may be the only way to maintain those traditions" (p. 21). She goes on to say, "Keeping people disconnected and ignorant may respect cultural difference, but shows moral indifference. It clearly blocks people from using their full potential, in whatever cultural, social, and material environment they live in" (p. 21). It seems Fallis (2007) would agree in terms of who should be knowledgeable. With knowledge at the core of the discussion, Fallis uses epistemic value theory to discuss whether one system of knowledge distribution is superior to another, which he refers to as 'epistemic betterment'. Within this discussion he claims,

it is clearly epistemically better for a person to have more knowledge rather than less. In other words, it is epistemically better to have knowledge on a particular topic than to be ignorant (or in error) on the topic. In addition, it follows that it is epistemically better for more people to have knowledge rather than fewer. (p. 32)

While Hellsten's and Fallis' statements certainly ring true in terms of how Gullah communities have been affected by their lack of technological engagement and subsequent lack of knowledge, we are very cognizant of the fact that thoughtless, culturally insensitive and even arbitrary promotion of digital inclusion projects in Gullah communities could also have negative and dire unintended consequences in terms of their culture and way of life.

While Smith (1991) speaks of grassroots resistance and the 'Gullah movement' to protect property and culture (p. 288-289), which occurred prior to the information age, he also notes that this social movement occurred after gentrification of Gullah communities was well underway. He states, "In an effort to save the land base, local leaders initiated public education drives to formalize land ownership, pay taxes, and create surveyed boundaries" (p. 294). Smith also points out that although the leadership came from inside the Gullah community it consisted primarily of middle class Blacks of Gullah descent, who had moved away but were returning as retirees or for other reasons (p. 295-296). In other words, these educated leaders were not reflective of those Gullah who had lived their lives in traditional closed communities while upholding and maintaining cultural values and practices. However, by enlisting the aid of local advocates, both Black and White, they were somewhat successful in promoting education around Gullah culture and issues and promoting policies to protect Gullah land (p. 296). However, we ultimately believe more research is necessary to learn how we can support the entrance of traditional Gullah communities into the digital age in ways *the communities* deem beneficial and without jeopardizing their culture, ways of knowing, or ways of being.

CONCLUSION

Hellsten (2007) contends, "In most cases and most areas people have not chosen their own isolation. The poor simply lack the options to live in any other way" (p. 21). As authors of this chapter, we certainly found that to be true during the enslavement of the ancestors of the Gullah, whose descendents are presently residing in the Sea Islands off the coast of South Carolina. We described this history of forced isolation and how it affected Gullah practices, customs, traditions, and ways of communicating within these closed communities, as they developed from slavery and thrived through emancipation into 21st century U.S. society. This isolation allowed for the preservation of Gullah culture and language and also gave rise to suspicions of those outside Gullah communities. Others might argue that since emancipation, contrary to Hellsten's statement, Gullah have chosen their own isolation. Yet we dispute this view in that, given their history of oppression and their lack of opportunities, there have been few 'options to live any other way'. Moreover, using narratives and critical race theory perspectives, we showcased the tenacious efforts of Gullah to resist their oppressors in their quest for education, while yet embracing and maintaining their distinct Gullah culture and language, all made possible through this isolation. In the end, we also pointed out how this isolation, whether forced or 'chosen', has negative implications for Gullah in a modern technological U. S. society.

Despite the Gullahs' tenacious quest for education after emancipation and even into the industrial age, they were rendered even further behind educationally, politically, and socially with the onset of the information age. We argue that the same isolation that allowed Gullah to maintain their culture is the same isolation, and value of it, that causes Gullah to resist engagement with the tools of this information age. The information communication technologies (ICTs) of the 20th and 21st centuries defy isolation, while allowing those who do engage with ICT networks to have access to important information and knowledge. Again, using a CRT perspective to interrogate constructs of property and property rights, we demonstrated how this lack of engagement with information networks threaten Gullah even more in terms of loss of land and culture through gentrification.

While Dixon (1997) admonishes Blacks to take control and own these computer networks, efforts in that direction have been minimal or absent for Gullahs who are maintaining their traditional culture. Dixon explains that traditional and digital media outlets typically tell communities what they need and what is good for them, even when the contrary is true. As new land development information technologies arise, those who own those technologies and have access to information can use their control, influence, and power to gain ownership of now valuable Gullah lands. Gullah customs and culture, which are intricately tied to the land, are also in jeopardy. As Jones-Jackson (1987) reports in her book, *When Roots Die: Endangered Traditions on the Sea Island*: "As the islander often put it, 'Everything change up now.' One of the major areas of change in recent years has been in the use and ownership of land in the islands" (p. 165).

REFERENCES

Bell, D. (1987). *And we are not saved: The elusive quest for racial justice.* New York, NY: BasicBooks.

Bell, D. (1992). *Faces at the bottom of the well: The permanence of racism.* New York, NY: BasicBooks.

Billingsley, A. (2007). *Yearning to breathe free: Robert Smalls of South Carolina and his families.* Columbia: University of South Carolina Press.

Blake, J. H. (2007, May). *We come from a distance: Education and Gullah culture.* Paper presented at The Original Gullah Festival, Beaufort, SC.

Blake, J. H., & Simmons, E. R. (2008). A Daufuskie Island lad in an academic community: An extraordinary journey of personal transformation. *Journal of College and Character, 10*(1), 1–14. doi:10.2202/1940-1639.1061

Burn, B. (2003). *An island named Daufuskie.* Spartanburg, SC: The Reprint Publishers.

Clark, S. P. (1986). *Ready from within: Septima Clark and the civil rights movement—A first person narrative.* Navarro, CA: Wild Trees Press.

Clark, S. P., & Blythe, L. (1962). *Echo in my soul.* New York, NY: E. P. Dutton.

Collins, P. H. (2009). *Black feminist thought: Knowledge, consciousness, and the politics of empowerment.* New York, NY: Routledge.

Creel, M. W. (1988). *A peculiar people: Slave religion and community-culture among the Gullahs.* New York, NY: New York University Press.

Davis, A. Y. (1983). *Women, race & class.* New, NY: Vintage Books.

Delgado, R. (Ed.). (1995). *Critical race theory: The cutting edge.* Philadelphia, PA: Temple University Press.

Dixon, B. R. (1997). Toting technology: Taking it to the streets. In Gordon, L. (Ed.), *Existence in Black: An anthology of Black existential philosophy* (pp. 135–147). New York, NY: Routledge.

Fallis, D. (2007). Epistemic value theory and the digital divide. In Rooksby, E., & Weckert, J. (Eds.), *Information Technology and Social Justice* (pp. 29–46). Hershey, PA: Information Science Publishing.

Forten, C. (1864). Life on the Sea Islands. *The Atlantic Monthly, 13*(79), Issue 79, 587-596.

Grabbatin, B. C. (2008). *Sweetgrass basketry: The political ecology of an African American art in the South Carolina Lowcountry.* Unpublished master's thesis, College of Charleston, South Carolina.

Hacker, K. L., Mason, S. M., & Morgan, E. L. (2007). Digital disempowerment. In Rooksby, E., & Weckert, J. (Eds.), *Information technology and social justice* (pp. 112–147). Hershey, PA: Information Science Publishing.

Hargrove, M. D. (2009). Mapping the "social field of Whiteness": White racism as habitus in the city where history lives. *Transforming Anthropology, 17*(2), 93–104. doi:10.1111/j.1548-7466.2009.01048.x

Harris, C. (1995). Whiteness a property. In Crenshaw, K., Gotanda, N., Peller, G., & Thomas, K. (Eds.), *Critical race theory: The key writings that formed the movement* (pp. 276–291). New York, NY: The New Press.

Hellsten, S. K. (2007). From information society to global village of wisdom? The role of ICT in realizing social justice in the developing world. In Rooksby, E., & Weckert, J. (Eds.), *Information Technology and Social Justice* (pp. 1–28). Hershey, PA: Information Science Publishing.

Housing Authority Council. (2005). *They paved paradise… Gentrification in rural areas*. Washington, DC: Author.

Jones-Jackson, P. (1987). *When roots die: Endangered traditions on the Sea Islands*. Athens, GA: The University of Georgia Press.

Jordan, L. W., & Stringfellow, E. H. (2000). *A Place Called St. John's*. Spartanburg, SC: The Reprint Publishers.

Ladson-Billings, G., & Tate, W. (1995). Toward a critical race theory of education. *Teachers College Record, 97*(1), 47–68.

Lerner, G. (Ed.). (1992). *Black women in White America: A documentary history*. New York, NY: Vintage Books.

Miller, E. A. Jr. (2008). *Gullah Statesman: Robert Smalls from Slavery to Congress, 1839-1915*. Columbia: University of South Carolina Press.

National Park Service. (2005). *Low Country Gullah culture special resource study and final environmental impact statement*. Atlanta, GA: NPS Southeast Regional Office.

Pollitzer, W. S. (2005). *The Gullah people and their African heritage*. Athens, GA: The University of Georgia.

Smith, J. P. (1991). Cultural preservation of the Sea Island Gullah: A Black social movement in the post-civil rights era. *Rural Sociology, 56*(2), 284–298. doi:10.1111/j.1549-0831.1991.tb00437.x

South Carolina. (1868). *Proceedings of the Constitutional Convention of South Carolina*. [Held at Charleston, SC, beginning January 14th and ending March 17th, 1868, reprinted by Arno Press and the New York Times, 1968.]

Spring, J. (2004). *Deculturalization and the struggle for equality: A brief history of the education of dominated cultures in the United States*. New York, NY: McGraw Hill.

Tate, W. (1996). Critical race theory and education: History, theory, and implications. *Review of Research in Education, 22*, 195–247. doi:10.3102/0091732X022001195

Taylor, K. A. (2005). Mary S. Peake and Charlotte L. Forten: Black teachers during the Civil War and Reconstruction. *The Journal of Negro Education, 74*(2), 124–137.

Thomas, J. M. (1980). The impact of corporate tourism on Gullah Blacks: Notes on issues of employment. *Phylon, 41*(1), 1–11. doi:10.2307/274663

Toepke, A., & Serrano, A. (1998). *The language you cry in* [videorecording]. San Francisco, CA: California Newsreel.

van Dijk, J. A. (2005). *The deepening divide: Inequality in the information society*. Thousand Oaks, CA: Sage Publications, Inc.

Williams, H. A. (2005). *Self-taught: African American education in slavery and freedom*. Chapel Hill, NC: The University of North Carolina Press.

ADDITIONAL READING

Branch, T. (1988). *Parting the waters: America in the King years: 1954-63*. New York: Simon and Schuster.

Carawan, G., & Carawan, C. (1966). *Ain't you got a right to the tree of life?* New York, NY: Simon and Schuster.

Coclanis, P. A. (1989). *Economic life & death in the South Carolina Low Country: 1670-1920*. New York: Oxford University Press.

Cross, W. (2008). *Gullah culture in America*. Westport, CN: Praeger Publishers.

Goodwine, M. L. (1995). *Gullah/Geechee: The Survival of Africa's seed in the winds of the diaspora*. Brooklyn, NY: Kinship Publicatons.

Goodwine, M. L. (1997). *Gawd dun smile pun we: Beaufort Isles*. Brooklyn, NY: Kinship Publications.

Goodwine, M. L. & Clarity Press Gullah Project. (Eds.). (1998). *The legacy of Ibo Landing: Gullah roots of African American culture*. Atlanta, GA: Clarity Press.

Goodwine, M. L. (1999). *Frum wi soul tuh de soil cotton, rice, and indigo*. Brooklyn, NY: Kinship Publications.

Guthrie, P. (1996). *Catching sense: African American communities on a South Carolina sea island*. Westport, CN: Bergin & Garvey.

Haynie, C. W. (2007). *Images of America: John's Island*. Charleston, SC: Arcadia Publishing.

Joyner, C. (1984). *Down by the riverside: A South Carolina slave community*. Urbana, IL: University of Illinois Press.

Kinlaw-Ross, E. (1996). *Dat Gullah and other Geechee traditons*. Atlanta, GA: Crick Edge Productions.

Leland, E. (1992). *The vanishing coast*. Salem, NC: John F. Blair.

Moutoussamy-Ashe, J. (2007). *Daufuskie Island*. Columbia, SC: The University of South Carolina Press.

Opala, J. A. (1987). *The Gullah: Rice, slavery, and the Sierra Leone-American connection*. Freetown, Sierra Leone: United States Information Service.

Robinson, C. S., & Dortch, W. R. (1985). *The Blacks in these Sea Islands: Then and now*. New York, NY: Vantage Press.

Schwalm, L. A. (1997). *A hard fight for we: Women's transition from slavery to freedom in South Carolina*. Urbana, IL: University of Illinois Press.

Simms, L. A. (1992). *Profiles of African American females in the Low Country of South Carolina*. Charleston, SC: College of Charleston.

Twining, M. A., & Baird, K. E. (1990). *Sea Island roots: African presence in the Carolinas and Georgia*. Trenton, NJ: Africa World Press.

ENDNOTES

[1] Located in Charleston County, Johns Island is the second largest island on the east coast of the United States. Horseshoe shaped it is ten miles across and thirty-two miles from point to point. See: Laylon W. Jordan and Elizabeth H. Stringfellow, *A Place Called St. John's*, Spartanburg, SC: The Reprint Publishers, 2000, pp. 4, 402.

[2] The second author—Blake—met Clark in 1967 while conducting an evaluation of the Citizenship Education Schools for the Emil Schwarzhaupt Foundation. He made frequent visits to her home in Charleston

and conducted extensive interviews with her. In addition he arranged for her to spend two periods of time at the University of California, Santa Cruz where she was a visiting resident. During this time she gave him several lengthy hand-written descriptions of her work throughout the South with Citizenship Education Schools and later with the Southern Christian Leadership Conference. See also: Taylor Branch, *Parting the Waters: America in the King Years 1954-63*, New York: Simon and Schuster, 1988, pp. 575-77.

[3] A barrier island in Beaufort County, SC, Daufuskie Island is south of Hilton Head Island and north of Savannah, GA. It is approximately 2.5 miles wide and 5 miles long. See: Billie Burn, *A Island Named Daufuskie*, Spartanburg, SC: The Reprint Publishers, 2003, p. xxi.

[4] Moore, E. L., Presentation, Society for Information Technology and Teacher Education International Conference, March 2009. See also: Moore, E. (2004) HIV/AIDS Prevention in a Rural African American Community: The Impact of Grassroots Women Leaders. [Extended Abstract]. Medimond S.r.l., Bologna, Italy, E710S4371, 207-211.

[5] In Patricia Hill Collins' (2009) work, *Black feminist thought: Knowledge, consciousness, and the politics of empowerment*, she discusses the positionality of Black domestic workers who care for White families. These workers have intimate knowledge of White family dynamics and, to some degree, may be considered a part of the family. However, Collins contends that these workers are very cognizant of the fact that they are 'outsiders' who can never be included in the families in the real sense, even though they do have some 'insider' status. Thus, she coined the term 'outsider within' to describe the status of these Black domestic workers. (pp. 13-15) In the case of the closed and isolated Gullah communities, gentrification has brought White and wealthy outsiders in as neighbors. However, these new residents can never be true neighbors even though they have powerful insider status. So the authors here contend that the White residents are the 'outsiders within' gentrified Gullah communities, while the power, in these cases, remains with these outsiders.

Chapter 10
Governing Digital Divides:
Power Structures and ICT Strategies in a Global Perspective

Francesco Amoretti
University of Salerno, Italy

Fortunato Musella
University of Naples "Federico II", Italy

ABSTRACT

A great part of the rhetoric accompanying the rapid diffusion of information and communication technologies (ICTs) in Western societies in recent decades has put the spotlight on their potential for generating economic growth and development in the socio-political arena. Yet mechanisms that generate disparities among citizens do not go away with the advent of electronic citizenship, as asymmetric access to economic and political resources limit access to new technologies. This contribution will be divided in three sections. In the first part, the concept of "digital divide" will be analysed by considering its first formulation in the US political debate during the Nineties, as well as the more recent efforts to consider the multidimensional nature of such category. In the second section significant quantitative measure of digital disparities between countries will be provided. Finally, it will show how developing countries adopting proprietary softwares are becoming dependent on the power of providers of ICT goods and services, which are mainly concentrated in the United States.

INTRODUCTION. UNIVERSAL PROMISES, DIGITAL INEQUALITIES

A great part of the rhetoric accompanying the rapid diffusion of information and communication technologies (ICTs) in Western societies in recent

decades has put the spotlight on their potential for generating economic growth and development in the socio-political arena. New forms of media are regarded to perform a similar role in countries with very different traditions and history. More important, new digital infrastructures are supposed to be capable of accelerating developmental prospects for the poorest regions of the world, as

DOI: 10.4018/978-1-61520-793-0.ch010

Copyright © 2011, IGI Global. Copying or distributing in print or electronic forms without written permission of IGI Global is prohibited.

they facilitate "all types of economies (developed, developing and in transition) to bring benefits of the emerging global information society to the largest possible part of their respective populations …." (World Bank, 2002, preface). Other universal promises regard the spread of socio-political rights. The myth of "information revolution" has shaped the Western social imagination by presenting a "timeless and borderless world" where also the inhabitants of territories located on the fringes and in backward areas would access a new form of citizenship (Mosco, 2004).

Electronic citizenship – given its immaterial nature – seems to open up new opportunities to citizens exercising those rights that have remained only in constitutional charters in the past, or to claim new kinds of rights (Barlow, 1996; Bollier, 2008). Yet, the history of citizenship shows, in different forms, the fundamentally exclusive nature of this category, replicating the causes of the gaps existing between and within countries (Wallerstein, 2006). Mechanisms that generate disparities among citizens do not go away with the advent of electronic citizenship: asymmetric access to economic and political resources limit access to new technologies; differences in levels of education exclude a relevant part of the population from the use of the Internet. Digital citizenship encounters obstacles quite similar to those met during the initial phase of (pre)-electronic citizenship (Mossberger, Tolbert & McNeal, 2007).

Research shows, indeed, that the diffusion of ICTs varies territorially, mainly in the wake of wealth distribution, among other variables1. This consideration would corroborate the view of those reading the rhetoric over IS as a facade covering the restructuring of the capitalist economy at the global level and arguing that the diffusion of ICTs, based on an unequal model of development, further strengthens rather than reduces territorial and socio-economic divides between the center and outer fringes. Notwithstanding the many promises of ICTs, a report published by the United Nations Conference on Trade and Development in 2006 shows that a person in a high-income country was 22 times more likely to use the Internet than someone in a low-income country (Unctad, 2006). In addition, a recent report by the United Nations (2008) has confirmed the limits of digital policies as a global project. The emergence of new technologies may provide limited reason for optimism, given the persistent disparities among countries. It has been calculated that 67% of Internet users were from Organisation for Economic Co-operation and Development (OECD) countries, where nearly 14% of world population lives. Adding to this, it can be observed that the lack of a global policy encouraging the development of poorest nations constitutes a key question. Digital policy at the global level represents an "opportunity for all", or a new way for liberalizing markets in less advantaged countries? Does it create a new international balance or simply reinforces age-old dependencies?

Newly found chasms can be observed between different geopolitical areas emerging following the diffusion of digital technologies. Although digital divide is often considered politically neutral, a map of powers structure in Internet governance will suggest a quite different interpretation, leading to consider a new geography of cyberspace (Amoretti, 2009). Disparities have to be accounted for as the product of the strategies of specific actors that continue to reproduce socio-economic and political divisions among developed and poorer countries, operating control and regulation on different levels of the Internet.

This chapter is divided into three sections. In the first part, the concept of "digital divide" is analysed by considering its first formulation in the US political debate during the Nineties, as well as the more recent efforts to consider the multidimensional nature of such category. This helps to overcome the bipolar meaning of "digital divide" based on the distinction between the "information haves" and the "information have-nots". In the second section, significant quantitative measure of disparities between countries is provided. More in particular, attention is devoted to the most diffused index elaborated by prominent International Organizations (Global E-Government Readiness Index, World Internet Usage and Population Index, Information Development Index) in order to "map" digital divides worldwide and underline differences between geopolitical areas. Yet, processes which occur in such areas are not independent of each other. The last part of the chapter shows how developing countries adopting proprietary softwares are becoming dependent on the power of providers of ICT goods and services, which are mainly concentrated in the United States. Starting from Lessig's perspective, the definition of the Internet code, i.e. the set of standards which defines the rules of access and digital behavior, calls to mind a long-established law-making activity.

BACKGROUND: THE CONCEPT OF DIGITAL DIVIDE

A new term has also been coined to express a new type of inequality created by the widespread diffusion of the information society: "digital divide" refers to the disadvantages of those who lack access or refrain from using ICTs in their everyday lives (Ibrahim, 2008). A larger analysis has come to consider the different access to ICT opportunities in a global perspective, between nations or larger areas.

The *Oxford English Dictionary* registered the first occurrence of "digital divide" in an article published in 1995 in the *Columbus Dispatch* (Ohio), giving the following definition: "the gulf between those who have ready access to current digital technology (esp. computers and the Internet) and those who do not; (also) the perceived social or educational inequality resulting from this" (Ranieri, 2008). Then the bipolar meaning of the notion of digital divide was made popular in a series of reports in the mid 1990's by Larry Irving, a former United States Assistant Secretary of Commerce and Technology adviser

to the Clinton Administration (Ferri, 2008). For instance, the various *Falling through the Net* studies conducted by the Department of Commerce's National Telecommunications and Information Administration (NTIA) adopt this view about the "divide" by simply considering users' chance to use the computer and Internet (NTIA, 1999). With clear implications either on the methodology followed for measuring the phenomenon2, and on the definition of policies for 'closing the gap" (Stewart et al, 2006).

Indeed, the diffusion of ICTs is interpreted as a 'natural path', already undertaken by other mass media (telephone, radio, TV): initially access to the new technology is restricted to an *élite*, with a large divide between the 'haves' and the 'have-nots', but with time its increasing penetration within society progressively reduces any gaps. For instance, according to Benjamin N. Compaine, Internet inequalities are closing rapidly, so rapidly that governments' attempt to intervene might mean spending billions to address needs that no longer exist (Compaine, 2001). The so called *"normalisation thesis"* is based on the idea that a series of factors linked with the development of technology (increasingly lower costs, user friendly access, differentiated contents) will create a saturation in the market allowing 'have-nots' groups to access innovation (Amoretti & Casula, 2009). The role played by processes of liberalization of the ICT sector, reducing the digital divide due to increased competition in the telecommunications market, is often praised in studies supporting such a thesis (OECD, 2001; Wef & Insea, 2007). Limited contribution to governmental action is also allowed as far as it enhances this path, through policy measures aiming to facilitate the development of the ICT sector (infrastructure building, facilitated connections in schools and other public institutions, introduction of digital alphabetization in education programs) and market liberalization (regulatory actions to grant free competition).

Although the first meaning of "digital divide" emphasizes technological aspects, by stressing the gap between different ICTs equipment within populations, such a concept was then enriched by other dimensions (Bertot, 2003). This leads to doubt that "information inequality in the use of digital technology or computer-mediated communication (CMC) is solved at the moment that everyone has the ability to obtain a personal computer and a connection to the Internet" (van Dijk & Hacker 2003, p. 316). In the remaining part of this paragraph, it will present some conceptions that try to broaden the simplistic and bipolar meaning of "digital divide", usually far too focused on the distinction between the "information haves" and the "information have-nots".

The first effort was to consider how different uses of new technology are influenced by contextual, cultural, and knowledge resources available to individuals and groups3. For instance, Paul DiMaggio and Eszter Hargittai suggest the shift from a conception of "digital divide" interpreted as the chasm between the "haves" and "have-nots" – differentiated by dichotomous measures of access to new technologies – to the larger category "digital inequality", including differences in equipment, autonomy of use, skills, social support, and the purposes for which the technology is employed (DiMaggio & Hargittai, 2001; DiMaggio et al, 2003). Going beyond the binary view of access, such authors propose to expand the focus of digital divide research by investigating the use of technology for different subgroups within the population. Questions concerning who is 'connected' to information and technology were gradually surpassed by questions regarding what is meant by 'access'. This leads to consider digital divide "as a hierarchy of access to various forms of technology in various contexts, resulting in differing levels of engagement and consequences" (Selwyn, 2004, p. 351), depending both on technological equipment and users skills in the use of the Internet. In the same manner, some classifications identify components of full social access, such as cognitive aspects, material resources, institutional structures (Warschauer,

2002; 2003; Hargittai, 2003; De Hann, 2004; Van Dijk, 2006). Moreover, the passage from the idea of 'digital divide' to the multidimensional conception of 'full social access' helps to focus attention on inequalities present also in developed countries, against the widespread hypothesis of a process tending to close the digital gap (Sartori, 2006).

Another approach aims at overcoming the methodological individualism frequently connected to studies on digital divides: a way to analyse digital divide regards the close relationship between inequality and industrial development (Bindé, 2005). For instance, according to Norris (2001), while "social divide" and "democratic divide" points to inequalities among the population within one nation, the notion of "global divide" encompasses differences among industrialized and lesser developed nations. Also, new measures of digital divide try to aggregate different indicators of digitalization, in order to measure the digital divide within a set of countries or geographical areas (Corrocher & Ordanini, 2002).

An additional dimension to consider on the macro level is the relationship between politics and the Internet. Such position suggests the recognition of 'embeddedness' of the use of new technologies in different socio-institutional contexts. Indeed, the Internet can be adopted as an instrument of freedom or as a mean of centralization and control. For instance, some governments enact preventive censuring measures by filtering the information resources available and reducing universal access to information. On the other hand, according to Milner (2006), democratic governments facilitate the spread of the Internet compared to autocratic ones, as "data from roughly 190 countries over the past decade (1991-2001) show that a country's regime type matters greatly, even when controlling for other economic, technological, political and sociological factors" (p. 176). The Internet does not erase national borders: it is so easy to remember Google's struggles with the French government and Yahoo's capitulation to the Chinese regime (Goldsmith & Wu, 2006). In more general terms, the position of the neo-institutional school seems reasonable, according to which "cognitive, cultural, social, and institutional structures influences the design, perceptions, and uses of the Internet and related [information technology]" (Fountain 2001, p. 88). Therefore, it can been argued that technology represents a social construct, because it shows that any technological application is strongly influenced by social aspects, such as cognitive frames, political culture, local traditions and so forth (Amoretti & Musella, 2009).

Yet, also such more complex perspectives usually do not take into consideration "the material impoverishment of large numbers of the world's population by those both better equipped to take advantage of ICTs and also use it for the protection of their privileged position; a social and economic process which has much in continuity with previous epochs" (Loader, 1998, p. 8). More advanced country and International organizations have assumed a leadership role in offering financial resources and guidelines for the development of digital capacities in less developed countries. Yet it would seem that these policies often have not only proven to be inefficient, but also that the new course does not take into due consideration the reasons for such failure; on the contrary, attempts to bridge the digital divide may have the effect of locking developing countries into a new form of dependency on the West. As Wade (2002), once a World Bank economist, said:

The technologies and 'regimes' (international standards governing ICTs) are designed by developed country entities for developed country conditions. As the developing countries participate in ICTs, they become more vulnerable to the increasing complexity of the hardware and software and to the quasi-monopolistic power of providers of key ICT services ... Much of the ICT-for-development literature talks about plans, intentions and opportunities provided—and blurs the distinction between these and verified actions on the ground. It talks about benefits and not costs.

And it explains cases of failure, when noted, in ways that protect the assumption that ICT investment is a top priority. (pp. 443-461)

Developing countries are in danger of locking themselves into a new form of e-dependency on the West as they introduce software and hardware technical and investment assistance. This process is accompanied by definitions of criteria of selectivity and evaluation that maximize the project development impact, and at the same time enhance the role of the organizations that are responsible (Putzel, 2005).

MAPPING DIGITAL DIVIDES: QUANTITATIVE EVIDENCE

Notwithstanding the recent diffusion of ICTs, developing countries remain far behind developed ones in the adoption of ICTs and their use by citizens and enterprises. It has been shown that by the end of 2008, the world had reached unprecedented ICT levels: over 4 billion mobile cellular subscriptions, 1.3 billion fixed telephone lines and close to a quarter of the world's population using the Internet, yet, despite overall high growth rates, record numbers, and all-high penetration rates, major differences in ICT levels between macro-regions and between the developed and developing economies remain (ITU, 2009, p. 71).

Note reports produced by the United Nations (2008) show significant disparities in the use of new technologies in different geo-political areas. They place countries of North America and Europe in the leadership position in the world in terms of "e-government readiness", a quantitative index that measures the capacity and willingness of countries to use e-government4 (Table 1). More particularly, in 2005 the United States reached an index of 0.8744, representing the world leader. In that year the World e-government readiness average was 0.43, while African countries present a very low value (0.27). Three years later, in

Table 1. Regional e-government readiness ranking

World Regions	2008	2005	2004	2003
North America	0.8408	0.8744	0.8751	0.8670
Europe	0.6490	0.6012	0.5866	0.5580
South and Eastern Asia	0.4290	0.4922	0.4603	0.4370
South and Central America	0.4838	0.4643	0.4558	0.4420
Western Asia	0.4857	0.4384	0.4093	0.4100
Caribbean	0.4480	0.4282	0.4106	0.4010
South & Central Asia	0,3628	0.3448	0.3213	0.2920
Oceania	0.4338	0.2888	0.3006	0.3510
Africa	0.2739	0.2642	0.2528	0.2460
World Average	*0.4514*	*0.4267*	*0.4130*	*0.4020*

Source: *Global E-Government Readiness Report 2005* (United Nations, 2005, p. 30) and *Global E-Government Survey. From E-Government to Connected Governance* (United Nations, 2008, p. 22).

2008, Scandinavian countries surpassed the United States as the leader: a result that probably reflects the reduction in ICTs investment during the Bush Administration. Nevertheless, most developing countries do not present significant improvements in e-government rankings.

Such disparities are based on relevant gaps in terms of infrastructures (Musella, 2009). The United States are at the top of the degree of penetration of new technologies in society (Table 2). As the Internet World States show, the North American population constitutes only 5.0% of world population. Yet, it represents 15.8% of world users, with an ICTs penetration rate of 73.9%. Furthermore, it can be calculated that US growth for the period 2000-2009 reaches 132.9%.

Asia provides very different results. The above four billion Asian inhabitants constitute 56.2% of world population, with a penetration of only 18.5%. This continent has seen an incredible usage growth during the period 2000-2009 (516,1%), and it shows potential for further development. Yet, the ranking position of this "big giant" is completely unsatisfactory. It can be inferred here

Table 2. World Internet Usage and Population Index

World Regions	Population (2009 est.)	Internet Users (December 30, 2000)	Internet Users (Latest Data)	Penetration (% Population)	Growth 2000-2009 (%)	Internet Users/ Population (Latest Data)
Africa	991.002.342	4.514.400	65.903.900	6,7	1.359,9	6,65
Asia	3.808.070.503	114.304.000	704.213.930	18,5	516,1	18,49
Europe	803.850.858	105.096.093	402.380.474	50,1	282,9	50,05
Midle East	202.687.005	3.284.800	47.964.146	23,7	1360,2	23,6
North America	340.831.831	108.096.800	251.735.500	73,9	132,9	73,8
Latin America/ Caribbean	586.662.468	18.068.919	175.834.439	30,0	873,1	29,97
Oceania/Australia	34.700.201	7.620.480	20.838.019	60,1	173,4	60,05
World Total	*6.767.805.208*	*360.985.492*	*1.668.870.408*	*24,7*	*362,3*	*24,65*

Source: Internet World Stats, http://www.Internetworldstats.com/stats.htm, November 30, 2007 and further elaboration.

Figure 1. Internet Usage per Population

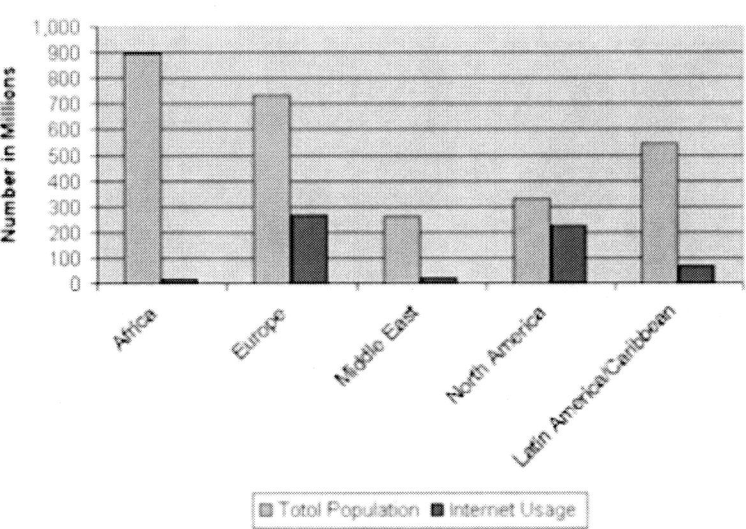

that there is a quite evident disparity in ICTs access in and among regions and countries of the world. Despite their initial efforts, the majority of developing countries are way behind achieving any meaningful economy-wide benefits from the information society.

Digital divides are so persistent that a long list of paradoxes can be presented. As Rahman points out: the total Internet bandwidth in Africa is equal to that of the Brazilian city of Sao Paulo; the total Internet bandwidth in all of Latin America is equal to that in Seoul Republic of Korea; as a proportion of monthly income, Internet access in the United States is 250 times cheaper than in Nepal and 50 times cheaper than in Sri Lanka; and in the United States, 54.3 per cent of citizens use the Internet, compared to a global average of 6.7 per cent. In the Indian subcontinent, the proportion is 0.4 per cent (Rahman, 2006, quoted in United States, 2008, p. 116).

In the following table (Table 3), countries are grouped on the basis of different ICT levels, in order to compare the magnitude of differences between them. It can be noted that 88 of 154 national cases present a medium or low ICT level, amounting to 63% of world population. Indeed, the group with a "medium level" includes economies that account for more than one-third of the total population, with countries like China and Indonesia, while the group with a "low level" includes most of the Southern-Asian countries along with most of the Sub-Saharan African countries. In the last columns of the table a measure (Information Development Index) is presented, which combines three subcomponents: ICT infrastructure and access, ICT use and the intensity of use; ICT skills (or ability necessary to use ICTs effectively).

Notwithstanding, e-government policy initiatives have gained international validity by the donor community as a catalyst for relevant reforms5, disparities among different geopolitical areas do not seem to be mitigated by the intensive use of new technologies. It may be argued that: "since differences in modern resources are of long standing, the distribution of collective Internet capital is a consequence rather than a cause of international differences" (Rose, 2005, p. 7).

Table 3. Country groups with different ICT levels

Group	Number of countries	Share in population (%)*	IDI 2007	
			Minimum	Maximum
High	33	15.1	5.29	7.50
Upper	33	11.9	3.41	5.25
Medium	44	37.4	2.05	3.34
Low	44	35.6	0.82	2.03
All countries	154	100.0	0.81	7.50

Source: ITU 2009

Indeed, such trends seem to contradict OECD assumptions, which in some official reports have asserted that ICTs are capable of contributing to development goals by reducing transaction costs and helping to deliver information services. Indeed, the concept of good governance, mainly developed in the framework of United Nations Development Program (UNDP) since the mid 1990s, has been strongly connected to the condition of liberalizing markets, on the basis of the statement that only a competitive and non discriminatory market order is conducive to economic growth. Thus ICTs programs typically carry a political agenda of *"government by the market"* (Wade, 2002, p. 449). Especially after the recent economic crisis, when countercyclical fiscal stimulus in both developed and developing countries in response to a downturn in global demand was linked to clear liberalization strategies, with the aim of *"making the market work first"* (World Bank, 2009a). Yet, is it enough to close the gap between geopolitical areas? The question is how and to what extent is it achieving a redefinition of power relationships among geo-political areas, with an erosion of US hegemony, or whether it is bringing about new forms of (digital) dependency. If the vitality of some countries or blocs of countries, such as China, India and South American states, seem to suggest unprecedented scenarios, the direction of such developments still remain to be seen.

INTERNET ARCHITECTURES

Often, developing countries have laid serious doubts on the significant role of new technology programs in reducing global gaps among different geo-political areas. In 2000 the Group of 77 (G-77), a coalition of 133 developing nations, complained that the huge income gap between rich and poor has been exacerbated by a North-South "digital divide"6. Moreover, the adoption of proprietary software make developing countries more vulnerable to the power of ICT providers of key goods and services, which are mainly concentrated in the United States and continue to dominate the global market. As Carmel notes (1997), the United States benefits from several factors that sustain its advantage in this industry. A long list allows to better understand how US firms have gained such a position: skilled labor, favorable capital conditions, sophisticated customers, close association with hardware vendors, a competitive marketplace, geographic concentrations, first-mover advantage, a strong intellectual property regime, and English as the *lingua franca* software. Conditions leading to a quasi-monopolistic regime brings developing countries very far from the idea of a digital "open architecture networking, in which any type of network anywhere can be included" (OECD, 2005, p. 29).

Indeed, even the distribution of software houses firmly places the United States in a leader position in forging global digital policy. As table 4

Table 4. Software house location

SOFTWARE HOUSE	HEADQUARTER	SOFTWARE	CODE
Microsoft	Redmond, Washington, USA	Microsoft Windows, Microsoft Office (Word, Excel, Power Point, Access, Publisher, etc.), Internet Explorer, Windows Live	Proprietary
Adobe	San José, California, USA	Acrobat, Photoshop, AfterEffects, Macromedia MX (Flash, Freehand, Dreamweaver, Fireworks)	Proprietary
Mozilla Foundation	Mountain View, California, USA	Firefox, Thunderbird, Seamonkey	Open Source
Novell	Waltham, Massachusetts, USA	Suse Linux	Open Source
Red Hat	Raleigh, Carolina del Nord, USA	Red Hat Linux, Fedora	Open Source
Autodesk	San Rafael, California, USA	Autocad	Proprietary
Electronic Arts	Redwood City, California, USA	Sim City, The Sims, Need for Speed, Battlefield, ecc.	Proprietary
Blizzard Entertainment	Irvine, California, USA	World of Warcraft, RPM racing, ecc.	Proprietary
Symantec	Cupertino, California, USA	Norton Antivirus	Proprietary

Source: our elaboration

clearly shows, the most important software houses in the world are located in the United States. As highlighted in the third table's column, the most widespread software applications in the world are produced by only three corporations (Microsoft, Adobe and Mozilla Foundation). Yet, software produced by the first two corporations are different from the Mozilla ones for their *proprietary code*, with restrictions on use or private modifications7. This means that there are strong restrictions on the part of those who purchase such software in reading and modifying sofware's instructions, as their source codes are almost always kept secret. From the point of view of power relationships, the code closeness of most diffused software lead to two relevant implications. First, software is not adaptable to users' requirements. Dominant software impose their cultural standards, starting from the imposition of English as *Internet lingua franca*. Local cultures must passively accept the interpretative schema and models of interaction developed in California. Second, thanks to proprietary codes software houses can choose to provide new parts to software's code (patch, frame work, upgrade, etc.) or to force users to purchase a new version of the software. Countries that create their digital infrastructures basing them on proprietary codes lose control on their technological investments.

The power of US companies in defining the condition of Internet access and use underlines the dependency of developing countries, becoming one of the most important basis for the global digital divide. Attention on the theme of access has usually led to a sort of "empirical reductionism" in the field of electronic citizenship: the main problems were considered to ensure the use of digital goods – hardware, software and contents – to a growing number of people. However, despite the common image of the Internet as a free territory, governments and firms of the most developed countries colonize it by controlling its underlying code and by shaping the legal environment in which it operates. Indeed, the question of inequalities in digital access is not only related to individual resources and behavior, as so many definitions of "digital divide" lead to assume, but it lies in a more systematic approach, regarding "the interactions of corporations' strategic choices, individual users' responses to these choices, programmers' decisions about code" (DiMaggio et al, 2003, p. 15; DiMaggio et al, 2001). In other words, cyberspace is not lawless and anarchic: there is a "politics of code" however, which regulates how the Internet and the web are used, by whom, and under what conditions (Lessig, 1999; Longford, 2005). More in particular, the definition of a computer-based infrastructure establishes a true code able to regulate, probably even better than the traditional regulatory mechanisms, digital behaviors, activities, choices and citizen's orientations. As Lawrence Lessig has clarified: "cyberspace has an architecture, [… that] embeds certain principles; its sets the terms on which one uses the space; it defines what's possible in the space" (Lessig, 2000, p. 1). For this reason, digital architectures can be assimilated to a sort of Constitution: "in the context of digital technology and new media, the technical architecture of the Internet and the various software codes and applications which run on it are analogous to legislative declarations and founding political documents which delimit the form, content and extent of citizen rights and obligations in a given polity" (Longford, 2005, p. 71).

In just a number of years cyberspace has become a relevant arena of political conflict and economic conquest, with two main actors: States and corporation, i.e. those institutions that were deeply underestimated by the first interpretations of digital phenomenon and by the egalitarian and utopian elaborations of cybercultures (Goldsmith & Wu, 2006). Two institutions that govern our lives in an invisible and shady manner. Examples of such practices are relevant and numerous, both in the growing commercialization of cyberspace and

in security policies: from the impressive activities of e-commerce – through *web browsers* and *cookies* – the "perpetual pedagogy of surveillance" (Luke, 2002), to the spread of the logic of mega-portals for accessing information and services; from the strengthening of proprietary rights to the diffusion of surveillance instruments, especially developed after the events of September 11th, 2001. Such examples show that "the colonization of cyberspace by commercial, proprietary code amounts to the declaration and enactment of a new constitution for cyberspace which lays down commercial terms and conditions of cyber-citizenship, including new rights (intellectual property) and obligations (compulsory visibility, identification, pay-per-view/play), and which also identifies and excludes non-citizens and outsiders (hackers, file-sharers, the unconnected). Secondly, proprietary code is designed through opaque processes of product-development and marketing by centralized, secretive corporations who conceal their source codes from the wider Internet public, this despite the fact that such codes have potentially profound implications for the production of users as subjects. Lastly, the production of new subjects and citizens of cyberspace through commercial code may spill over into and shape processes of subjectification in the off-line world as well, with troubling consequences for the cultivation of democratic citizens" (Longford, 2005, p. 81; Luke, 2002).

FUTURE RESEARCH DIRECTIONS

The prevalence of proprietary models is one of the most important causes of disparities in digital resources at the global level. Such considerations lead to developing a more complex conception of digital divide, on the basis of a different territorial distribution of powers to regulate cyberspace, and to overcome the established procedure of measuring digital inequalities by analyzing the number of individual access lines per inhabitants.

A useful direction for future research regards implications of the use of open softwares to overcome digital divides (May, 2006; von Krogh & Spaeth, 2007). For instance James (2003a; 2003b) underlines the advantages of Gnu/Lunux operating system for developing countries seeking to bridge global disparities. Also development organizations and international non-governmental organizations have been emphasizing the high potential of free and open source software for the less developed countries, with particular reference to the themes of cost reduction and less vendor dependency. According to data presented at a meeting on e-development organized by the World Bank (2009b), within the last decade more than 60 countries and international organizations have published nearly 275 policy documents related with the use of open source in public sector. Yet, first experiments of free and open source software adoption seems unsatisfactory, especially in the African continent (Van Reijswoud, 2008). Thus, although the use of open source presents many opportunities for developing countries, it is not clear what the impact of open source applications will be in providing all citizens with equal possibilities and reducing the digital gap among countries.

Digital divide has been usually analyzed through a self-centered lens. Much of the research focuses on the question of technological access, and more in particular on disadvantages of individuals with lower social-economic status, less education or disabled (Rogers, 2001). In recent years many scholars have underlined that it encompasses not only the availability of computers, but also the required skills for using them and what users do with computers, trying to go beyond the simplistic bipolar split between the "have" and the "have-nots" (Warshauer, 2003; Selwyn, 2004). Although it has been recognized that access to the Internet depends on full individual social capital, little attention was dedicated to those systematic aspects that affect inequalities within and among countries. Indeed, initial US official reports presenting problems connected to disparities in digital equipment in the population until now, studies on digital divide have been reluctant in considering the role of geo-political factors in technological development.

CONCLUSION: AN HOLISTIC APPROACH

According to the usual image of digital revolution, societies can "leapfrog" over entire stages of economic and social reorganization followed by developed countries in the past, on the basis of the simple assumption that countries that take steps to create a competitive market environment for ICT generally have a larger share of people using ICT services than those that have not. As a United Nations report underlined in 2001, the world seems to "evolve—in fact, is evolving—toward a seamless "information society", organized in global networks, in which individuals and countries can escape the confines of poverty or underdevelopment, simply through exploiting new access to information" (p. 3). Yet the passage from digital investments to effective development remains related to the definition of a complex structure of opportunities at the global level, as developed countries possess either more resources in managing digital revolution and the power to define rules of access and use of new technologies. For instance, decisions on *global ICT issues*, including intellectual property rights, management of the electromagnetic spectrum, use of information gained through remote sensing, and so forth, are taken in the United States. While "cross-national differences in IT resources and capabilities are sharpening global inequalities and, moreover, setting a new geography of global centrality/marginality" (Drori & Jang, 2003, p. 155), developing countries are locking themselves into a new form of e-dependency on the West (Wade, 2002; Ciborra, 2005). Considerations on different distribution of software and hardware resources, as well as of the power to define standards and rules, seem to damp enthusiasm in the first stages of the diffusion of new technologies, (re)presented the question of digital divide in a new – and more alarming – form.

REFERENCES

Amoretti, F. (Ed.). (2009). *Electronic Constitution: Social, Cultural and Politcal Implications.* Hershey-New York: Information Science Reference, IGI Global.

Amoretti, F., & Casula, C. (2009). From Digital Divides to Digital Inequalities. In M. Khosrow-Pour (Ed.), *Encyclopedia of Information Science and Technology, 2nd edition* (pp. 1114–1119). Hershey-New York: Information Science Reference, IGI Global.

Amoretti, F., & Musella, F. (2009). Institutional Isomorphism and New Technologies. In M. Khosrow-Pour (Ed.), *Encyclopedia of Information Science and Technology, 2nd edition* (pp. 2066–2071). Hershey-New York: Information Science Reference, IGI Global.

Barlow, J.P. (1996). Declaration of the independence of cyberspace. *Cyber-Electronic List, 8.*

Bertot, J. C. (2003). The Multiple Dimensions of the Digital Divide: More than Technology "Haves" and "Have Nots". *Government Information Quarterly, 20,* 182–191. doi:10.1016/S0740-624X(03)00036-4

Bindé, J. (2005). *Towards Knowledge Societies: Unesco World Report.* Paris: Unesco Publishing.

Bollier, D. (2008). *Viral Spiral. How the Commoners Built a Digital Republic of Their Own.* New York: The New Press.

Chen, W., & Wellman, B. (2004). The Global Digital Divide. Within and Between Countries. *IT&Society, 1*(7), 39–45.

Ciborra, C. (2005). Interpreting E-government and Development: Efficiency, Transparency or Governance at a Distance? *Information Technology & People, 18*(3), 260–279. doi:10.1108/09593840510615879

Compaine, B. (Ed.). (2001). *The Digital Divide: Facing a Crisis or Creating a Myth?* Cambridge, MA: MIT Press.

Corrocher, N., & Ordanini, A. (2002). Measuring the digital divide: a framework for the analysis of cross-country differences. *Journal of Information Technology, 17*(1), 9–19. doi:10.1080/02683960210132061

De Haan, J. (2004). A Multifaceted Dynamic Model of the Digital Divide. *IT&Society, 1*(7), 68–88.

DiMaggio, P., & Hargittai, E. (2001). From the 'Digital Divide' to 'Digital Inequality': Studying Internet Use as Penetration Increases. *Centre for Arts and Cultural Policy Study, Princeton University, Working papers 15.*

DiMaggio, P., Hargittai, E., Celeste, C., & Shafer, S. (2003). From Unequal Access to Differentiated Use: A Literature Review and Agenda for Research on Digital Inequality. *Centre for Arts and Cultural Policy Study, Princeton University, Working Paper 29.*

DiMaggio, P., Hargittai, E., & Rusell Neuman, W. (2001). Social Implications of the Internet. *Annual Review of Sociology, 27,* 307–336. doi:10.1146/annurev.soc.27.1.307

Drori, G. S., & Jang, Y. S. (2003). The Global Digital Divide: A Sociological Assessment of Trends and Causes. *Social Science Computer Review, 21,* 144–161. doi:10.1177/0894439303021002002

Ferri, P. (2008), Children and computers. In A. Cartelli & M. Palma (Eds.), *Encyclopedia of Information and Communication Technology, 2nd edition* (pp. 75–83). Hershey-New York: Information Science Reference, IGI Global.

Fountain, J. E. (2001). *Building the Virtual State: Information Technology and Institutional Change.* Washington: Brookings Institution.

Fuchs, C. (2009). The Role of Income Inequality in a Multivariate Cross-National Analysis of the Digital Divide. *Social Science Computer Review, 27,* 41–58. doi:10.1177/0894439308321628

Goldsmith, J., & Wu, G. (2006). *Who Controls the Internet? Illusions of a Borderless World.* Oxford: Oxford University Press.

Hargittai, E. (2003). The Digital Divide and What To Do About It. In Economy Handbook, N. (Ed.), *D.C. Jones* (pp. 822–841). San Diego: Academic Press.

Hemphill, T. A. (2005). Government Technology Acquisition Policy: The Case of Proprietary Versus Open Source Software. *Bulletin of Science, Technology & Society, 25,* 484–490. doi:10.1177/0270467605282245

Ibrahim, Y. (2008). Contemporary Concerns of Digital Divide in an Information Society. In A. Cartelli & M. Palma (Eds.), *Encyclopedia of Information Science and Technology, 2nd edition* (722–727). Hershey-New York: Information Science Reference, IGI Global.

ITU, International Telecommunication Union. (2009). *Measuring the Information Society.* Geneva: The Ict Development Index.

James, J. (2003). Free Software and the Digital Divide: Opportunities and Constraints for Developing Countries. *Journal of Information Science, 29*(1), 25–33. doi:10.1177/016555150302900103

Jung, J., Qiu, J., & Kim, Y. (2001). Internet Connectedness and Inequality: Beyond the "Divide". *Communication Research, 28,* 507–535. doi:10.1177/009365001028004006

Lessig, L. (1999). *Code and Other Laws of Cyberspace.* New York: Basic Books.

Lessig, L. (2000). *The Code in Law, and the Law in Code.* Retrieved November 10th, 2009, from www.lessig.org/content/articles/works/pcforum.pdf

Lessig, L. (2001). *The future of Ideas: The Fate of the Commons in a Connected World.* New York: Vintage.

Loader, B. D. (1998). *Cyberspace Divide. Equality, Agency and Policy in the Information Society.* New York: Routledge. doi:10.4324/9780203169537

Longford, G. (2005). Pedagogies of Digital Citizenship and the Politics of Code. *Technè, 9*(1), 68–96.

Luke, R. (2002). Habit@line: Web Portals as Purchasing Ideology. *Topia: A Canadian Journal of Cultural Studies, 8.* Retrieved November 10th, 2009, from http://pi.library.yorku.ca/ojs/index.php/topia/issue/view/19/showToc

May, C. (2006). The FLOSS alternative: TRIPs, non-proprietary software and development. *Knowledge, Technology, and Policy, 18*(4), 142–163. doi:10.1007/s12130-006-1008-4

Milner, H. V. (2006). The Digital Divide: the Role of Political Institutions in Technology Diffusion. *Comparative Political Studies, 29*(2), 176–199. doi:10.1177/0010414005282983

Mosco, V. (2004). *The Digital Sublime. Myth, Power, and Cyberspace.* Cambridge, MA: MIT Press.

Mossberger, K., Tolbert, C., & McNeal, R. (2007) (Eds.). *Digital Citizenship.* Cambridge, MA: MIT Press.

Musella, F. (2009). American electronic constitution. Reinventing government and neo-liberal corporatism. In F. Amoretti (Ed.), *Electronic Constitution: Social, Cultural, and Political Implications* (pp. 54–70). Hershey-New York: Information Science Reference, IGI Global.

Norris, P. (2001). *Digital Divide, Civic Engagement, Information Poverty and the Internet Worldwide*. Cambridge: Cambridge University Press.

Ntia, National Telecommunications and Information Administration (1999). *Falling through the Net: Defining the Digital Divide*. Retrieved November 10th, 2009, from www.ntia.doc.gov

OECD. (2001). *Understanding the Digital Divide. Paris. OECD (2005). Science, Technology, and Industry Scoreboard 2005. Paris. OECD (2009). OECD*. Paris: Communications Outlook.

Putzel, J. (2005). Globalization, Liberalization, and Prospects for the State. *International Political Science Review*, *26*(1), 5–16. doi:10.1177/0192512105047893

Ranieri, M. (2008). Cyberspace's Ethical and Social Challenges in Knowledge Society. In A. Cartelli & M. Palma (Eds.), *Encyclopedia of Information and Communication Technology, 2nd edition* (132–137). Hershey-New York: Information Science Reference, IGI Global.

Rogers, E. M. (2001). The Digital Divide. *Convergence*, *7*, 96–111.

Rose, R. (2005). A Global Diffusion Model of E-Governance. *Journal of Public Policy*, *25*(1), 5–27. doi:10.1017/S0143814X05000279

Sartori, L. (2006). *Il Divario Digitale: Internet e le Nuove Disuguaglianze Sociali*. Bologna: Il Mulino.

Selwyn, N. (2004). Reconsidering Political and Popular Understandings of the Digital Divide. *New Media & Society*, *6*(3), 341–362. doi:10.1177/1461444804042519

Stewart, C. M., Gil-Egui, G., Tian, Y., & Pileggi, M. I. (2006). Framing the Digital Divide: a Comparison of US and EU Policy Approaches. *New Media & Society*, *8*(5), 731–751. doi:10.1177/1461444806067585

Strover, S. (2003). Remapping the Digital Divide. *The Information Society*, *19*, 275–277. doi:10.1080/01972240309481

Unctad, United Nations Conference on Trade and Development (2006). *Information Economy Report: The Development Perspective*. New York.

United Nations. (2001). The Development Divide in a Digital Age. An Issues Paper, (Ed. C. Hewitt de Alcántara). *Technology, Business and Society Programme Paper Number 4*. New York.

United Nations. (2008). *Global E-Government Survey*. New York.

(2009). *United Nations*. New York: The Millennium Development Goals Reports.

Van Dijk, J. (2006). Digital Divide Research, Achievements and Shortcomings. *Poetics*, *34*(4-5), 221–235. doi:10.1016/j.poetic.2006.05.004

Van Dijk, J., & Hacker, K. (2003). The Digital Divide as a Complex and Dynamic Phenomenon. *The Information Society*, *19*, 315–326. doi:10.1080/01972240309487

Van Reijswoud, V. (2008). *Free and open source software for development: exploring expectations, achievements and the future*. Monza: Polimetrica.

von Krogh, G., & Spaeth, S. (2007). The open source software phenomenon: Characteristics that promote research. *The Journal of Strategic Information Systems*, *16*, 236–253. doi:10.1016/j.jsis.2007.06.001

Wade, R. H. (2002). Bridging the Digital Divide: New Route to Development or New Form of Dependency? *Global Governance*, *8*, 443–466.

Wallerstein, I. (2006). *European Universalism. The Rhetoric of Power*. New York: The New Press.

Warschauer, M. (2002). Reconceptualizing the Digital Divide. *First Monday*, *7*(7).

Warschauer, M. (2003). *Technology and Social Inclusion: Rethinking the Digital Divide*. Cambridge, MA: MIT Press.

Wef & Insead. (2007). *The Global Information Technology Report 2006-2007: Connecting to the Networked Economy* (Dutta, S., & Mia, I., Eds.). New York: Palgrave Macmillan.

(2002). *World Bank*. Washington: The E-Government Handbook for Developing Countries.

World Bank. (2008). *Key Trends in ICT Development* (Cieslikowski, D. A., Halewood, N. J., Kimura, K., & Qiang, C. Z., Eds.). Washington.

World Bank. (2009a). *Broadband Infrastructure Investment in Stimulus Packages: Relevance for Developing Countries* (Qiang, C. Z., Ed.). Washington.

World Bank (2009b). *Global Dialogue on Exploring the Results of Governmental Open Source Software Policies: Brazil Experience*. Event desciption, Dec. 17.

ADDITIONAL READING

Benkler, Y. (2006). *The Wealth of Networks: How Social Production Transforms Markets and Freedom*. New Haven, CT: Yale University Press.

Budd, L., & Morris, L. (Eds.). (2009). *E-Governance: Managing or Governing*. New York: Routledge.

Cawkell, T. (2001). Sociotechnology: The digital Divide. *Journal of Information Science, 27*(1), 55–60.

Chadwick, A., & Howard, P. N. (Eds.). (2008). *Routledge Handbook of Internet Politics*. New York: Routledge.

Cotten, S. R., Anderson, W. A., & Tufekci, Z. (2009). Old Wine in a New Technology, or a Different Type of Digital Divide? *New Media & Society, 11*(7), 1–24. doi:10.1177/1461444809342056

Crenshaw, E. M., & Robison, K. K. (2006). Globalization and the Digital Divide: The Roles of Structural Conduciveness and Global Connection in Internet Diffusion. *Social Science Quarterly, 87*, 190–207. doi:10.1111/j.0038-4941.2006.00376.x

Ferro, E., Dwivedi, K. Y., Gil-Garcia, R., & Williams, M. D. (2009). *Handbook of Research on Overcoming Digital Divides: Constructing an Equitable and Competitive Information Society*. Hershey-New York: Idea Group, IGI Global.

Fuchs, C. (2008). *Internet and Society: Social Theory in the Information Age*. New York: Routledge.

Guillén, M. F., & Suárez, S. L. (2005). Explaining the Global Digital Divide: Economic, Political and Sociological Drivers of Cross-National Internet Use. *Social Forces, 84*(2), 681–708. doi:10.1353/sof.2006.0015

James, J. (2003). *Bridging the Global Digital Divide*. Cheltenham: Edward Elgar Publishing.

Jorgensen, R. F. (Ed.). (2006). *Human Rights in the Global Information Society*. Cambridge, MA: The MIT Press.

Kogut, M. B. (Ed.). (2003). *The Global Internet Economy*. Cambridge, MA: MIT Press.

Korupp, S. E., & Szydlik, M. (2005). Causes and Trends of the Digital Divide. *European Sociological Review, 21*, 409–422. doi:10.1093/esr/jci030

Main, L. (2001). The Global Information Infrastructure: Empowerment or Imperialism? *Third World Development, 22*(1), 83–97. doi:10.1080/713701143

Mossberger, K., Tolbert, C. J., & Stansbury, M. (2003). *Virtual Inequality: Beyond the Digital Divide*. Washington, DC: Georgetown University Press.

Servon, L. J. (2002). *Bridging the Digital Divide: Technology, Community, and Public Policy*. Oxford: Wiley-Blackwell. doi:10.1002/9780470773529

Shelley, M., Thrane, L., Shulman, S., Lang, E., Beisser, S., Larson, T., & Mutiti, J. (2004). Digital Citizenship: Parameters of the Digital Divide. *Social Science Computer Review, 22*, 256–269. doi:10.1177/0894439303262580

Stewart, C. M., Gil-Egui, G., Tian, Y., & Pileggi, M. I. (2006). Framing the Digital divide: a Comparison of US and EU Policy Approaches. *New Media & Society, 8*(5), 731–751. doi:10.1177/1461444806067585

Toulouse, C., & Luke, T. W. (Eds.), *The Politics of Cyberspace*. New York: Routledge.

Van Dijk, J. A. (2005). *The Deepening Divide: Inequality in the Information Society*. London: Sage.

Wilson, E. J. (2006). *The Information Revolution and Developing Countries*. Cambridge, MA: MIT Press.

ENDNOTES

[1] Marginalization is influenced by socio-economic status, gender, life stage, and geographic place (Norris, 2001; Chen & Wellman, 2004), so that it is possible to find statistical correlation of socioeconomic, political, cultural, social, and technological factors shaping access to ICTs (Fuchs, 2009).

[2] As indicated by an OECD report: "At international level, the most basic, and the most important, indicator of the digital divide is the number of access lines per 100 inhabitants. It is the leading indicator for the level of universal service in telecommunications and a fundamental measure of the international digital divide" (2001, p. 6). Indeed, most studies on digital divide use binary measures (access/no-access) or time-based measures (number of hours spent online) (Jung, Qiu & Kim, 2001).

[3] As Strovers notes (2003, p. 275) the activity of 'mapping' digital divides is a true necessity as "technologies never exist in isolation. Social, political, and economic environments condition the scope of imagination that assigns technology certain roles in our lives, as well as its use, acceptance, integration, and utility. [...] However, the focus on technologies as discrete systems and their users as isolated individuals masks some of the contradictory ideas that routinely accompany information technology policy and programs".

[4] The term indicates the application of new information and communication technology for the restructuring of public administration and the renewal of the relationship between public institutions and citizen-users.

[5] The use of new technologies has been considered as part of the broader program of new public management, which in a few years has modified modalities of actions of public agencies in the new as well as in the old Continent. Digitalization and back-office restructuring have represented instruments for the "reinventing government", an expression used to indicate the attempt to focus the public sector on results in terms of efficiency, effectiveness, and quality of service.

[6] See Unesco Observatory, Newsletter No 45, March 30, 2000. Available at: http://www.apnic.net/mailing-lists/s-asia-it/archive/2000/04/msg00015.html, accessed January 30th, 2008.

[7] *Proprietary code* is typical of a technology that is owned by an individual or business entity, and for such feature it possesses "a wide range of rights conferred by the legal system in relation to discrete items of information that have resulted from some form of human intellectual activity" (Hemphill, 2005, p. 484).

Compilation of References

(2002). *World Bank*. Washington: The E-Government Handbook for Developing Countries.

(2009). *United Nations*. New York: The Millennium Development Goals Reports.

Ablett, J., Baijal, A., Beinhocker, E., Bose, A., Farrell, D., Gersch, U., et al. (2007, May). *The "bird of gold": The rise of India's consumer market. McKinsey Global Institute*. Retrieved December 14, 2008, from http://www.mckinsey.com/mgi/publications/india_consumer_market/index.asp

Addison, T., & Laakso, L. (2003). The Political Economy of Zimbabwe's Descent into Conflict. *Journal of International Development . Journal of International Development, 15*, 457–470. doi:10.1002/jid.996

Africa Research Bulletin (2008). *Social and Cultural: Education Zimbabwe: Education has fallen victim as teachers flee or are killed.* Africa Research Bulletin – 17551, May 1st – 31st 2008: Blackwell Publishing Ltd.

Afrobarometer (2002). Key Findings About Public Opinion in Africa: What do Africans think about democracy and development? *Afrobarometer Briefing Paper No. 1: The Afrobarometer Network.* Accessed 2 May 2009: http://www.afrobarometer.org/papers/AfrobriefNo1.pdf

Afrobarometer (2006). The Status of Democracy, 2005-2006: Findings from Afrobarometer Round 3 for 18 Countries. *Briefing Paper No. 40: The Afrobarometer Network.* Accessed 2 May 2009: http://www.afrobarometer.org/papers/AfrobriefNo40_revised16nov06.pdf

Alexander, C. (2001). Wiring the nation! Including first nations? *Journal of Canadian Studies. Revue d'Etudes Canadiennes, 35*(4), 227–239.

Alexander, C., & Pal, L. (Eds.). (1998). *Digital democracy: Policy and politics in the wired world.* Toronto: Oxford University Press.

Aloysius, G. (1997). *Nationalism without a nation in India.* New Delhi: Oxford University Press.

Al-Samarrai, S., & Bennell, P. (2007). Where has all the education gone in sub-Saharan Africa? Employment and other outcomes among secondary school and university leavers. *The Journal of Development Studies, 43*(7), 1270–1300. doi:10.1080/00220380701526592

Ambedkar, B. R. (1945). What Congress and Gandhi Have Done to the Untouchables. Bombay, Thacker & Co. Reprinted in Vasant Moon, (Ed.), *Dr. Babasaheb Ambedkar Writings and Speeches,* 9. Bombay, Government of Maharashtra, 1990.

Amoretti, F. (Ed.). (2009). *Electronic Constitution: Social, Cultural and Politcal Implications.* Hershey-New York: Information Science Reference, IGI Global.

Copyright © 2011, IGI Global. Copying or distributing in print or electronic forms without written permission of IGI Global is prohibited.

Amoretti, F., & Casula, C. (2009). From Digital Divides to Digital Inequalities. In M. Khosrow-Pour (Ed.), *Encyclopedia of Information Science and Technology, 2nd edition* (pp. 1114–1119). Hershey-New York: Information Science Reference, IGI Global.

Amoretti, F., & Musella, F. (2009). Institutional Isomorphism and New Technologies. In M. Khosrow-Pour (Ed.), *Encyclopedia of Information Science and Technology, 2nd edition* (pp. 2066–2071). Hershey-New York: Information Science Reference, IGI Global.

Andersen, D. A. (2001). *The internet and web design for teachers: A step by step guide to creating a virtual classroom*. New York: Longman.

Anderson, B. (1983). *Imagined communities: Reflections on the origin and spread of nationalism*. London: Verso.

Andreotti, V. (2008). *Development vs poverty: Notions of cultural supremacy in development education policy*. In D. Bourn (Ed), Development Education: Debates and Dialogues. (45-63). London: Institute of Education, University of London.

Andrews, G. (1992). Racial inequality in Brazil and the United States: A statistical comparison. *Journal of Social History*, *26*(2), 229–263.

Apple, M. W. (2003). Freire and the politics of race in education. *International Journal of Leadership in Education*, *6*(2), 107–108. doi:10.1080/13603120304821

Arrows, F. (Jacobs, D.) (2008). The Authentic Dissertation: Alternative ways of knowing, research, and representation. New York: Routledge.

Ashcroft, B., Griffith, G., & Tiffin, H. (1989). *The empire writes back: Theory and practice in post-colonial literature*. London, New York: Routledge. doi:10.4324/9780203426081

Attewell, P. (2001). The First and Second Digital Divides. *Sociology of Education*, *74*(3), 252–259. doi:10.2307/2673277

Bandopadhyay, S., & Von Eschen, D. (1991). Agricultural failure: Caste, class and power in rural West Bengal . In Gupta, D. (Ed.), *Social stratification* (pp. 353–368). New Delhi: Oxford University Press.

Barlow, J.P. (1996). Declaration of the independence of cyberspace. *Cyber-Electronic List, 8*.

Battiste, M. (25 October 2005). *Reconciling aboriginal diversity in learning systems*. Keynote presentation. URL: www.usask.ca/education/people/battistem/keynote-OSSTF-2008.pdf

BBC. (2002). *BBC News Talking Point: Can we narrow the digital divide?* Online discussion, accessed 15 January, 2010: http://news.bbc.co.uk/1/hi/talking_point/2369155.stm

Belich, J. (1996). *Making peoples: A history of the New Zealanders: From Polynesian settlement to the end of the nineteenth century*. London: Allen Lane.

Bell, D. (1987). *And we are not saved: The elusive quest for racial justice*. New York, NY: BasicBooks.

Bell, D. (1992). *Faces at the bottom of the well: The permanence of racism*. New York, NY: BasicBooks.

Bell, D. A. (1980). Brown and the interest-convergence dilemma . In Bell, D. (Ed.), *Shades of Brown: New perspectives on school desegregation* (pp. 90–106). New York: Teachers College Press.

Bertot, J. C. (2003). The Multiple Dimensions of the Digital Divide: More than Technology "Haves" and "Have Nots". *Government Information Quarterly*, *20*, 182–191. doi:10.1016/S0740-624X(03)00036-4

Béteille, A. (1991). Caste, class, and power . In Gupta, D. (Ed.), *Social stratification* (pp. 339–352). New Delhi: Oxford University Press.

Billingsley, A. (2007). *Yearning to breathe free: Robert Smalls of South Carolina and his families.* Columbia: University of South Carolina Press.

Bindé, J. (2005). *Towards Knowledge Societies: Unesco World Report.* Paris: Unesco Publishing.

Bishop, R. (2005). Freeing ourselves from neocolonial domination in research: A kaupapa Māori approach to knowledge creation . In Denzin, N. K., & Lincoln, Y. S. (Eds.), *The Sage handbook of qualitative research* (3rd ed.). Los Angeles, London: Sage.

Bishop, R., & Glynn, T. (1999). *Culture counts: Changing power relations in education.* Palmerston North: Dunmore Press.

Blake, J. H., & Simmons, E. R. (2008). A Daufuskie Island lad in an academic community: An extraordinary journey of personal transformation. *Journal of College and Character, 10*(1), 1–14. doi:10.2202/1940-1639.1061

Blake, J. H. (2007, May). *We come from a distance: Education and Gullah culture.* Paper presented at The Original Gullah Festival, Beaufort, SC.

Bloch, A. S. (2006). Emigration from Zimbabwe: Migrant Perspectives. *Social Policy and Administration, 40*(1), 67–87. doi:10.1111/j.1467-9515.2006.00477.x

Block, A. A. (1994). Marxism and education . In Martusewicz, R. A., & Reynolds, W. M. (Eds.), *Inside out: Contemporary Critical perspectives in education* (pp. 61–78). Mahwah, NJ: Lawrence Erlbaum Associates, Publishers.

Bollier, D. (2008). *Viral Spiral. How the Commoners Built a Digital Republic of Their Own.* New York: The New Press.

Boulou E. D'beri. (2005). *The new practices of memory.* URL: mokk.bme.hu/centre/conferences/reactivism/FP/fpBB

Brender, A. (2004). Asia's new high-tech tiger. *The Chronicle of Higher Education, 50*(4).

Broadbent, K., & Ford, M. (2007). *Women and labour organizing in Asia: Diversity, autonomy and activism. Retrieved from Ebook Library database.* Boca Raton, FL: Taylor & Francis.

Brookfield, S. D. (2005). *The Power of Critical Theory for Adult Learning and Teaching.* Maidenhead, England: Open University Press.

Burn, B. (2003). *An island named Daufuskie.* Spartanburg, SC: The Reprint Publishers.

Canada. Department of Indian Affairs and Northern Development. (11 June 2008). *Statement of apology to former students of indian residential schools by prime minister stephen sarper.* [http://www.ainc-inac.gc.ca/ai/rqpi/apo/index-eng.asp]

Cant, G. (2005). Ngāi Tahu and its eighteen papatipu rūnanga in a contested post-colonial New Zealand . In Cant, G., Goodall, A., & Inns, J. (Eds.), *Discourses and silences: Indigenous peoples, risks and resistance* (pp. 199–208). Christchurch: University of Canterbury.

Cartier, C., Castells, M., & Qiu, J. L. (2005). The information have-less: Inequality, mobility and translocal networks in Chinese cities. *Studies in Comparative International Development, 40*(2), 9–34. doi:10.1007/BF02686292

Chan, S., & Primorac, R. (2004). The Imagination of Land and the Reality of Seizure: Zimbabwe's Complex Reinventions. [Columbia University School of International Public Affairs: Gale Group.]. *Journal of International Affairs, 57*(2), 63.

Chandra, S. (2002, Sept.). Information in a networked world: The Indian perspective. *The International Information & Library Review, 34*(3), 235–246. doi:10.1006/iilr.2002.0202

Chandrasekhar, G. P. (2006). The political economy of IT-driven outsourcing . In Parayil, G. (Ed.), *Political economy and information capitalism in India: Digital divide and equity* (pp. 35–60). New York: Palgrave Macmillan.

Chen, W., & Wellman, B. (2003). *Charting and bridging digital divides: Comparing socio-economic, gender, life stage, and rural-urban Internet access and use in eight countries: GCAB.* The AMD Global Consumer Advisory Board.

Chen, W., & Wellman, B. (2004). The Global Digital Divide. Within and Between Countries. *IT&Society, 1*(7), 39–45.

Chikwanha, A. Sithole, T.& Bratton, M. (2004). *Working Paper No. 42: The Power of Propaganda: Public Opinion in Zimbabwe.* Afrobarometer Working Papers, accessed 2 May, 2009: http://www.afrobarometer.org/papers/AfropaperNo42.pdf

Cho, J. (2001). 정보격차 해소 종합계획의 내용과 향후과제[The digital divide plan's content and future path]. The path to informationalization, 56.

Choi, D. (2003) 우리나라 정보격차의 특성 및 정보격차 해소를 위한 정책 과제[Korea's characteristics of digital divide and its solution]. Korean Information Culture Center Publication, 1-19.

Choudrie, J., Papazafeiropoulou, A., & Lee, H. (2003). A web of stakeholders and strategies: A broadband diffusion in South Korea. *Journal of Information Technology, 18*, 281–290. doi:10.1080/0268396032000150816

Ciborra, C. (2005). Interpreting E-government and Development: Efficiency, Transparency or Governance at a Distance? *Information Technology & People, 18*(3), 260–279. doi:10.1108/09593840510615879

World Bank. (2008). *Key Trends in ICT Development* (Cieslikowski, D. A., Halewood, N. J., Kimura, K., & Qiang, C. Z., Eds.). Washington.

Clark, A. W., & Sekher, T. V. (2007). Can career-minded young women reverse gender discrimination? A view from Bangalore's high-tech sector. *Gender, Technology and Development, 11*(3), 285–319. doi:10.1177/097185240701100301

Clark, S. P. (1986). *Ready from within: Septima Clark and the civil rights movement—A first person narrative.* Navarro, CA: Wild Trees Press.

Clark, S. P., & Blythe, L. (1962). *Echo in my soul.* New York, NY: E. P. Dutton.

CNNIC (China Internet Network Information Center). (2003) *Semiannual Survey on the Development of China's Internet.* Retrieved September 15, 2006 from http://www.ccnic.org.cn

CNNIC (China Internet Network Information Center). (2004) *Semiannual Survey on the Development of China's Internet.* Retrieved September 15, 2007 from http://www.ccnic.org.cn

Coates, K. S., Lackenbauer, P. W., Morrison, W. R., & Poelzer, G. (2008). *Arctic front: Defending canada in the far north.* Toronto: Thomas Allen Publishers.

Cohen, J. & Fung, A. (2004) Radical Democracy. *Swiss Journal of Political Science, 10*(4).

Collins, P. H. (2009). *Black feminist thought: Knowledge, consciousness, and the politics of empowerment.* New York, NY: Routledge.

Compaine, B. (Ed.). (2001). *The Digital Divide: Facing a Crisis or Creating a Myth?* Cambridge, MA: MIT Press.

Corrocher, N., & Ordanini, A. (2002). Measuring the digital divide: a framework for the analysis of cross-country differences. *Journal of Information Technology, 17*(1), 9–19. doi:10.1080/02683960210132061

Creel, M. W. (1988). *A peculiar people: Slave religion and community-culture among the Gullahs.* New York, NY: New York University Press.

D'Costa, A. (2006). ICTs and decoupled development: Theories, trajectories, and transitions . In Parayil, G. (Ed.), *Political economy and information capitalism in India: Digital divide and equity* (pp. 11–34). New York: Palgrave Macmillan.

Damarin, S. K. (2000). The 'digital divide' versus differences: principles of equitable use of technology in education. *Educational Technology, 40*(4), 17–22.

Davis, N. E., & Fletcher, J. (2009in preparation). *E-learning for adults in New Zealand with literacy, numeracy and/or ESOL needs. Report for the New Zealand Ministry of Education.* Christchurch: University of Canterbury College of Education.

Davis, A. Y. (1983). *Women, race & class.* New, NY: Vintage Books.

Davis, N. E. (2008). How may teacher learning be promoted for educational renewal with IT? In Voogt, J., & Knezek, G. (Eds.), *International handbook of information technology in primary and secondary education.* Amsterdam: Springer. doi:10.1007/978-0-387-73315-9_31

de Almeida, J. (2003). Unveiling the mirror: Afro-Brazilian identity and the emergence of a community school movement. *Comparative Education, 1*(47), 41–63.

De Haan, J. (2004). A Multifaceted Dynamic Model of the Digital Divide. *IT&Society, 1*(7), 68–88.

Deer, K., & Hahanasson, K. A. (2005). Towards an indigenous vision for the information society. In T.J. van Weert (ed). *Education and the knowledge society: Information technology supporting human development.* pp 67-80. Boston/Dortrecht/London: Kluwer Academic Publishers.

Delgado, R. (Ed.). (1995). *Critical race theory: The cutting edge.* Philadelphia, PA: Temple University Press.

Deshpande, A., & Darity, W. (2003). Boundaries of clan and color: An introduction . In Deshpande, A., & Darity, W. (Eds.), *Boundaries of clan and color: Transitional comparisons of inter-group disparity* (pp. 1–13). London: Routledge.

DiMaggio, P., Hargittai, E., Neuman, W. R., & Robinson, J. P. (2001). Social implications of the internet . *Annual Review of Sociology, 27,* 307–336. doi:10.1146/annurev.soc.27.1.307

DiMaggio, P., Hargittai, E., & Rusell Neuman, W. (2001). Social Implications of the Internet. *Annual Review of Sociology, 27,* 307–336. doi:10.1146/annurev.soc.27.1.307

DiMaggio, P., Hargittai, E., Celeste, C., & Shafer, S. (2004). Digital inequality: From unequal access to differentiated use . In Neckerman, K. (Ed.), *Social inequality* (pp. 355–400). New York: Russell Sage.

DiMaggio, P., & Hargittai, E. (2001). From the 'Digital Divide' to 'Digital Inequality': Studying Internet Use as Penetration Increases. *Centre for Arts and Cultural Policy Study, Princeton University, Working papers 15.*

DiMaggio, P., Hargittai, E., Celeste, C., & Shafer, S. (2003). From Unequal Access to Differentiated Use: A Literature Review and Agenda for Research on Digital Inequality. *Centre for Arts and Cultural Policy Study, Princeton University, Working Paper 29.*

Dixon, B. R. (1997). Toting technology: Taking it to the streets . In Gordon, L. (Ed.), *Existence in Black: An anthology of Black existential philosophy* (pp. 135–147). New York, NY: Routledge.

Doubleday, N. (2005). *Sustaining arctic visions, values and ecosystems: Writing inuit identity, reading inuit art in cape dorset, nunavut. Presenting and representing environments.* The Netherlands: Springer.

Drori, G. S., & Jang, Y. S. (2003). The Global Digital Divide: A Sociological Assessment of Trends and Causes. *Social Science Computer Review, 21*, 144–161. doi:10.1177/0894439303021002002

Durie, M. (2001). *A framework for considering Māori educational advancement. Opening address*. Hui Taumata Mātauranga.

Durie, M. H. (1998). *Te mana te kawanatanga: The politics of Māori self-determination*. Auckland: Oxford University Press.

Wef & Insead. (2007). *The Global Information Technology Report 2006-2007: Connecting to the Networked Economy* (Dutta, S., & Mia, I., Eds.). New York: Palgrave Macmillan.

Dutton, W. (2004). *Social Transformation in the Information Society*. Paris: UNESCO [online] http://portal.unesco.org/ci/en/ev.php-URL_ID=12848&URL_DO=DO_TOPIC&URL_SECTION=201.html

Dzidonu, C. K., Rodrigues, T., & Okot-Uma, R. (1989). The Emerging Global Electronic Messaging and Networking Technologies: An Analysis of their Potential Developmental Impact in Africa. *African Development Review, 10*(1), 189–210. doi:10.1111/j.1467-8268.1998.tb00104.x

E-Knowledge for Women in Southern Africa (EKOWISA). (2009). *EKOWISA-OKN High Glen Community Centre Information Needs Analysis Report by Zunguze, M.* Harare, Zimbabwe: EKOWISA.

Engler, M. (2008, May-June). The world is not flat: how Thomas Friedman gets it wrong about globalization. *Dollars and Sense, 276*(6), 20.

Evison, H. (1993). *Te Waipounamu: The greenstone island*. Wellington: Aoraki Press.

Fallis, D. (2007). Epistemic value theory and the digital divide . In Rooksby, E., & Weckert, J. (Eds.), *Information Technology and Social Justice* (pp. 29–46). Hershey, PA: Information Science Publishing.

Ferri, P. (2008), Children and computers. In A. Cartelli & M. Palma (Eds.), *Encyclopedia of Information and Communication Technology, 2nd edition* (pp. 75–83). Hershey-New York: Information Science Reference, IGI Global.

Forten, C. (1864). Life on the Sea Islands. *The Atlantic Monthly, 13*(79), Issue 79, 587-596.

Fountain, J. E. (2001). *Building the Virtual State: Information Technology and Institutional Change*. Washington: Brookings Institution.

Freire, P. (1970). *Pedagogy of the Oppressed. Harmondsworth*. Middlesex, England: Penguin.

Freire, P. (1972). *Pedagogy of the oppressed*. New York: Continuum.

Freire, P. (1974). *Education for critical consciousness*. London: Continuum.

Freire, P., & Macedo, D. (1987). *Literacy: reading the word and the world*. Massachusetts: Bergin & Garvey.

Friedman, T. (2005). *The world is flat: A brief history of the twenty-first century*. New York: Farrar, Straus and Giroux.

Fuchs, C. (2009). The Role of Income Inequality in a Multivariate Cross-National Analysis of the Digital Divide. *Social Science Computer Review, 27*, 41–58. doi:10.1177/0894439308321628

Fuchs, C. (2008a). *Internet and Society: Social Theory in the Information Age*. New York: Routledge, Routledge Research in Information Technology and Society Series, Number 8, Fuchs, C. (2008b). Critical Theory in Age of the Internet, presentation at the *1st World Forum of the International Sociological Association*. September 6, 2008, Barcelona, accessed 2 May, 2009: http://fuchs.icts.sbg.ac.at/i&s.html

Fukuda-Parr, S., & Birdsall, N. (2001). *Human development report 2001: Making new technologies work for human development*. New York: Oxford University Press for the United Nations Development Programme. Retrieved August 11, 2008, from http://hdr.undp.org/en/reports/global/hdr2001/

Fuller, C. (1991). Kerala Christians and the caste system . In Gupta, D. (Ed.), *Social stratification* (pp. 195–212). New Delhi: Oxford University Press.

Galabuzi, G. (2006). Canada's *economic epartheid: The social exclusion of racialized groups in the new century*. Toronto: Canadian Scholars' Press.

Gale, T., & Densmore, K. (2003). *Engaging teachers: Towards a radical democratic agenda for schooling*. Philadelphia, PA: Open University Press.

Ganguly, D. (2005). *Caste and Dalit Life Worlds: Postcolonial Perspectives*. New Delhi: Orient Longman.

GAPWUZ. (2009). *If something is wrong…The invisible suffering of commercial farm workers and their families due to "Land Reform"*. Report produced for the General Agricultural & Plantation. Workers Union of Zimbabwe [GAPWUZ] by the Research and Advocacy Unit [RAU] and the Justice For Agriculture [JAG] Trust: Report available on the kubatana.net website. Accessed 15 January, 2010: http://www.kubatana.net/docs/.../gapwuz_suffering_farm_workers_091111.pdf

Gee, J. P. (1992). *The social mind: Language, ideology and social practice*. New York: Bergin and Garvey.

Giroux, H. (1988). *Teachers as intellectuals: Towards a critical pedagogy of learning*. Granby, MA: Bergin & Garvey.

Giroux, H. (1979). Review: Paulo Freire's approach to radical educational reform. *Curriculum Inquiry, 9*(3), 257–272. doi:10.2307/3202124

Giroux, H. (1992). *Border crossings: Cultural workers and the politics of education*. New York: Routlege.

Giroux, H. (2000). Insurgent multiculturalism and the promise of pedagogy . In Duarte, E., & Smith, S. (Eds.), *Foundational Perspectives in Multicultural Education* (pp. 195–212). New York: Longman.

Global Information Infrastructure Commission. (2001). *Global Information Infrastructure Commission Survey*. Retrieved October 14, 2008 from http://www.giic.org/#survey

Goldsmith, J., & Wu, G. (2006). *Who Controls the Internet? Illusions of a Borderless World*. Oxford: Oxford University Press.

Gomez, M. (2006). Contemporary spheres for the teaching education: Freire's principles. *Turkish Online Journal of Distance Education, 7*(2), 52–65.

Goncalves e Silva, P. B. (2005). A new millennium research agenda in Black education: Some points to be considered for discussion and decisions . In King, J. E. (Ed.), *Black education: A transformative research and action agenda for the new century* (pp. 301–308). Mahwah, NJ: Lawrence Erlbaum Associates.

Gorski, P. (2005). Education and the digital divide. *Association for the advancement of computing in education journal, 13*(1), 3-45

Gough, K. (1991). Class and economic structure in Thanjavur . In Gupta, D. (Ed.), *Social stratification* (pp. 276–287). New Delhi: Oxford University Press.

Government of Nunavut. Department of Human Resources. (2005). *Inuit qaujimajatuqangit*. [http: http://www.gov.nu.ca/hr/site/beliefsystem.htm].

Govindan, P. (2006). Introduction: Information capitalism . In Parayil, G. (Ed.), *Political economy and information capitalism in India: Digital divide and equity* (pp. 1–10). New York: Palgrave Macmillan.

Grabbatin, B. C. (2008). *Sweetgrass basketry: The political ecology of an African American art in the South Carolina Lowcountry.* Unpublished master's thesis, College of Charleston, South Carolina.

Greenwood, J. (2002). *History of bicultural theatre: Mapping the terrain.* Christchurch: Christchurch College of Education.

Greenwood, J., & Wilson, A. M. (2006). *Te Mauri Pakeaka: A journey into the third space.* Auckland: Auckland University Press.

Greenwood, J, & Te Aika, Lynne Harata (2009). *Final report of the Hei Tauira project: Teaching and learning for success for Māori in tertiary settings.* Wellington: Ako Aotearoa.

Greenwood, J. (1999). *Journeys into a third space: a study of how theatre enables us to interpret the emergent space between cultures.* Doctoral thesis, Griffith University, Brisbane, www.gu.edu.au/ins/lils/adt/

Gupta, D. (1991). Caste, class and conflict . In Gupta, D. (Ed.), *Social stratification* (pp. 303–306). New Delhi: Oxford University Press.

Hacker, K. L., Mason, S. M., & Morgan, E. L. (2007). Digital disempowerment . In Rooksby, E., & Weckert, J. (Eds.), *Information technology and social justice* (pp. 112–147). Hershey, PA: Information Science Publishing.

Han, G. (2003). Broadband Adoption in the United States and Korea: Business Driven Rational Model Versus Culture Sensitive Policy Model. *Trend sin Communication. 11*(1), 3-25.

Harbison, F. H., & Myers, C. A. (1964). *Education manpower and economic growth: Strategies of human resource development.* New York: McGraw Hill.

Hargittai, E. (2003). The Digital Divide and What To Do About It . In Jones, D. C. (Ed.), *New Economy Handbook*. San Diego, CA: Academic Press.

Hargittai, E. (2002). Second level digital divide: Differences in people's online skills. *FirstMonday*, http://www.firstmonday.org/issues/issue7_4/hargittai/

Hargrove, M. D. (2009). Mapping the "social field of Whiteness": White racism as habitus in the city where history lives. *Transforming Anthropology, 17*(2), 93–104. doi:10.1111/j.1548-7466.2009.01048.x

Harris, C. (1995). Whiteness a property . In Crenshaw, K., Gotanda, N., Peller, G., & Thomas, K. (Eds.), *Critical race theory: The key writings that formed the movement* (pp. 276–291). New York, NY: The New Press.

Hawkins, J. N., & Su, Z. (2003). Asian education . In Arnove, R. F., & Torres, C. A. (Eds.), *Comparative education* (pp. 338–356). Lanham, Maryland: Rowman & Littlefield Publisher, Inc.

Haythornthwaite, C. (2007). Digital divide and e-learning . In Andrews, R., & Haythronthwaite, C. (Eds.), *The Sage handbook of e-learning research* (pp. 97–118). Los Angeles, London: Sage.

Hellsten, S. K. (2007). From information society to global village of wisdom? The role of ICT in realizing social justice in the developing world . In Rooksby, E., & Weckert, J. (Eds.), *Information Technology and Social Justice* (pp. 1–28). Hershey, PA: Information Science Publishing.

Hemphill, T. A. (2005). Government Technology Acquisition Policy: The Case of Proprietary Versus Open Source Software. *Bulletin of Science, Technology & Society, 25*, 484–490. doi:10.1177/0270467605282245

Horkheimer, M. (1982). *Critical Theory*. New York: Seabury Press.

Housing Authority Council. (2005). *They paved paradise... Gentrification in rural areas.* Washington, DC: Author.

How India voted: Verdict 2004. (2004, May 20). *The Hindu.* Retrieved December 14, 2009, from www.hindu.com/elections2004/verdict2004/index.htm

Huh, W., & Kim, H. (2003). Information flows on the Internet of Korea. *Journal of Urban Technology, 10*(1), 61–87. doi:10.1080/1063073032000086335

Human Rights Tribune. (8 January 2008). *Native suicide surge, a post-colonial trauma.* URL: http://www.humanrights-geneva.info/Native-suicide-surge-a-post,2630

Ibrahim, Y. (2008). Contemporary Concerns of Digital Divide in an Information Society. In A. Cartelli & M. Palma (Eds.), *Encyclopedia of Information Science and Technology, 2nd edition* (722–727). Hershey-New York: Information Science Reference, IGI Global.

India. (2009). The World Factbook [online]. Retrieved April 23, 2009, from U.S. Central Intelligence Agency: https://www.cia.gov/library/publications/the-world-factbook/geos/in.html

Inflation at 10-year high: Food prices to spill over to other areas. *The Economic Times.* (2009, December 11). Retrieved December 11, 2009 from http://economictimes.indiatimes.com/news/economy/indicators/Inflation-at-10-year-high-Food-prices-to-spill-over-to-other-areas/articleshow/5324792.cms

Inglehart, R., Foa, R., Peterson, C., & Welzel, C. (2008). Development, Freedom, and Rising Happiness: A Global Perspective (1981–2007). *Perspectives on Psychological Science, 3*(4), 264–285. doi:10.1111/j.1745-6924.2008.00078.x

International Monetary Fund (IMF). (2006) *World Economic Outlook Database.* Retrieved September 26, 2006 from http://www.imf.org/external/pubs/ft/weo/2006/02/data/index.aspx

International Telecommunication Union (ITU). (2003). *Broadband Korea: Internet Case Study.* Retrieved February 10, 2002 from http://www.itu.int/net/home/index.aspx

IOM (International Organization for Migration) (Ed.). (2005). *World Migration 2005: Costs and Benefits of International Migration.* Geneva: IOM.

Isaacs, S. (2007). *ICT in Education in Zimbabwe,* Survey of ICT and Education in Africa: Zimbabwe Country Report. Washington DC: InfoDev, World Bank, accessed 2 May, 2009: http://www.infodev.org/en/Publication.437.html

ITU, International Telecommunication Union. (2009). *Measuring the Information Society.* Geneva: The Ict Development Index.

ITU. (2007a). ICT-D Statistics online, reported from the Information and Telecommunications Union (ITU) *World Telecommunication/ICT Indicators Database 2006.* Online, accessed 2 May 2009: http://www.itu.int/ITU-D/ict/statistics/ict/index.html

ITU. (2007b). World Information Society Report 2007: Beyond WSIS - Access to ICTs: Executive Summary. *International Telecommunications Union United Nations Conference on Trade and Development,* Geneva 2003 and Tunis, 2005, online, accessed 2 May, 2009: http://www.itu.int/osg/spu/publications/worldinformationsociety/2007/WISR07-summary.pdf

ITU. (2007c). *Telecommunication/ICT Markets and Trends in Africa,* ITU Statistical Report, Geneva, Switzerland: International Telecommunications Union. Accessed 2 May, 2009: http://www.itu.int/ITU-D/ict/statistics/material/af_report07.pdf

ITU. (2007d). *Connect Africa: Facts and Figures,* statistical report online at ITU website. Accessed 2 May, 2009: http://www.itu.int/ITU-D/connect/africa/2007/bgdmaterial/figures.html

ITU. (2009). *Measuring the Information Society: the ICT Development Index*, ITU Report, Geneva, Switzerland: International Telecommunications Union. Accessed 15 January, 2010: http://www.itu.int/ITU-D/ict/publications/idi/2009/index.html

Jaffrelot, C. (2005). *Dr. Ambedkar and untouchability: fighting the Indian caste system*. New York: Columbia University Press.

James, J. (2003). Free Software and the Digital Divide: Opportunities and Constraints for Developing Countries. *Journal of Information Science*, *29*(1), 25–33. doi:10.1177/016555150302900103

Jameson, J. (2008). *Leadership: Professional Communities of Leadership Practice in Post-Compulsory Education:* Discussion in Education Series. Bristol: ESCalate, accessed 2 May, 2009: http://escalate.ac.uk/5130

Jeffrey, R. (2003). *India's newspaper revolution: Capitalism, politics and the Indian-language press*. New Delhi: Oxford University Press.

JISC infoNet (2006) *The CAMEL project: collaborative approaches to the management of e-learning*. Online: retrieved 2 May, 2009: http://www.jiscinfonet.ac.uk/publications

Jones-Jackson, P. (1987). *When roots die: Endangered traditions on the Sea Islands*. Athens, GA: The University of Georgia Press.

Jordan, L. W., & Stringfellow, E. H. (2000). *A Place Called St. John's*. Spartanburg, SC: The Reprint Publishers.

Jung, J., Qiu, J., & Kim, Y. (2001). Internet Connectedness and Inequality: Beyond the "Divide". *Communication Research*, *28*, 507–535. doi:10.1177/009365001028004006

JuxtConsult. (2009). *Internet usage behaviour and preferences of Indians: Snapshot 2009*. Retrieved April 16, 2009, from http://www.juxtconsult.com/download.asp

Kahn, R., & Kellner, D. (2007). Paulo Freire and Ivan Illich: technology politics and the reconstruction of education. *Policy Futures in Education*, *5*(4), 431–448. doi:10.2304/pfie.2007.5.4.431

Kawharu, I. H. (Ed.). (1989). *Waitangi: Māori and Pakeha perspectives of the Treaty*. Auckland, New Zealand: Oxford University Press.

Kay, J. (3 February 2009). "The left's aboriginal blind spot." *National Post*. URL: http://www.nationalpost.com/opinion/columnists/story.html?id=f80486d4-3866-4932-a3a5-4ec927795139&p=3

Keniston, K. (2004). Introduction: The four digital divides . In Keniston, K., & Kumar, D. (Eds.), *IT experience in India: Bridging the digital divide* (pp. 11–36). New Delhi: Sage Publications.

Kikkawa, T. (2004). Effect of educational expansion on educational inequality in post-industrialized societies: A cross-cultural comparison of Japan and the United States of America. *International Journal of Japanese Sociology*, *13*.

Kim, C., & Santiago, R. (2005). Construction of e-learning environments in Korea. *Educational Technology Research and Development*, *53*(4), 108–115. doi:10.1007/BF02504690

Kim, S. (2001). Korea's e-commerce: Present and future. *Asia-Pacific Review*, *8*(1), 75–85.

Kincheloe, J. K., & Steinberg, S. R. (2002). *Changing Multiculturalism*. Buckingham, UK: Open University Press.

King, M. (2004). *The Penguin History of New Zealand*. Auckland, New Zealand: Penguin.

Kipling, R. (1993). *The Elephant's Child, from the Just So Stories. (Republished from 1902 original), Hanworth*. London: Kenago Books.

Korea Network Information Center. (2002a). *A Survey on the Number of Internet Users and Internet Behavior in Korea: Summary*. Seoul: Korea Network Information Center.

Korea Network Information Center. (2002b). *Internet statistics*. Retrieved February 14, 2002 from http://stat.nic.or.kr/sdata.html

Korean Ministry of Information and Communication. (2001). *2001 Master Plan for Closing the Digital Divide Solution*. Retrieved March 23, 2008 from www.mic.go.kr/eng/index

Korean Ministry of Information and Communication. (2006). *2006 Master Plan for Closing the Digital Divide Solution*. Retrieved March 23, 2008 from www.mic.go.kr/eng/index

Kozma, R., McGhee, R., & Quellmalz, E. (2004). Closing the digital divide: Evaluation of the World Links program. *International Journal of Educational Development, 24*(4), 361–381. doi:10.1016/j.ijedudev.2003.11.014

Ladson-Billings, G., & Tate, W. (1995). Toward a critical race theory of education. *Teachers College Record, 97*(1), 47–68.

Lambert, J. (2007). *Digital Storytelling: Capturing lives, Creating Community*. Berkeley, CA: Digital Diner Press.

Leigh, P. R. (2008). Historical perspectives on analog and digital equity: A critical race theory approach . In Kidd, T., & Chen, I. L. (Eds.), *Social information technology: Connecting society and cultural Issues* (pp. 1–11). Hershey, PA: IGI Global.

Lelyveld, D. (1994). Upon the subdominant: Administering music on All-India Radio. *Social Text, 39*, 111–127. doi:10.2307/466366

Lerner, G. (Ed.). (1992). *Black women in White America: A documentary history*. New York, NY: Vintage Books.

Lessig, L. (1999). *Code and Other Laws of Cyberspace*. New York: Basic Books.

Lessig, L. (2001). *The future of Ideas: The Fate of the Commons in a Connected World*. New York: Vintage.

Lessig, L. (2000). *The Code in Law, and the Law in Code*. Retrieved November 10th, 2009, from www.lessig.org/content/articles/works/pcforum.pdf

Light, A., & Luckin, R. (2008). *Designing for social justice: people, technology, learning*. Retrieved December 15, 2008 from www.futurelab.org.uk/openingeducation

Loader, B. D. (1998). *Cyberspace Divide. Equality, Agency and Policy in the Information Society*. New York: Routledge. doi:10.4324/9780203169537

Logan, B. I. (1999). The Reverse Transfer of Technology from Sub-Saharan Africa: The Case of Zimbabwe. *International Migration (Geneva, Switzerland), 37*(2), 437–463. doi:10.1111/1468-2435.00079

Longford, G. (2005). Pedagogies of Digital Citizenship and the Politics of Code. *Technè, 9*(1), 68–96.

Lovell, P. (1994). Race, gender, and development in Brazil. *Latin American Research Review, 29*(3), 7–35.

Luke, R. (2002). Habit@line: Web Portals as Purchasing Ideology. *Topia: A Canadian Journal of Cultural Studies, 8*. Retrieved November 10th, 2009, from http://pi.library.yorku.ca/ojs/index.php/topia/issue/view/19/showToc

Macfarlane, A. (2007). *Discipline, democracy and diversity: Working with students with behaviuour difficulties*. Wellington: NZCER Press.

Machado da Silva, T. J. (2005). Black people and Brazilian education . In King, J. E. (Ed.), *Black education: A transformative research and action agenda for the new century* (pp. 297–300). Mahwah, NJ: Lawrence Erlbaum Associates, Publishers.

Mack, R. L. (2001). *The digital divide: Standing at the intersection of race & technology*. Durham, NC: Carolina Academic Press.

Martusewicz, R. A., & Reynolds, W. M. (1994). *Inside Out: Contemporary critical perspectives in education*. Mahwah, NJ: Lawrence Erlbaum Associates, Publishers.

Masunungure, E., Ndapwadza, A., & Sibanda, N. (2006). Support for Democracy and Democratic Institutions in Zimbabwe. *Afrobarometer Briefing Paper No.27*. http://www.afrobarometer.org/papers/AfrobriefNo27.pdf

Matsika, K. (2007). Intellectual Property, Libraries and Access to Information in Zimbabwe. *IFLA Journal, 33*(2), 160–167. doi:10.1177/0340035207080556

Mattoso, K. M. de Queiros (1986). *To be a slave in Brazil*. New Brunswick, NJ: Rutgers University Press.

May, C. (2006). The FLOSS alternative: TRIPs, non-proprietary software and development. *Knowledge, Technology, and Policy, 18*(4), 142–163. doi:10.1007/s12130-006-1008-4

McLaren, P. (1997). *Revolutionary multiculturalism: Pedagogies of dissent for the new millenium*. New York: Routledge.

McLaren, P. (2000). White terror and oppositional agency: Towards a critical multiculturalism . In Duarte, E., & Smith, S. (Eds.), *Foundational Perspectives in Multicultural Education* (pp. 195–212). New York: Longman.

Meisenhelder, T. (1994). The Decline of Socialism in Zimbabwe. [Crime and Social Justice Associates: Gale Group.]. *Social Justice (San Francisco, Calif.), 21*(4), 83.

Meldrum, A. (2007). Refugees flood from Zimbabwe, newspaper article. *The Observer,* 1 July 2007, accessed 2 May, 2009: http://www.guardian.co.uk/world/2007/jul/01/zimbabwe.southafrica

Merchant, K. (2006, May 1). The women behind closed digital doors: Technology development: Social conservatism in India is harming efforts to increase computer literacy - and wasting talent. *Financial Times (North American Edition)*, 8.

Miller, E. A. Jr. (2008). *Gullah Statesman: Robert Smalls from Slavery to Congress, 1839-1915*. Columbia: University of South Carolina Press.

Milner, H. V. (2006). The Digital Divide: the Role of Political Institutions in Technology Diffusion. *Comparative Political Studies, 29*(2), 176–199. doi:10.1177/0010414005282983

Ministry of Education. (2003). Masterplan 2 for IT in Education. Retrieved May 11, 2008 from http://www.moe.gov.sg/edumall/mp2/mp2.htm

Ministry of Public Management. Home Affairs, Posts and Telecommunications, Japan. (2002). *Information and Communications in Japan White Paper 2002: Stirring of the IT-prevalent Society*. Tokyo: Ministry of Public Management, Home Affairs, Posts and Telecommunications, Japan.

Ministry of Public Management. Home Affairs, Posts and Telecommunications, Japan. (2003). *Information and Communication in Japan White Paper 2003: Building a "New, Japan-inspired IT Society"*. Tokyo: Ministry of Public Management, Home Affairs, Posts and Telecommunications, Japan.

MISAZimbabwe. (2008). *Foreign currency billing system, deprivation of right to communicate*. Article on Media Institute of Southern Africa website, accessed 2 May 2009: http://www.misazim.co.zw/index.php?option=com_content&task=view&id=429&Itemid=5

MISAZimbabwe. (2009). *MISA-Zimbabwe statement on the State of the telecommunications sector in Zimbabwe*. Article on Media Institute of Southern Africa website, accessed 2 May 2009: http://www.misazim.co.zw/index.php?option=com_content&task=view&id=382&Itemid=5

Miyata, K., Boase, J., Wellman, B., & Ikeda, K. (2004). The Mobile-izing Japanese: Connecting to the Internet by PC and Webphone in Yamanashi . In Ito, M. (Ed.), *Portable, Personal, Intimate: Mobile Phones in Japanese Life*. Cambridge, MA: MIT Press.

Morris, P. (1996). Asia's four little tigers: A comparison of the role of education in their development. *Comparative Education, 21*(1), 95–109. doi:10.1080/03050069628948

Morse, T. (2004). Ensuring equality of educational opportunity in the digital age. *Education and Urban Society, 36*(3). doi:10.1177/0013124504264103

Mosco, V. (2004). *The Digital Sublime. Myth, Power, and Cyberspace*. Cambridge, MA: MIT Press.

Mossberger, K., Tolbert, C., & McNeal, R. (2007) (Eds.). *Digital Citizenship*. Cambridge, MA: MIT Press.

Mukherjee, S. (1991). The Bhadraloks of Bengal . In Gupta, D. (Ed.), *Social stratification* (pp. 176–182). New Delhi: Oxford University Press.

Murthy, N. R. N. (2008, Sept 25). Caste away: Great ideas, great minds – state accountability. *India Today*. Retrieved October 23, 2008, from http://indiatoday.digitaltoday.in/index.php?option=com_content&issueid=72&task=view&id=16150&acc=high

Musella, F. (2009). American electronic constitution. Reinventing government and neo-liberal corporatism. In F. Amoretti (Ed.), *Electronic Constitution: Social, Cultural, and Political Implications* (pp. 54–70). Hershey-New York: Information Science Reference, IGI Global.

Naito, S., & Hausman, B. (2005). *Information and communications technology in Japan. A general overview of the current Japanese initiatives and trends in the area of ICT*. Sweden: VINNOVA. Retrieved January 12, 2009 from http://www.vinnova.se/upload/EPiStorePDF/vr-05-04.pdf

Nakayama, M., & Santiago, R. (2004). Two categories of e-learning in Japan. *Educational Technology Research and Development, 52*(3), 91–111. doi:10.1007/BF02504680

Nambisan, V. (2000). *Bihar is in the eye of the beholder*. New Delhi: Penguin Books India.

Narula, S. (2008). Equal by law, unequal by caste: The "untouchable" condition in critical race perspective. *Wisconsin International Law Journal, 26*, 255.

NASSCOM. (2008). *Indian IT-BPO industry factsheet: NASSCOM analysis*. New Delhi: National Association of Software and Service Companies. Retrieved December 12, 2008, from http://www.nasscom.in/Nasscom/templates/NormalPage.aspx?id=53615

National Computerization Agency and Ministry of Information and Communications. (2003). *2003 White Paper Internet Korea, 72*, NCA & MIC, Seoul.

National Council of Applied Economic Research. (2007). *India rural infrastructure report*. SAGE. Retrieved December 16, 2008, from http://books.google.com/books?id=xjpqUSei38kC&printsec=frontcover#PPP16,M1

National Park Service. (2005). *Low Country Gullah culture special resource study and final environmental impact statement*. Atlanta, GA: NPS Southeast Regional Office.

New Economics Foundation (NEF). (2009) *The (un)Happy Index Planet 2.0*. Research report produced by the New Economics Foundation. Written by Saamah Abdallah, Sam Thompson, Juliet Michaelson, Nic Marks and Nicola Steuer. London, UK. Accessed 18 January, 2010: http://www.happyplanetindex.org/

Nherera, C. M. (2000). Globalisation, Qualifications and Livelihoods: the case of Zimbabwe. *Assessment in Education, 7*(3), 335–362. doi:10.1080/09695940050201343

Nieto, S., & Bode, P. (2008). *Affirming diversity: The sociopolitical context of multicultural education*. Boston: Pearson Education.

Norris, P. (2001). *Digital Divide, Civic Engagement, Information Poverty and the Internet Worldwide*. Cambridge: Cambridge University Press.

Ntia, National Telecommunications and Information Administration (1999). *Falling through the Net: Defining the Digital Divide*. Retrieved November 10th, 2009, from www.ntia.doc.gov

OECD. (2001). *Understanding the Digital Divide. Paris. OECD (2005). Science, Technology, and Industry Scoreboard 2005. Paris. OECD (2009). OECD*. Paris: Communications Outlook.

OECD. (2004). *OECD African Economic Outlook 2003/2004: 22 Country Studies: Zimbabwe*. Paris: Organisation for Economic Co-operation and Development, accessed 2 May, 2009: http://www.oecd.org/document/44/0,3343, en_2649_15162846_32404716_1_1_1_1,00.html

Ohia, M. (2004). *Introduction: Critical success factors for effective use of e-learning with Māori learners*. Retrieved from http://elearning.itpnz. ac.nz/files/Hui_Report_final1_Effective_use_ of_Māori_eLearning.pdf

Ohler, J. (2008). *Digital story telling in the classroom: New media pathways to literacy, learning, and creativity*. Thousand Oaks, CA: Corwin Press.

Orange, C. (2004). *An illustrated history of the Treaty of Waitangi*. Wellington: Bridget Williams Books.

Organization for Economic Co-operation and Development (OECD). (2001). *The development of broadband access in OECD countries*. Paris: OECD.

Organization for Economic Co-operation and Development (OECD). (2006). The development of broadband access in OECD countries. *OECD Composite Leading Indicators (CLIs) for OECD Countries and Major Non-Member Economies*. Retrieved September 14, 2007 from http://www.oecd.org/statisticsdata/0,2643, en_2649_37443_1_119656_1_1_37443,00.htm

Pande, R. (2005). Looking at information technology from a gender perspective: The call centers in India. *Asian Journal of Women's Studies, 11*(1), 58–82.

Pandian, M. S. S. (2007). *Brahmin Non Brahmin: Geneologies of the Tamil political present*. New Delhi: Permanent Black.

Pang, V. (2001). *Multicultural Education: A Caring-centered, Reflective Approach*. Boston: McGraw Hill.

Parayil, G. (2006). The political economy of informational development: A normative appraisal. In Parayil, G. (Ed.), *Political economy and information capitalism in India: Digital divide and equity* (pp. 196–217). New York: Palgrave Macmillan.

Pauktuutit, Inuit Women of Canada. (1 October 2006). Keepers of the Light: Inuit Women's Action Plan, Ottawa. http://www.pauktuutit.ca/pdf/publications/pauktuutit/KeepersOfTheLight_e.pdf

Pere, R. (1991). *Te wheke*. Wellington: Ao Ako Global Learning Ltd.

Peresuh, M., & Ndawi, O. P. (1998). Education for All—the challenges for a developing country: the Zimbabwe experience'. *International Journal of Inclusive Education, 2*(3), 209–224. doi:10.1080/1360311980020302

Picot, A., & Wenicek, C. (2007). The role of government in broadband access. *Telecommunications Policy, 31*, 660–674. doi:10.1016/j. telpol.2007.08.002

Poll results verdict against globalisation: CPI(ML) (2004, May 17). *The Times of India*. Retrieved December 14, 2009 from http://timesofindia.indiatimes.com/city/thirupuram/Poll-results-verdict-against-globalisation-CPIML/articleshow/681922.cms

Pollitzer, W. S. (2005). *The Gullah people and their African heritage*. Athens, GA: The University of Georgia.

Pomerantz, L. (2001). Bridging the digital divide: Reflections on "Teaching and learning in the digital age" . *The History Teacher, 34*(4), 509–522. doi:10.2307/3054203

Power, S. (2003). How to Kill a Country: Turning a breadbasket into a basket case in ten easy steps—the Robert Mugabe way. *The Atlantic Magazine*, December, 2003, accessed 2 May 2009: http://www.theatlantic.com/doc/200312/power

Prasad, M. (2004). Reigning stars: The political career of south Indian cinema . In Fischer, L., & Landy, M. (Eds.), *Stars: The film reader* (pp. 97–114). New York: Routledge.

Putzel, J. (2005). Globalization, Liberalization, and Prospects for the State. *International Political Science Review, 26*(1), 5–16. doi:10.1177/0192512105047893

World Bank. (2009a). *Broadband Infrastructure Investment in Stimulus Packages: Relevance for Developing Countries* (Qiang, C. Z., Ed.). Washington.

Rajagopal, A. (2001). *Politics after television: Hindu nationalism and the reshaping of the public in India*. Cambridge: Cambridge University Press. doi:10.1017/CBO9780511489051

Randall, C. H., Reichgelt, H., & Price, B. A. (2003). *Demography and IT/IS students: Is this digital divide widening?* Paper presented at the 17th Annual Conference of the International Academy for Information Management, Barcelona, Spain.

Ranganathan, C. (2009, December 10). IT companies will tweak HR policies to remain competitive: Lakshmi Narayanan. *The Economic Times*. Retrieved December 10, 2009 from http://economictimes.indiatimes.com/opinion/interviews/IT-companies-will-tweak-HR-policies-to-remain-competitive-Lakshmi-Narayanan/articleshow/5323977.cms

Ranieri, M. (2008). Cyberspace's Ethical and Social Challenges in Knowledge Society. In A. Cartelli & M. Palma (Eds.), *Encyclopedia of Information and Communication Technology, 2nd edition* (132–137). Hershey-New York: Information Science Reference, IGI Global.

Rawat, V. B. (2009). *Fire incident in Deoria, Mushhar families hurt: Please help*. Retrieved August 11, 2009, from http://awareness-2009.blogspot.com/2009/03/fire-incident-in-deoria-mushhar.html

Reed, A. W. (1935). *The Māori and his first printed books*. Dunedin: Reed.

Reedy, T. (2000). Te reo Māori: The past 20 years and looking forward. *Oceanic linguistic, 39*(1), 157–169.

Resta, P., Christal, M., & Roy, L. (2004). Digital technology to empower indigenous culture and education . In Brown, A., & Davis, N. E. (Eds.), *Digital technology, communities and education. World Yearbook of Education 2004* (pp. 179–195). London: Routledge Falmer. doi:10.4324/9780203416174_chapter_11

Rho, K. (2002). Uses of Internet in Korea. *Educational Technology Research and Development, 50*(1), 84–88. doi:10.1007/BF02504964

Robin, B. (2008). Digital storytelling: A powerful technology tool for the 21st century classroom. *Theory into Practice, 47*, 220–228. doi:10.1080/00405840802153916

Rogers, E. M. (2001). The Digital Divide. *Convergence*, *7*, 96–111.

Roller, L.-H., & Waverman, L. (2001). Telecommunications Infrastructure and Economic Development: A Simultaneous Approach, *American Economic Review, American Economic Association*, 91(4):909-23, September. Accessed 14 January, 2010: http://www.e-aer.org/archive/9104/91040909.pdf

Rong, X. L., & Shi, T. (2001). Inequality in Chinese education. *Journal of Contemporary China*, *10*(26), 107–124. doi:10.1080/10670560124330

Rose, R. (2005). A Global Diffusion Model of E-Governance. *Journal of Public Policy*, *25*(1), 5–27. doi:10.1017/S0143814X05000279

Said, E. (1978). *Orientalism*. New York: Vintage Books.

Sartori, L. (2006). *Il Divario Digitale: Internet e le Nuove Disuguaglianze Sociali*. Bologna: Il Mulino.

Saul, J. R. (2008). *A fair country: Telling truths about canada*. Toronto: Viking Canada.

Schwartz, S. B. (1985). *Sugar plantations in the formation of Brazilian society: Bahia, 1550-1835*. Cambridge, MA: Cambridge University Press.

Selka, S. (2007). *Religion and the politics of ethnic identity in Bahia, Brazil*. Gainesville, FL: University Press of Florida.

Selwyn, N. (2004). Reconsidering Political and Popular Understandings of the Digital Divide. *New Media & Society*, *6*(3), 341–362. doi:10.1177/1461444804042519

Seo, Y. (2001). 디지털 시대의 평등사회 구현을 위한 정보격차 해소방안 [A digital divide solution for an equitable digital society]. Policy and Computers, *23*(1).

Shah, G., Mander, H., Thorat, S., Deshpande, S., & Baviskar, A. (Eds.). (2006). *Untouchability in rural India*. New Delhi: Sage.

Shah, G. (1991). Tribal identity and class differentiations: A case study of the Chaudhri tribe . In Gupta, D. (Ed.), *Social stratification* (pp. 288–302). New Delhi: Oxford University Press.

Shimo, A. (25 February 2009). *Tough critique or hate speech*? Macleans.ca URL: http://www2.macleans.ca/2009/02/25/tough-critique-or-hate-speech/print/

Shiva, V. (2005, May). The polarised world of globalization (A response to Friedman's flat earth hypothesis). *Z-Net*. Retrieved May 7, 2009, from http://www.zmag.org/zspace/commentaries/2299

Shor, I. (1987). *Pedagogy for liberation*. MA: Bergin & Garvey.

Simon, J. (1986). *Ideology in schooling of Māori children: Delta Research Monograph, No 7*. Palmerston North: Massey University Press.

Simon, M. (13 February 2009). Assimilation is no solution. *National Post*. URL: http://www.nationalpost.com/life/footprint/story.html?id=1285763

Singh, J. (2002). From atoms to bits: Consequences of the emerging digital divide in India. *The International Information & Library Review*, *34*(2), 187–200. doi:10.1006/iilr.2002.0194

Singh, T. (2008, Dec 11). Only 13% of rural India has access to telephone. *The Times of India*. Retrieved December 20, 2008, from http://timesofindia.indiatimes.com/articleshow/3820808.cms

Sleeter, C., & Grant, C. (2003). *Making choices for multicultural education: Five approaches to race, class, and gender*. New York: Merrill.

Sleeter, C., & Bernal, D. (2004). Critical pedagogy, critical race theory, and antiracist education: Implications for multicultural education . In Banks, J. A., & Banks, C. (Eds.), *Handbook of Research on Multicultural Education*. San Francisco, CA: Jossey-Bass.

Slemon, S. (1987). Monuments of empire: Allegory/counter-discourse/post-colonial writing. *Kunapipi* 9.3, p. i-i6.

Smith, D. (2007). *Hinduism and modernity.* Chichester, England: John Wiley & Sons, Ltd.

Smith, L. (1999). *Decolonizing methodologies: Research and indigenous peoples.* Dunedin: University of Otago Press.

Smith, C., & Ward, G. (Eds.). (2000). *Indigenous cultures in an interconnected world.* Toronto: University of British Columbia Press.

Smith, J. P. (1991). Cultural preservation of the Sea Island Gullah: A Black social movement in the post-civil rights era. *Rural Sociology, 56*(2), 284–298. doi:10.1111/j.1549-0831.1991.tb00437.x

Smith, G. (2000). Protecting and respecting indigenous knowledge . In Battiste, M. (Ed.), *Reclaiming indigenous voice and vision* (pp. 209–225). Vancouver, BC: UBC Press.

Smith, L. (2005). On tricky ground: Researching the native in the age of uncertainty . In Denzin, N., & Lincoln, Y. (Eds.), *The Sage handbook of qualitative research* (3rd ed., pp. 85–108). Thousand Oaks, CA: Sage.

Smith, L. T. (2005). On tricky ground: Researching the native in an age of uncertainty . In Denzin, N. K., & Lincoln, Y. S. (Eds.), *The Sage Handbook of Qualitative Research* (3rd ed.). Los Angeles, London: Sage.

Smith, M. K. (1997, 2002) Paulo Freire and informal education, *The Encyclopaedia of Informal Education.* Accessed 15 January, 2010: http://www.infed.org/thinkers/et-freir.htm

Solorzano, D. G., & Bernal, D. D. (2001). Examining transformational resistance through a Critical Race and LatCrit Theory framework: Chicana and Chicano students in an urban context . *Urban Education, 36*(3), 308–342. doi:10.1177/0042085901363002

South Carolina. (1868). *Proceedings of the Constitutional Convention of South Carolina.* [Held at Charleston, SC, beginning January 14th and ending March 17th, 1868, reprinted by Arno Press and the New York Times, 1968.]

Spener, D. (1990, 1992) *The Freirean Approach to Adult Literacy Education*, ESL Resources: Digests, National Center for ESL Literacy Education. April 1990, Revised November 1992. Accessed 16 January, 2010: http://www.cal.org/caela/esl_resources/digests/FREIREQA.html

Spivak, G. (1996). Explanation and culture: Marginalia . In Landry, D., & MacLean, G. (Eds.), *The Spivak reader* (pp. 29–51). New York: Routledge.

Spring, J. (2007). *Deculturalization and the struggle for equality: A brief history of the education of dominated cultures in the United States* (5th ed.). Boston: McGraw Hill.

Spring, J. (2004). *Deculturalization and the struggle for equality: A brief history of the education of dominated cultures in the United States.* New York, NY: McGraw Hill.

Sreekumar, T. (2006). ICTs for the rural poor: Civil society and cyberlibertarian developmentalism in India . In Parayil, G. (Ed.), *Political economy and information capitalism in India: Digital divide and equity* (pp. 61–87). New York: Palgrave Macmillan.

Srinivas, M. N. (1991a). The dominant caste in Rampura . In Gupta, D. (Ed.), *Social stratification* (pp. 307–311). New Delhi: Oxford University Press.

Srinivas, M. N. (1991b). Mobility in the caste system . In Gupta, D. (Ed.), *Social stratification* (pp. 312–325). New Delhi: Oxford University Press.

Stewart, C. M., Gil-Egui, G., Tian, Y., & Pileggi, M. I. (2006). Framing the Digital Divide: a Comparison of US and EU Policy Approaches. *New Media & Society, 8*(5), 731–751. doi:10.1177/1461444806067585

Strover, S. (2003). Remapping the Digital Divide. *The Information Society, 19*, 275–277. doi:10.1080/01972240309481

Tahu, K. (2006) *Kotahi mano kāika, kotahi, mano wawata: One thousand homes, one thousand aspirations*. Retrieved from http://www.kmk.maori.nz/ Accessed 16 February 2009

Talmon-Chvaicer, M. (2008). *The hidden history of capoeira: A collision of cultures in the Brazilian battle dance*. Austin, TX: University of Texas Press.

Tartakov, G. M. (2009). Why compare Dalits and African Americans? They are neither unique nor alone . In Natrajan, B., & Greenough, P. (Eds.), *Against stigma: Studies in caste, race and justice since Durban* (pp. 95–140). Hyderabad: Orient Black Swan.

Tate, W. (1996). Critical race theory and education: History, theory, and implications. *Review of Research in Education, 22*, 195–247. doi:10.3102/0091732X022001195

Taylor, K. A. (2005). Mary S. Peake and Charlotte L. Forten: Black teachers during the Civil War and Reconstruction. *The Journal of Negro Education, 74*(2), 124–137.

Te Aika, L. H., & Greenwood, J. (2009). Ko tātou te rangahau, ko te rangahau, ko tātou: A Māori approach to participatory action research. In D. Kapoor & S. Jordan (Eds) *Education, participatory action research, and social change: International perspectives. Palgrave Macmillan*.

Television, M. (2009a). *Mission statement*. Retrieved from http://www.maoritelevision.com/

Television, M. (2009b). *Website: frequently asked questions*. Retrieved from http://corporate.maoritelevision.com/faq.htm

Teltumbde, A. (2008). *Khairlanji: A strange and bitter crop*. New Delhi: Navayana.

Thirumal, P. (2008). Situating the new media: Reformulating the Dalit question . In Gajjala, R., & Gajjala, V. (Eds.), *South Asian Technospaces* (pp. 97–122). New York: Peter Lang.

Thomas, J. M. (1980). The impact of corporate tourism on Gullah Blacks: Notes on issues of employment. *Phylon, 41*(1), 1–11. doi:10.2307/274663

Thomas, J. J. (2006). Informational development in rural areas: Some evidence from Andhra Pradesh and Kerala . In Parayil, G. (Ed.), *Political economy and information capitalism in India: Digital divide and equity* (pp. 109–132). New York: Palgrave Macmillan.

Thorat, S. (2009). Caste, race and United Nations' perspective on discrimination: Coping with challenges from Asia and Africa . In Natrajan, B., & Greenough, P. (Eds.), *Against Stigma: Studies in Caste, Race and Justice since Durban* (pp. 141–167). Hyderabad: Orient Longman.

Tiene, D. (2004). Bridging the digital divide in the schools of developing countries. *International Journal of Instructional Media, 31*(1), 89–97.

Tiene, D. (2002). Addressing the global digital divide and its impact on educational opportunity. *Educational Media International, 39*(3-4), 211–222. doi:10.1080/09523980210166440

Times, Z. (2009). *Zimbabwe scores badly on ICT*. Newspaper article published in Harare: March 26, 2009, accessed 2 May 2009: http://www.thezimbabwetimes.com/?p=14058

Toepke, A., & Serrano, A. (1998). *The language you cry in* [videorecording]. San Francisco, CA: California Newsreel.

Tribunal, W. (1986). *Report of the Waitangi Tribunal on the Te Reo Māori claim*. Wellington: Government Printer.

Uchida, H. (2004). Information Technology-Driven Education in Japan: Problems and Solutions . *Educational Technology Research and Development, 52*(3), 91–111. doi:10.1007/BF02504679

UN General Assembly. (2007*). Declaration on the rights of indigenous peoples.*http://www.iwgia.org/sw248.asp

Unctad, United Nations Conference on Trade and Development (2006). *Information Economy Report: The Development Perspective.* New York.

UNESCO. (2003). *Meta-survey on the Use of Technologies in Education in Asia and the Pacific 2003-2004.* Retrieved January 12, 2009 from www.iosn.net/education/Metasurvey

UNESCO. (2007). *ICT in education: Japan.* Retrieved May 12, 2008. http://www.unescobkk.org/index.php?id=1381

UNICEF. (2008). India statistics. Retrieved October 23, 2008, from http://www.unicef.org/infobycountry/india_india_statistics.html

United Nations. (2001). The Development Divide in a Digital Age. An Issues Paper, (Ed. C. Hewitt de Alcántara). *Technology, Business and Society Programme Paper Number 4.* New York.

United Nations. (2008). *Global E-Government Survey.* New York.

Universal Declaration of Human Rights. (1948) Retrieved January 19, 2009 from http://www.unhchr.ch/udhr/lang/eng.htm

Van Deusen, K. (2009). *Kiviuq: An inuit hero and his siberian cousins.* Montreal: McGill-Queen's University Press.

van Dijk, J. A. (2005). *The deepening divide: Inequality in the information society.* Thousand Oaks, CA: Sage Publications, Inc.

van Dijk, J. A. (2005). *The deepening divide: Inequality in the information society.* Thousand Oaks, CA: Sage Publications, Inc.

Van Dijk, J. (2006). Digital Divide Research, Achievements and Shortcomings. *Poetics, 34*(4-5), 221–235. doi:10.1016/j.poetic.2006.05.004

Van Dijk, J., & Hacker, K. (2003). The Digital Divide as a Complex and Dynamic Phenomenon. *The Information Society, 19*, 315–326. doi:10.1080/01972240309487

Van Reijswoud, V. (2008). *Free and open source software for development: exploring expectations, achievements and the future.* Monza: Polimetrica.

Varma, P. (2007). *The great Indian middle class* (2nd ed.). New Delhi: Penguin Books.

von Krogh, G., & Spaeth, S. (2007). The open source software phenomenon: Characteristics that promote research. *The Journal of Strategic Information Systems, 16*, 236–253. doi:10.1016/j.jsis.2007.06.001

Wade, R. H. (2002). Bridging the Digital Divide: New Route to Development or New Form of Dependency? *Global Governance, 8*, 443–466.

Waldon, W. (2004) Official launch of Māori television http://www.maoritelevision.com/newsletter/issue5/index.htm

Walker, R. (1990). *Ka whawhai tonu matou: Struggle without end.* Auckland: Penguin.

Wallerstein, I. (2006). *European Universalism. The Rhetoric of Power.* New York: The New Press.

Warschauer, M. (2004). *Technology and Social Inclusion: Rethinking the Digital Divide.* Cambridge, MA: The MIT Press.

Warschauer, M. (2002). Reconceptualizing the Digital Divide. *First Monday, 7*(7).

Warschauer, M. (2003). *Technology and Social Inclusion: Rethinking the Digital Divide.* Cambridge, MA: MIT Press.

Warshauer, M., Knobel, M., & Stone, L. (2004). Technology and equity in schooling: Deconstructing the digital divide. *Educational Policy, 18*(4), 562–588. doi:10.1177/0895904804266469

Weber, M. (1946). Class, status, party . In Gerth, H., & Wright Mills, C. (Eds.), *From Max Weber: Essays in Sociology* (pp. 114–124). New York: Oxford University Press.

WEF. (2009). *The Global Information Technology Report 2008-2009*. INSEAD Business School: World Economic Forum, available online at http://www.weforum.org accessed 15 January 2010: http://www.weforum.org/en/initiatives/gcp/Global%20Information%20Technology%20Report/index.htm

Weisskopf, T. (2004). *Affirmative action in the United States and India; A comparative perspective*. London: Routledge.

Widdowson, F., & Howard, A. (2008). *Disrobing the aboriginal industry*. Toronto: McGill-Queen's University Press.

Williams, H. A. (2005). *Self-taught: African American education in slavery and freedom*. Chapel Hill, NC: The University of North Carolina Press.

Williams, M. (2008). *Broadband for Africa: Policy for Promoting the Development of Backbone Networks*. Washington, DC: World Bank Report funded by InfoDev, accessed 2 May 2009: http://www.infodev.org/en/Publication.526.html

Women and Men of Zimbabwe Arise (WOZA). (2009). *The State of Education in Zimbabwe - a Dream Shattered: A Women and Men of Zimbabwe Arise (WOZA) perspective, February 2009*. Famona, Bulawayo, Zimbabwe, accessed 15 January, 2010: http://wozazimbabwe.org/?p=315

World Bank (2009b). *Global Dialogue on Exploring the Results of Governmental Open Source Software Policies: Brazil Experience*. Event desciption, Dec. 17.

World Internet Statistics. (2007). Internet Usage and Statistics - The Big Picture. World Internet *Users and Population Stats*. Retrieved September 30, 2007 from http://www.Internetworldstats.com/stats.htm

Worst over for Indian economy, may grow at 6.5% in FY10: EIU. (December 10, 2009). *The Economic Times*. Retrieved December 10, 2009 from http://economictimes.indiatimes.com/news/economy/indicators/Worst-over-for-Indian-economy-may-grow-at-65-in-FY10-EIU/articleshow/5324299.cms

WOZA. (2010) *Looking back to look forward - education in Zimbabwe: a WOZA perspective – January 2010*. WOZA Report on Education: Bulawayo, Zimbabwe, accessed 15 January, 2010: http://wozazimbabwe.org/?p=607

Yu, S., Wang, M., & Che, H. (2005). An exposition of the crucial issues in china's educational informatization. *Educational Technology Research and Development, 53*(4), 88–101. doi:10.1007/BF02504688

Zelliot, E. (1992). *From Untouchable to Dalit: Essays on the Ambedkar Movement*. New Delhi: Manohar.

Zhang, W., & Perris, K. (2004). Researching the efficacy of online learning: A collaborative effort amongst scholars in Asian open universities. *Open Learning, 19*(3), 247–264. doi:10.1080/0268051042000280110

Zhu, J., & Wang, E. (2005). Diffusion, use, and effect of the Internet in china. *Communications of the ACM, 48*(4). doi:10.1145/1053291.1053317

Zimbizi, G. (2000). *Scenario Planning for Farm Worker Displacement.* Harare: Zimbabwe Network for Informal Settlement Action (ZINISA).

Zimbizi, G. (2001). *Study on Socio-Economic Status of Children and Women on Commercial Farms, Mining and Peri-Urban Areas.* Report prepared for UNICEF, accessed 15 January, 2010: http://www.unicef.org/evaldatabase/files/ZIM_01-800_Part1.pdf

About the Contributors

Patricia Randolph Leigh is Associate Professor of Curriculum & Instruction and is affiliated with the Center for Technology in Learning and Teaching in the College of Human Sciences at Iowa State University. Dr. Leigh also provides leadership to the George Washington Carver Academy, a university undergraduate scholarship program. She teaches courses in the areas of educational foundations, multicultural education, and instructional technology, in which she embeds her social justice and equity scholarship. Dr. Leigh's research is also informed by this expertise as she focuses on the impact of historical discrimination on technology equity in the digital age, the equality of educational opportunities historically afforded underserved children, and the impact that economic discrimination and residential segregation has had upon the public schooling of ethnic/racial minority children in the U.S. Dr. Leigh's more recent work centers on globalization and social justice, particularly issues affecting those in the Africa Diaspora.

* * *

Francesco Amoretti is Professor of Political Communication, and of E-democracy and E-Government Policies, University of Salerno, Graduate Degree Course in Communication Science. Since 1999 he has been a member of the Directive Committee of the Political Communication Review. He has published journal articles in several areas, including social policies, administrative reforms, and mass media and political systems. Currently his interests focus broadly on new technologies and politics – e-democracy and e-government - communication policy, European public space, and cyberspace. Recent publications are in the Encyclopedia of Digital Government, Encyclopedia of Information Science and Technology, and in a issue of Review of Policy Research. He is also editor of the volume "Electronic Constitution: Social, Cultural and Political Implications" (Hershey, IGI Global Publications, 2009).

J. Herman Blake is the inaugural Humanities Scholar in Residence at the Medical University of South Carolina. He served as founding Provost of Oakes College at the University of California, Santa Cruz, and President of Tougaloo College in Mississippi. He was also the Eugene M. Lang Visiting Professor for Social Change at Swarthmore College; Vice Chancellor for Undergraduate Education at Indiana University Purdue University Indianapolis; and Director of African American Studies at Iowa State University. He is Professor of Sociology-Emeritus at Iowa State University. His research focuses on minority students in higher education; urban militants in the African American community and social change and community development in rural and urban African American communities. His publications

Copyright © 2011, IGI Global. Copying or distributing in print or electronic forms without written permission of IGI Global is prohibited.

include the book *Revolutionary Suicide,* the autobiography of Huey P. Newton, and founder of the Black Panther Party for Self-Defense. He has been awarded six honorary degrees and two presidential medals.

Niki Davis is Professor of E-Learning at the University of Canterbury College of Education, New Zealand and the director of the e-Learning Lab, with current research projects including online education, and e-learning for adults with needs in literacy, language and numeracy. She is an international leader in information and communication technologies (ICT) in teacher education and distance education, plus related organizational change with an ecological perspective. Sought by UNESCO, European Commission, national agencies, companies, scholarly societies and institutions for her expertise, Niki has hundreds of publications. Niki is a very recent immigrant to New Zealand from Northern Ireland by way of the UK and USA. She has decades of teaching and research experience in secondary and tertiary education. In 2009 Niki became the inaugural coordinator of New Zealand Collaborative Action and Research Network (NZCA&RN), which is linked within both CARN and NZARE.

Janinka Greenwood's research grows out of her work as a teacher and teacher educator and as an artist. Therefore it is based in a group of interconnected areas: education, theatre and the intercultural space where these take place. Some projects have been sited in one or more of these separate areas, and she also keenly interested in where they overlap and impact one another other, and in the ways in which they may inform each other in extending our conceptualisations of aesthetics, semiotics, scholarship and knowledge. Research work in the fields of theatre, drama in education, critical literacies, education and cross-cultural perspectives have also lead to further examination of particular methodologies such as practitioner research, action research creative work, and to epistemological questions about the cultural locatedness of knowledge including Māori and indigenous perspectives.

Lynne Harata Te Aika leads the School of Māori, Social and Cultural Studies in Education. Her research interests are primarily in the area of Maori education, including all aspects of te reo Maori, bilingual and immersion education from birth, early childhood and compulsory schooling to tertiary and lifelong learning. She is also interested in indigenous education and providing for the needs of diverse learners. She has led and been involved in Ngai Tahu tribal education and te reo initiatives over the past ten years. She is also involved in pan-tribal education initiatives at the local, regional and national level as well as developing links with indigenous communities in Pacific Rim countries.

Jill Jameson, Director of Research & Enterprise, University of Greenwich; Chair SRHE Conference 2010, Co-Chair, ALT-C 2008; Director JISC eLIDA CAMEL; Director JISC eLISA, Convenor SRHE HE-FE Network; AACE Journal Editorial Board; E-LEARN, BERA, BELMAS & ALT-C presenter; Biographee, Marquis Who's Who in the World; Special Editor, BJET (2006) & Alt-J (2000).

Elizabeth Langran has been an assistant professor and director of the Educational Technology program in the Graduate School of Education and Allied Professions at Fairfield University since 2006. She received her Ph.D. in Instructional Technology and a certificate in International Leadership in Educational Technology from the University of Virginia after nine years teaching middle and high school social studies, French, and English as a Second Language and two years teaching English at a Moroccan university as a Peace Corps volunteer. At Fairfield University she is engaged in research and partnerships with schools in Nicaragua, India, Senegal, Kyrgyzstan and Kazakhstan.

James C. McShay currently directs the Office of Multicultural Involvement and Community Advocacy and teaches courses in Leadership and Identity Development within the College of Education at the University of Maryland, College Park. He is the former director of undergraduate education for the Department of Curriculum and Instruction at Iowa State University, USA. During his tenure at ISU, he taught courses in multicultural education, ethnicity and learning, and antiracist education. His research has a special focus on how technology can be used to support liberatory pedagogies in K-16 education. His work can be found in the Journal of Multicultural Perspectives; Multicultural Education and Technology Journal; Contemporary Issues in Technology and Teacher Education; and Critical Multiculturalism: From theory to practice.

Emily L. Moore, Ed. D, is professor at the Department of Health Sciences and Research as well as Director of the Master in Health Administration (MHA) - Global program in the College of Health Professions, Medical University of South Carolina. She is Professor Emerita in the Department of Educational Leadership and Policy Studies, Iowa State University. She served as Provost and Vice President of Academic Affairs at Dillard University in New Orleans, Louisiana. She has also served as Vice-President of Academic Affairs and Dean of Faculty at Concordia University and Professor of Educational Leadership and Policy Studies at Iowa State University. She is President of Scholars for Educational Excellence and Diversity Inc., a higher education consulting firm. Her research interests include health education intervention in HIV/AIDS in sub-Saharan Africa, China and rural populations in the Sea Islands of South Carolina; Health Behaviors among the rural Black elderly; current issues in global health and high education.

Thalia M. Mulvihill currently serves as a Professor in Higher Education and Social Foundations of Education, and the Asst. Dept. Chair/Director of Doctoral Programs in the Dept. of Educational Studies at Ball State University's Teachers College. Dr. Mulvihill holds a Ph.D. in Cultural Foundations of Education and Curriculum from Syracuse University, an interdisciplinary degree with a focus in History and Sociology of Education/Higher Education and a Concentration in Gender and Education Studies. Her research agenda focuses on the history and sociology of higher education with a focus on women and gender issues. Some areas of special interest include critical theory and pedagogies that focus on democracy and social justice issues, and internationalizing the curriculum for educators.

Fortunato Musella has a Ph.D. in Political Science at the University of Florence, with a dissertation thesis dedicated to the Italian regional governments, and he is currently researcher at the University of Naples. His main research interests embrace a) The study of government, b) Elections and personalization of vote, c) New technologies and politics. His recent publications include the volume "Governi monocratici. La svolta presidenziale nelle regioni italiane" (Bologna, il Mulino, 2009), as well as several articles and book's chapters appearing in Quaderni di Scienza Politica, Quaderni dell'Osservatorio Elettorale, Polis, Comunicazione Politica. Recently he has contributed to the volume "Electronic Constitution: Social, Cultural and Political Implications" (Hershey, IGI Global Publications, 2009).

Gary Michael Tartakov is social historian and comparativist specializing in the visual arts and design of South Asia and the European and American tradition. He is a Professor Emeritus of Art and Design at Iowa State University. He has been working in and around India since 1963. Beginning with the temple culture of the ancient period he has moved in recent years through studies of Orientalism, and the impact of European and American scholars on the study of India, to a concentration on contemporary visual imagery in India concerned with the portrayal of Dalits and with Dalit uses of visual imagery. Samples of this later work can be seen in his essay "Why compare Dalits and African Americans? They are neither unique nor alone" in B. Natrajan & P. Greenough (Eds.), Against stigma: Studies in caste, race and justice since Durban (Hyderabad: Orient Black Swan, 2009) and the forthcoming volume Dalit Art and Visual Imagery.

P. Thirumal is a cultural theorist at the Department of Communication, University of Hyderabad. He has been writing and researching in the area of social and cultural history of media in India. Further, he has been associated with Dalit movements in the country. Thirumal's doctoral work engaged with the formation of the regional community and the imbrication of the region in the print culture that it produced during the post-colonial period. He has been teaching history of languages, literary cultures and media for post graduate students for over a decade now. Thirumal started his career at the Centre for Science and Technology Communication, Pondicherry University. Apart from designing courses, he was closely involved in popularization of science through Non Governmental Organizations. It was at his initiative that the special paper on Science and Communication was introduced at Department of Communication, University of Hyderabad.

Sunnie Lee Watson is an Assistant Professor in the Department of Educational Studies at Ball State University. She received her dual major doctorate in Educational Policy studies with a concentration in International and Comparative Education, and Instructional Systems Technology at Indiana University, Bloomington. She has worked as a high school teacher of English as a Second Language in South Korea and in human resources development for various Fortune 100 companies. Some of her research interests include, Critical Systems Theory for educational change and research, internationalization of K-12 education through virtual schooling, the creation of learner-centered environments for marginalized and disadvantaged students, and international technology policies for digital equity and social justice.

Index

Copyright © 2011, IGI Global. Copying or distributing in print or electronic forms without written permission of IGI Global is prohibited.

T

U

V

W

Z

Breinigsville, PA USA
15 November 2010
249270BV00004B/7/P